Human Rights and the Environment under African Union Law

Michael Addaney
Ademola Oluborode Jegede
Editors

Human Rights and the Environment under African Union Law

palgrave
macmillan

Editors
Michael Addaney
Research Institute of Environmental Law
Wuhan University
Wuhan, China

University of Energy and Natural Resources
Sunyani, Ghana

Ademola Oluborode Jegede
Department of Public Law
University of Venda
Thohoyandou, South Africa

ISBN 978-3-030-46522-3 ISBN 978-3-030-46523-0 (eBook)
https://doi.org/10.1007/978-3-030-46523-0

Cover illustration: kenkuza_shutterstock.com

This Palgrave Macmillan imprint is published by the registered company Springer Nature Switzerland AG.
The registered company address is: Gewerbestrasse 11, 6330 Cham, Switzerland

Foreword

This book on Human Rights and the Environment under African Union Law is a timely publication. Africa and its people depend on the environment for their well-being and survival. The African continent's biodiversity and cultures are unique in the world and Africa includes some of the last wilderness areas in the world. Unfortunately, that is also Africa's downfall as its rich mineral and biological resources are exploited by non-Africans and multi-nationals. The Global North prescribes measures for the Global South that are not always implementable within an African context. In the end it is the people that suffer and their human rights that are trampled upon. Africa has other challenges too, such as failed governments, increased urbanization, poaching, climate change as well as climate-induced migration, to mention a few. This is a book on African environmental governance by Africans, providing a Global South perspective. The authors provide a unique and novel discussion on how and whether African Union Law, sub-regional law and specific selected countries address environmental issues in relation to human rights. The African Union was the first region to include an environmental right in its constituting document. The African Union and the sub-regions have adopted protocols in relation to environmental and human rights matters. Many African countries include an environmental right or environmental obligation in their constitutions. Most of the African countries have environmental framework legislation. There is unique case law explaining what an environmental right or the protection of human rights within an African context entail. All African countries have a responsibility to ensure that both human rights and the environment are protected. It is now the time for action and for African countries to ensure, within their regions as well as within their own jurisdictions, that environmental governance addresses human rights in relation to the environment. This book provides both scholarly and practical solutions in this regard.

Potchefstroom, South Africa Willemien du Plessis

Acknowledgments

The editors would like to thank the authors for their time, resources and determination toward completing this book project. It is a big volume to bring to fruition, and without their strong contributions and commitment, it simply would not have been possible. The editorial team at Palgrave Macmillan, including Rachael Ballard and Joanna O'Neill, was very patient and supportive along the way, as were the copyeditors. They helped in shaping the concept of the book and refined the final output to ensure that the publication was on schedule. Michael acknowledges the support of the Research Unit on Law, Justice and Sustainability at the Faculty of Law of North-West University, Potchefstroom, South Africa, for granting him six weeks visiting fellowship to enhance the completion of this project. The support of University of Venda which, through the research account of the co-editor, Professor AO Jegede, made available partial travel funds to support the execution of the book project is also acknowledged. The editors are also grateful to the two anonymous reviewers for the thorough and insightful reviews.

About the Book

This book combines contributions from emerging African and internationally recognized scholars in the field of human rights law, environmental law and policy, climate change law and African studies. The chapters in the book have undergone double blind peer-reviewed and revised by the authors; the editors particularly ensured that recent scholarship and distinctions are engaged by the authors. The primary aim of the book is to contribute to the development of African scholarship and to reinforce the international discourse on the future of human rights law, climate change and the African environment. The key contributions of the book are its analytic framework and enunciation of the specificities of the African context and (re)theorization of the conceptual legal links between human rights, climate change and the environment. Combining specialisms in human rights law and environmental law and politics, this book is a truly inter-disciplinary resource that will be essential for advanced students of human rights, environmental studies, international law, international relations, environmental politics and philosophy in the context of Africa. This book thus provides comprehensive examination of the environment and human rights under the African regional huvman rights system, focusing on normative developments, nature and content of their interactions and overlaps, including their normative unity with other rights, the correlating obligations and enforcement in situations of violations. Besides being enriching in its academic features, the book charts avenues for further research and discussion questions for beyond Africa. Due to the dearth of knowledge on the chosen thematic areas in Africa, the publication will remain relevant for now and several years into the future.

Wuhan, China/Sunyani, Ghana Michael Addaney
Thohoyandou, South Africa Ademola Oluborode Jegede

Contents

Abbreviations

AAEPA	Addis Ababa Environment Protection Agency
ACHPR	African Charter on Human and Peoples' Rights
ACHPR	African Commission on Human and Peoples' Rights
ACRWC	African Charter on the Rights and Welfare of the Child
ACtHPR	African Court of Human and Peoples' Rights
ADLI	Agriculture Development Led Industrialization
AMCEN	African Ministerial Conference on the Environment
ASEAN	Association of South East Asian Nations
AU	African Union
BIAT	Action Plan for Boosting Intra-Africa Trade
CBD	Convention on Biological Biodiversity
CEBR	Centre for Economics and Business Research
CEDAW	Convention on the Elimination of All Forms of Discrimination Against Women
CRC	Convention on the Rights of the Child
COMESA	Common Market for Eastern and Southern Africa
COP	Conference of Parties
CRPD	Convention on the Rights of Persons with Disabilities
CRGE	Climate-Resilient Green Economy
DEFRA	Department for Environment, Food and Rural Affairs
DPSP	Fundamental Objectives and Directive Principles of State Policy
DRC	Democratic Republic of the Congo
EAC	East African Community

EC	Environmental Council
ECOWAS	Economic Community of West African States
EIA	Environmental Impact Assessment
EIC	Ethiopian Investment Commission
EITAG	Extractive Industries Transparency and Accountability Guidelines
EITI	Extractive Industries Transparency Initiative
ELC	Environmental and Land Court
EMCA	Environmental Management and Co-ordination Act
EPA	Environmental Protection Authority
EROL	Environmental Rule of Law
ESIA	Environmental and Social Impact Assessment
EU	European Union
FEDEV	Fund for Environment and Development
FMP	Forest Management Plans
FPIC	Free, Prior and Informed Consent
FRELIMO	Frente de Libertação de Moçambique
FTAA	Free Trade Agreement of the Americas
FTA	Free Trade Area
GAP	Gender Action Plan
GMOs	Genetically Modified Organisms
GOU	Government of Uganda
GTP	Growth and Transformation Plan
HRBA	Human Rights Based Approach
ILA	International Law Association
ICCPR	International Covenant on Civil and Political Rights
ICESCR	International Covenant on Economic, Social and Cultural Rights
IUCN	International Union for Conservation of Nature
IPCC	Inter-governmental Panel on Climate Change
JCRA	J.C. Rathbone Associates Limited
LCB	Land Consultative Board
MINFOF	Ministry of Forestry
MNCs	Multinational Corporations
NAFTA	North American Free Trade Area
NBSAPs	National Biodiversity Strategies and Action Plans
NEA	National Environment Act
NEIPAs	National Extractive Industry Programmes of Action

NEITI	Nigerian Extractive Industries Transparency Initiative
NEMA	National Environmental Management Authority
NGOs	Non-governmental Organizations
OHCHR	Office of the High Commissioner for Human Rights
RACCNN	Revised African Convention on the Conservation of Nature and Natural Resources
RECs	Regional Economic Communities
SADC	Southern African Development Community
SAHRC	South African Human Rights Commission
SDGs	Sustainable Development Goals
SERAC	The Social and Economic Rights Action Centre
UDD	Universal Declaration of Human Rights
UDHR	Universal Declaration of Human Rights
UK	United Kingdom
UN	United Nations
UNCCD	United Nations Convention to Combat Desertification
UNDP	United Nations Development Programme
UNDRIP	UN Declaration on the Rights of Indigenous Peoples
UNEP	United Nations Environment Programme
UNFCCC	United Nations Framework Convention on Climate Change
UNICEF	United Nations Children's Fund
UNGA	United Nations General Assembly
UNHRC	United Nations Human Rights Council
UNPFII	UN Permanent Forum on Indigenous Issues
VPA	Voluntary Partnership Agreement
WGC	Women and Gender Constituency
WGEI	Working Group on Extractive Industries, Environment and Human Rights
WTO	World Trade Organisation
ZACPLAN	Zambezi Action Plan
ZIMASSET	Zimbabwe Agenda for Sustainable Socio-Economic Transformation

Notes on Contributors

Funmi Abioye is Associate Professor of Law in the Department of Jurisprudence at the University of South Africa (UNISA), South Africa. She obtained her LL.M. and LL.D. degrees from the University of Pretoria (UP), South Africa, and has variously taught public international law, private international law, comparative law and social welfare law in the past. Her research interests include the rule of law in Africa, the African Union, good governance, legitimacy and accountability of governments. She has written in these fields. She teaches Comparative Law and Private International Law, and is an editor of the *Comparative and International Law Journal of Southern Africa* (*CILSA*) and has a number of administrative posts within the College of Law.

Michael Addaney recently graduated from a Doctor of Law program in Science of Environment and Natural Resources Protection Law from the Research Institute of Environmental Law of Wuhan University, Wuhan, China, and also affiliated to the Department of Planning and Sustainability at the University of Energy and Natural Resources, Sunyani, Ghana. Addaney's research interests broadly focus on the role of international human rights law in framing and implementing responses to sustainability challenges in sub-Saharan Africa. Addaney was a visiting doctoral researcher at the Research Unit on Law, Justice and Sustainability of the Faculty of Law, North-West University, Potchefstroom, South Africa, and the Grotius

Centre for International Legal Studies of Leiden Law School, Leiden University, The Netherlands.

Desalegn Amsalu has been a researcher at Addis Ababa University, Ethiopia, since 2008. He holds LL.B. (Hons.), M.A. and Ph.D. He has a wide range of research interests in legal pluralism, environmental rights, minorities and minority rights. His recent publications include "Environmental Rights in Ethiopia: Shifting from Theory to Practical Realization". He is the research head of the Institute of Ethiopian Studies, Addis Ababa University.

Jean-Claude N. Ashukem holds LL.D. and LL.M. from the North-West University, and Maitrise en Droit and Licence en Droit from the University of Yaoundé II. Ashukem is a post-doctoral research fellow at the University of the Free State.

Elsabé Boshoff worked as a technical assistant at the African Commission on Human and Peoples' Rights, Banjul, the Gambia. She has written and co-written articles in the *African Human Rights Yearbook* and the *Australian Journal of Human Rights*, among other articles and book chapters, and co-edited the book entitled *Governance, Human Rights, and Political Transformation in Africa* published by Palgrave Macmillan in 2020. Her research interests include human rights, in particular women and children's rights, environmental law and climate change.

Lenin Tinashe Chisaira holds an LL.M. in International Human Rights Law and Public Policy from the University College Cork, Ireland; an LL.M. in Economic Regulation and a Postgraduate Certificate of Laws in Environmental and Natural Resources Law from the University of London, UK; and an LL.B. Hons. from the University of Zimbabwe, Zimbabwe. A registered lawyer and activist, Lenin is the founder and director of Advocates4Earth, a non-profit environmental law, climate and wildlife justice organization campaigning on Global South issues and based in Harare, Zimbabwe. Lenin's research interests include environmental law, climate and wildlife justice, economic regulation, eco-socialism, labour and human rights. His recent work is a book chapter entitled

'Climate Change and Labour Law in Zimbabwe: A Critical Perspective' appearing in the book, *Climate Change Law in Zimbabwe: Concepts and Insights* published by KAS and Weaver Press in 2019.

Eghosa Osa Ekhator joined the University of Chester Law School, UK, in January 2017 from the University of Hull where he had taught on numerous undergraduate and postgraduate modules. Ekhator's main research area lies in environmental justice/rights in developing countries—Nigeria and African Union in particular. Ekhator holds an LL.B. in Law from the University of Benin, and an LL.M. and Ph.D. in Law from the University of Hull. His works have appeared in the *Journal of African Law*, *African Journal of International & Comparative Law*, *Environmental Law Review*, *Commonwealth Law Bulletin*, *International Community Law Review*, *Annual Survey of International and Comparative Law* and *International Energy Law Review*, among others. He has presented academic papers at conferences in the UK, USA, South Africa, Nigeria, Germany and Portugal. Some of these include Society of Legal Scholars and Socio-Legal Studies Association annual conferences.

Charlotte Kabaseke is an advocate of the Courts of Judicature of Uganda. Kabaseke obtained a Bachelor of Laws Degree (LL.B.) from Uganda Christian University and Postgraduate Diploma in Legal Practice from the Law Development Centre, Kampala, following which she obtained a Master of Laws Degree (LL.M.) from Makerere University, Kampala, Uganda. She is a 2019 fellow of the UN International Law Fellowship Programme and an alumnus of the Hague Academy of International Law having attended the Summer course on Public International Law in July 2019. Kabaseke worked as a legal assistant (advocate) at the Justice Centres Uganda, a legal aid service provider. She was appointed Lecturer of Laws at Bishop Stuart University, Mbarara, Uganda, and later became an acting Dean of the Faculty of Law in the same institution. At the same time, she practiced as an advocate with Nakuya and Co. Advocates, Kampala, Uganda. Kabaseke recently graduated with a Doctor of Law in Environment and Natural Resources Protection Law from the School of Law, Wuhan University, Wuhan,

China. Kabaseke's research is focused on the intersectionality between gender and environment in East Africa.

Emma Charlene Lubaale is Associate Professor of Law at the Faculty of Law, Rhodes University, Grahamstown, South Africa. Prior to this, she was a senior lecturer at the University of Venda where she taught and researched various aspects of law. She has also taught and researched law in the capacity of a post-doctoral research fellow at the University of Pretoria's Institute of International and Comparative Law in Africa. She holds a Doctorate and Master's degree in Law from the University of Pretoria; a Bachelor's degree in Law from Makerere University and a Post-Graduate Diploma in Legal Practice from Uganda's Law Development Center. She is currently obtaining a Post-Graduate Diploma in Higher Education at the University of KwaZulu-Natal. Her current areas of interest are international human rights law, international criminal law, constitutional law and criminal law in the domestic perspective.

Mulesa Lumina is a researcher in the human rights field as well as a blogger and women's rights activist. She is the Chief Executive Officer and a senior consultant at the Human Rights & Governance Consultancy (HURICON)—an independent human rights consulting firm. She holds a Bachelor of Laws from the University of Cape Town and a Master of Laws in Human Rights and Democratization in Africa from the University of Pretoria, South Africa. She is also a Women Deliver Young Leader Alumni, volunteer with the International Youth Alliance for Family Planning—Zambia and former blog and media editor for the *Business and Human Rights Journal.*

Chantelle Gloria Moyo is an academic and registered legal practitioner with the Law Society of Zimbabwe. Moyo is also a doctoral researcher at the South African Research Chair in Cities, Law and Environmental Sustainability (CLES) at North-West University, Potchefstroom, South Africa. Her research experience ranges from human rights and the environment to Wildlife Law, Water Law, Pollution Control Law, Mining Law, Energy Law (renewable energy), Environmental Governance and

Enforcement Procedures. She was a recipient of a research grant from the Raoul Wallenberg Institute for International Humanitarian Law in 2019.

Collins Odote is a senior lecturer at the Centre for Advanced Studies in Environmental Law and Policy (CASELAP), University of Nairobi, Kenya, and an advocate of the High Court of Kenya. He has a Ph.D. in Law from the University of Nairobi with a specialization in land and environmental law. His research interests span the areas of governance, elections, property theory, natural resource management and extractives. He has written several articles and co-authored books in his areas of expertise. His most recent publications include "The Role of the Environment and Land Court in Governing Natural Resources in Kenya" in Patricia Kameri Mbote, et al (eds) *Law, Environment Africa* (2019); and "Implications of the Ecosystem-Based Approach to Wetlands Management to the Kenyan Coast in David Langlet and Rosemary Rayfuse (eds) *The Ecosystem Approach in Ocean Planning and Governance: Perspectives from Europe and Beyond* (2019).

Ademola Oluborode Jegede is Professor of Law in the Department of Public Law, University of Venda, Thohoyandou, South Africa. He holds degrees from Obafemi Awolowo University Ile-Ife, University of Ibadan and the Centre for Human Rights, Faculty of Law, University of Pretoria. He has been a research visitor to the Centre for International Environmental Law, USA; Human Rights Institute at Abo Akademi, Finland; and a fellow of Salzburg Global Seminar, Austria. His research focuses on the interface of climate change with human rights of vulnerable groups and general international human rights law.

Thabang Ramakhula is an LL.D. candidate at the Free State Centre for Human Rights, of the University of the Free State, South Africa. Her research interests center around women's political participation. She is also a women's rights and good governance activist and development worker.

Omphemetse S. Sibanda Snr is Professor of Law in the Department of Public and Environmental Law at the School of Law, Faculty of Management and Law, University of Limpopo, South Africa. He holds a Doctor of Laws in International Economic Law from the North-West University. Other qualifications include Master of Laws from Georgetown University Law Centre, USA; LL.B. (Hons.) and B.JURIS from the then Vista University, Soweto Campus. He has numerous publications in national and international academic journals. His research focus is on trade and consumer protections, trade-related aspects of human rights and the environment and appropriate dispute settlement approaches.

Brewsters Caiphas Soyapi is a senior lecturer at the North-West University, South Africa. His areas of interest are international environmental law, environmental rights and aspects of African human rights systems.

Oluwabusayo Temitope Wuraola is a PhD researcher at the School of Law, University of Leicester, UK. Her research focus is on "The Implications of the Grant of Legal Personhood to Natural Entities". Wuraola is an expert member of the United Nations Harmony with Nature Program and also a lecturer in the Department of Private and Property Law, Faculty of Law, University of Ilorin, Nigeria. Wuraola holds a Bachelor of Law degree (2010) and a Master of Law degree (2015) from the University of Ilorin, Nigeria. She was admitted as a Barrister and Solicitor of the Supreme Court of Nigeria in 2012.

Legal Materials

International Treaties

UNECE Convention On Access to Information, Public Participation in Decision-Making and Access to Justice in Environmental Matters, usually known as the Aarhus Convention, (adopted on 25 June 1998, entered into force 30 October 2001) 38 ILM 517.

African Charter on Human and Peoples' Rights (adopted 27 June 1981, entered into force 21 October 1986) (1982) 21 ILM 58.

African Charter on the Rights and Welfare of the Child (adopted 11 July 1990, entered into force 29 November 1999) OAU Doc CAB/LEG/24.9/49 (1990).

African Convention on the Conservation of Nature and Natural Resources, (adopted 11 July 2003, entered into force 23 July 2016) 7782 AU Treaties 0029.

African Convention on the Conservation of Nature and Natural Resources, September 25, 1968, in force June 16, 1969, OAU Doc. CAB/LEG/24.1.

Bamako Convention on the Ban of Import into Africa and the Control of Transboundary Movement and Management of Hazardous Waste within Africa, January 30, 1991, in force April 22, 1998.

Charter of the United Nations (24 October 1945) 1 UNTS XVI.

Constitutive Act of the African Union, July 11, 2000, in force May 26, 2001, OAU Doc. CAB/ LEG/23.15

Convention on the Elimination of All Forms of Discrimination against Women (adopted 18 December 1979, entered into force 3 September 1981) 1249 UNTS 13.

Convention on the Preservation of Wild Animals, Birds, and Fish in Africa, May 19, 1900, not in force.

Convention Relative to the Preservation of Fauna and Flora in Their Natural State, adopted on November 8, 1933, in force January 14, 1936.

International Covenant on Civil and Political Rights (adopted 16 December 1966, entered into force 23 March 1976) 999 UNTS 171.

International Covenant on Economic, Social and Cultural Rights (adopted 16 December 1966, entered into force 3 January 1976) 993 UNTS 3.

Kyoto Protocol to the United Nations Framework Convention on Climate Change, (adopted 11 December 1997, entered into force 16 February 2005) 2303 UNTS 148.

Protocol to the ACHPR on the Rights of Women in Africa (adopted by the 2nd Ord Sess of the Assembly of the African Union, Maputo on 13 September 2000, entered into force 25 November 2005) AU Doc CAB/LEG/66.6.

Protocol to the African Charter on Human and Peoples' Rights on the Establishment of an African Court on Human and Peoples' Rights, July 10, 1998, in force January 25, 2004, OAU Doc. OAU/LEG/EXP/AFCHPR/PROT (III).

United Nations Framework Convention on Climate Change (adopted 9 May 1992, entered into force 19 June 1993) 1771 UNTS 107.

United Nations Convention to Combat Desertification in those Countries experiencing Serious Drought and/or Desertification, particularly in Africa, October 14, 1994, 1994 UNTS 3.

UNFCCC, Paris Agreement (adopted 12 December 2015, entered into force 4 November 2016) UN Doc FCCC/CP/2015/L.9/Rev.1.

United Nations Convention on Biodiversity, June 5, 1992, 2760 UNTS 89.

Universal Declaration of Human Rights, General Assembly Resolution 217 A(III) (adopted 10 December 1948) UN Doc A/810 at 71 (1948).

UN/AU Resolutions and Other Official Documents

ACHPR, Guidelines and Principles on Articles 21 and 24 of the African Charter on Human and Peoples Rights related to the Extractive Industries, Environment and Human Rights.

African Commission on Human and Peoples' Rights, 'Climate Change and Human Rights in Africa', ACHPR/Res342 (LVIII), adopted at the 58th Ordinary Session (20 April 2016).

African Commission on Human and Peoples' Rights, 'Climate Change and the Need to Study Its Impacts on Africa', ACHPR/ Res153 (XLVI)09, adopted at the 46th Ordinary Session (25 November 2009).

African Commission on Human and Peoples' Rights, 'Climate Change in Africa', ACHPR/Res271, adopted at the 55th Ordinary Session (12 May 2014).

CEDAW, Statement of the CEDAW Committee on Gender and Climate Change, adopted at 44th Sess held in New York, USA, from 20 July–7 August 2009.

CEDAW Committee, Concluding Observations: Uganda A/57/38 (23 August 2002).

CEDAW Committee, Concluding Observations: Mali CEDAW /C/ MLI/CO/5 (3 February 2006).

CEDAW *General Recommendation No. 23: Political and Public Life*, 1997, A/52/38.

CEDAW Committee, General recommendation No. 27 on older women and protection of their human rights, CEDAW/C/GC/27, 16 December 2010.

CEDAW Committee, *General recommendation No. 34 on the rights of rural women*. CEDAW/C/GC/34, General. 4 March 2016.

CESCR Concluding Observations: Zambia E/C12/1/Add106 (23 June 2005).

CESCR Committee, *General Comment No. 12: The Right to Adequate Food (Art. 11 of the Covenant)*, 12 May 1999.

CESCR Committee, *General Comment No. 15: The Right to Water* (Arts. 11 and 12 of the Covenant) 20 January 2003.

CESCR Committee, *General Comment No. 14: The Right to the Highest Attainable Standard of Health* (Art. 12 of the Covenant), 11 August 2000, E/C.12/2000/4.

Gro Harlem Brundtland, "Chairman's Foreword" Our Common Future (1978).

Human Rights Council Resolution 7/23, Human Rights and Climate Change (UN Doc A/HRC/7/78, 14 July 2008).

Knox, JH, Report of the Special Rapporteur on the Issue of Human Rights Obligations Relating to the Enjoyment of a Safe, Clean, Healthy and Sustainable Environment, UN Doc A/HRC/31/52 (1 February 2016).

Rio Declaration on Environment and Development, United Nations Conference on Environment and Development, UN Doc A/CONF.151/1 (3–14 June 1992).

Stockholm Declaration of the United Nations Conference on the Human Environment (Stockholm, 16 June 1972), printed in ILM (1972), 1416.

UNFCCC, COP 23 (Decision 3/CP.23 FCCC/CP/2017/11/Add.1 8 February 2018). Report of the Conference of the Parties on its twenty-third session, held in Bonn from 6 to 18 November 2017.

UNFCCC, 'Gender Composition Report by the Secretariat' Conference of the Parties Twenty-fourth session Katowice, 2–14 December 2018, FCCC/CP/2018/3 21 September 2018. <https://unfccc.int/sites/default/files/resource/03e.pdf> accessed 18 October, 2019.

UNFCCC, 'Intended Nationally Determined Contribution (INDC) for the Republic of Chad' (UNFCCC 2015).

UNFCCC, 'Intended Nationally Determined Contribution of the Gambia' (UNFCCC 2015).

UNFCCC, 'Kenya's Intended Nationally Determined Contribution' (UNFCCC 2015).

UNFCCC, 'Morocco Nationally Determined Contribution under the UNFCCC' (UNFCCC 2015).

UNFCCC, 'Republic of Malawi Intended Nationally Determined Contribution' (UNFCCC 2015).

UNFCCC, 'Republic of Tunisia Intended Nationally Determined Contribution' (UNFCC 2015).

UNFCCC, 'South Africa's Intended Nationally Determined Contribution '(UNFCCC 2015).
UNFCCC, 'Zimbabwe's Intended Nationally Determined Contribution' (UNFCCCN 2015).
UNFCCC, The State of Eritrea Nationally Determined Contributions Report to UNFCCC' (UNFCC 2015).
United Nations General Assembly resolution 45/94 on the need to ensure a healthy environment for the well-being of individuals.
UNHRC, 'Climate Change and Human Rights,' A/HRC/32/L.34 (30 June 2016);
UNHRC, 'Human Rights and Climate Change,' A/HRC/26/L.33 (23 June 2014).
UNHRC, 'Human Rights and Climate Change,' A/HRC/RES/18/22 (24 March 2011)
UNHRC, 'Human Rights and Climate Change,' Res. 10/4, 41st Meeting, A/HRC/RES10/4 (25 March 2009).
UN HRC, 'Human Rights and Climate Change,' Res. 7/23, UN Doc. A/HRC/7/78 (2 July 2009).

National Legislation and Other Official Reports

Constitution of the Republic of Cameroon, 1996 (Cameroon).
Constitution of the Republic of Eritrea of 1997.
Constitution of the Republic of Kenya, 2010 (Kenya).
Constitution of the Republic of Morocco of 2011.
Constitution of the Republic of Tunisia of 2014 (as translated by UNDP).
Constitution of the Republic of Uganda of 1995 (as amended in 2005 and 2015).
Environmental and Land Court Act of 2011 (Kenya).
Environmental and Strategic Impact Assessments Decree No. 2013/0171/PM of 2013 (Cameroon).
Environmental Management Law No 96/12/5 of 1996 (Cameroon).
Equal Opportunities Commission Act of 2007 (Uganda).
Fisheries and Wildlife Regulations Law No 94/01 of 1994 (Cameroon).

Forestry Act no. 8 of 2003 (Uganda).
Forestry Policy of 2001 (Uganda).
Management of National Lands Ordinance No 76/166 of 27 April 1976 (Cameroon).
Modern Biotechnology Law No 2003/006/21 of 2003 (Cameroon).
National Climate Change Bill No. 17 of 2017 (Uganda).
National Environment Act 153 of 2000 (Uganda) (repealed).
National Environment Act 5 of 2019 (Uganda).
Ratification of Treaties Act (Chapter 204 Laws of 2000) (Uganda).
Water Act no. 152 of 2000 (Uganda).

Cases

International Cases

African Commission on Human and Peoples' Rights v. Kenya, Application 006/2012, Judgment (African Court of Human and Peoples' Rights, 2017) (Ogiek case or ACHPR v. Kenya (Judgment)).

Centre for Minority Rights Development and Minority Rights Group (on behalf of the Endorois v Kenya AHRLR 75 (ACHPR 2009).

Democratic Republic of Congo v Burundi, Rwanda and Uganda, Communication 227/99 (2004).

Front for the Liberation of the State of Cabinda v Republic of Angola, ACHPR Communication 328/06 (2013).

Gabcikovo-Nagymaros Project (Hungary v Slovakia) ICJ Report 7 (1997).

Michelot Yogogombaye v *Senegal* Appl No 1/2008, ACtHPR.

Peter Ngoge v Republic of Kenya, Communication 524/15.

Socio-Economic Rights and Accountability Project v *Nigeria* (*SERAP case*) ECW/CCJ/APP/08/09.

Social and Economic Rights Action Center (SERAC) and Center for Economic and Social Rights (CESR) v. Nigeria, ACHPR Communication 155/96 (2001).

Sudan Human Rights Organisation v Sudan, Communication 297/05 (2009).

Welfare Council v *Kenya,* Communication 276/03 (2009).

National Cases

African Centre for Rights and Governance (ACRAG) &3 others v Municipal Council of Naivasha Petition No. 50 of 2012 [2017] EKLR.

Chandra Bhavan Boarding and v The State of Mysore and Another 1970 SCR (2) 600.

Earthlife Africa Johannesburg v. Minister of Environmental Affairs and others [2017] JOL 37526 (GP).

FEDEV v China Road and Bridge Corporation [2009] Unreported decision No CFIB/004M/09.

Festo Balegele v Dar es Salaam City Council, Misc. Civil Case No. 90 of 1991, High Court of Tanzania.

Gbemre v. *Shell Petroleum Development Company Nigeria Limited and Others* [2005]) AHRLR 151 (NgHC 2005).

Gouriet v the National Union of Post Office Workers (1977) 3 All ER.

Greenwatch v Attorney General and National Environment Management Authority Miscellaneous Application No. 371 of 2002, High Court of Uganda.

Joseph Leboo & 2 others v Director Kenya Forest Services & another Environment and Land No. 273 of 2013 eKLR.

Minerva Mills Ltd. & Others v Union of India & Others 1981 SCR (1) 206.

Moffat Kamau & 9 others v Aelous Kenya Limited & 9 others Constitutional Petition No. 13 of 2015 (2016) eKLR.

Mtikila v The Attorney General, Tanzanian Civil Suit No. 5 of 1993.

New Patriotic Party v Attorney-General (1996–97) Supreme Court of Ghana Law Reports

Sajjan Singh v State of Rajasthan 1965 SCR (1) 933.

Rodgers Muema Nzioka & 2 Others v Tiomin Kenya Limited (*Muema* case) HCCC 97 of 2001.

Uganda Electricity Transmission Co Ltd v De Samaline Incorporation Ltd, Miscellaneous Cause No. 181 of 2004, High Court of Uganda.

Waman Rao and Ors v Union of India and Others (1981) 2 SCC 362.

Wangari Maathai v Kenya Times Media Trust Ltd (1989) 1 KLR (E&L).

Part I

New Frontiers in Human Rights and Environment: Environmental Protection Under African Union Law

1

Human Rights, Regional Law and the Environment in Africa: Legal and Conceptual Foundations

Michael Addaney, Chantelle Gloria Moyo, and Thabang Ramakhula

1 Introduction

Scholarship on human rights and the environment has gained prominence over the last three decades as scholars, practitioners and advocates examine how these distinctive areas of law interact and reinforce each other. However, most existing scholarship focuses on the Global North, with limited attention given to the subject in the Global South, particularly Africa. This book attempts to offer original and novel perspectives to the discourse by focusing on the African region, especially regional

M. Addaney (✉)
Research Institute of Environmental Law, Wuhan University, Wuhan, China

University of Energy and Natural Resources, Sunyani, Ghana
e-mail: michael.addaney@uenr.edu.gh

C. G. Moyo
North-West University, Potchefstroom, South Africa

T. Ramakhula
University of the Free State, Bloemfontein, South Africa

M. Addaney, A. Oluborode Jegede (eds.), *Human Rights and the Environment under African Union Law*, https://doi.org/10.1007/978-3-030-46523-0_1

environmental law and the African human rights system. What makes this edited collection different from other works is both its analytical framework that focuses on African Union Law as well as its geographical scope. The concept of an African Union Law or the African Union legal order originates from the efforts of African States to integrate economically and politically beginning from pre-independence Pan-Africanism.[1] This is symbolized by the formation of the Organisation of African Unity (OAU) on 25 May 1966 in Addis Ababa, Ethiopia, with 32 signatory governments. The OAU was succeeded by the African Union (AU) in 2002.[2] The establishment of the AU officially marked the beginning of continental legal order as different from the OAU, created through a Constitutive Act which is a recognized social pact and legal code that currently fortifies the continental legal order.[3] This development implies that the African region currently has the prerequisite basis for a continent-wide legal order.

In the area of human rights and the environment, the AU Constitutive Act has consolidated and advanced the framework for promoting and protecting human rights in Africa by giving it constitutional status, thereby making it binding on all member states. Since its formation, the AU has made significant progress in meeting its ambitious goal of political integration. For instance, the AU General Assembly commissioned the Kagame report which endorsed a number of institutional and legal reforms including the improvement of the effectiveness of the AU Court system, clarification of the legislative powers of the Pan-African Parliament and the speedy implementation of the AU continental passport.[4] Similar to the legal and institutional design of the European Union (EU), the laws and institutions of the AU, a supranational organization is evolving progressively. This is not to say that there is no need for certain reforms.

---

[1] See, generally, Olufemi Amao, *African Union Law: The Emergence of a Sui Generis Legal Order,* 1st Edition (Routledge 2020).

[2] The African Union (AU) was officially launched in July 2002 in Durban, South Africa, following a decision in September 1999 by its predecessor, the OAU, to create a new continental organisation to build on its work. See, AU, 'About the African Union', https://au.int/en/overview (accessed on 18th January 2020).

[3] Amao, (n 1 above).

[4] See, 'AU Commission Chairperson in Kigali for consultative meeting on AU institutional reforms' 08 May 2017, https://au.int/en/pressreleases/20170508/au-commission-chairperson-kigali-consultative---meeting-au-institutional (accessed on 12th January 2020).

As it is generally recognized that the AU legal order is in its infancy, the protection of the environment has been an essential part of African social, cultural and religious life for many generations.[5] It is also an essential part of human rights protection in Africa. As observed by the African Commission on Human and Peoples' Rights (African Commission), 'collective rights, environmental rights, and economic and social rights are essential elements of human rights in Africa'.[6] It is of significance that at the African regional level, environmental rights are recognized as explicit treaty norms, with normative harmony with other rights and corresponding obligations. However, Africa continues to be characterized by, among other problems, resource exploitation, water pollution, loss of biodiversity and continuous degradation of the environment,[7] with the latter having negative effects on human rights.[8] The emergence of climate change, globalization and increased urbanization are key factors contributing to Africa's environmental challenges.[9] Although Africa depends on the extraction of natural resources to foster economic growth, achieve sustainable development and reduce poverty, extractive activities also contribute significantly to environmental damage on the continent. As observed by the African Commission, 'the increasing rate of the destruction of the African environment and ecosystem by extractive industrial activities with impunity' is an issue of concern.[10] Consequently, the African Commission has elevated attention to the impact of extractive activities on environmental protection and environmental rights with the

---

[5] Emeka Polycarp Amechi, 'Enhancing Environmental Protection and Socio-Economic Development in Africa: A Fresh Look at the Right to a General Satisfactory Environment under the African Charter on Human and Peoples' Rights,' Law, Environment and Development Journal 5(1) (2009), pp. 58–72, 62.

[6] Social and Economic Rights Action Centre and the Centre for Economic and Social Rights v. Nigeria, Communication No. 155/96 (2001) (SERAC case or SERAC v. Nigeria), para. 68 (emphasis added).

[7] Werner Scholtz, 'Human Rights and the Environment in the African Union Context,' in Anna Grear and Louis J. Kotzé (eds.), *Research Handbook on Human Rights and the Environment* (Edward Elgar Publishing, 2015), p. 401; Achim Steiner, An Introduction to the African Convention on the Conservation of Nature and Natural Resources (International Union for Conservation of Nature Environmental Law Centre, 2004), p. 1.

[8] Scholtz, "Human Rights and the Environment," p. 401.

[9] New Partnership for Africa's Development (NEPAD), Review of the Implementation of the Action Plan of the AU/NEPAD Environment (2012), p. 10.

[10] African Commission on Human and Peoples' Rights res. 321 (November 18, 2015) pmbl.

establishment of the Working Group on Extractive Industries, Environment and Human Rights Violations.[11] In both national and regional priorities, concerns relating to the environment rank highly,[12] with the AU having developed various policy responses aimed at addressing Africa's environmental concerns.[13]

2 Integrating Human Rights in Regional Environmental Governance in Africa

The connection between human rights and the environment emerges in several human rights approaches to environmental issues, which contain the extension and sometimes, re-interpretation of existing human rights, dependence on procedural rights and the conceptualization of the right to environment.[14] To successfully achieve environmental conservation and protection, there must be a trend of innovative and refined legal analysis of human rights. Questions on whether human rights are an appropriate tool in addressing contemporary environmental concerns are pertinent for a number of reasons. Firstly, current developments in treaty law tend to recognize the entitlements of individuals, communities and groups to be a part of environmental decision-making processes as well as access to justice when these rights have been violated in terms of human rights.[15] This phenomenon is popularly known as the procedualization of human rights because of its individual and social empowerment to participate in a deliberate process of decision-making and to have rights of

[11] African Commission on Human and Peoples' Rights res. 148 (November 25, 2009).

[12] See Steiner, Introduction to the African Convention, p. 1.

[13] Constitutive Act of the African Union, July 11, 2000, in force May 26, 2001, OAU Doc. CAB/LEG/23.15, art 13(1). As at 22 January 2020, all 55 African states have ratified/acceded to the Constitutive Act (see ratification/accession status list (updated June 15, 2017) available at www.au.int/web/sites/default/files/treaties/7758-sl-constitutive_act_of_the_african_union_2.pdf).

[14] Malgosia Fitzmaurice, David M. Ong and Panos Merkouris (eds) *Human Rights and the Environment* (Martinus/Nijhof Publishers 2011) 71.

[15] Francesco Francioni 'International Human Rights in an Environmental Horizon' (2010) 1 *European Journal of International Law* 42.

recourse against environmental harm.[16] Secondly, the reason behind revisiting the characteristics of environmental rights is substantive. Current practice prominently features socio-cultural contexts which show that it is a sine qua non for the full enjoyment of human rights by individuals, communities or particular groups of people.[17] A good example of this trend can be found in the UN Declaration on the Rights of Indigenous Peoples.[18] Furthermore, the African Charter on Human and Peoples' Rights proclaims, in Article 24, the right of African peoples to 'a general satisfactory environment favourable to their development'.[19]

The Intersection of Human Rights and the Environment in Africa

Human rights theorists aver that since the reason for environmental conservation and protection is the enhancement of quality of human life, it falls squarely within the bounds of human rights.[20] On the other hand, environmental lawyers argue that a human-centered approach to environmental concerns runs the risk of reducing environmental values to mere roles that they play in improving the quality of life of human beings to the exclusion of everything else.[21] According to this view, the protection of the environment would be inspired by what human beings can benefit out of it thereby sacrificing other aspects such as ecological growth, restoration and function. A critical look at the nature and content of environmental rights makes it clear that these are *sui generis* and do not fit perfectly into any category under the traditional formulation of

---

[16] Ibid. See, also, Catherine Redgwell 'Access to Environmental Justice', in Francesco Francioni (ed) *Access to Justice as a Human Right* (2007) 153; James Cameron and Ruth Mackenzie 'Access to Environmental Justice and Procedural Rights in International Institutions' in Alan Boyle and Michael Anderson (eds) *Human Rights Approaches to Environmental Protection* (1996) 129.

[17] See, Frencioni (n 15 above) 43.

[18] UN Declaration on the Rights of Indigenous Peoples Adopted by GA Res 61/295 of 13 Sept. 2007. See Article 29 of the Declaration of the Rights of Indigenous Peoples.

[19] African Charter on Human and Peoples' Rights, adopted 27 June 1981, entered into force 21 October 1986.

[20] Takele Soboka Bulto, 'The Environment and Human Rights' in Anja Mihr (ed) *The Sage Handbook of Human Rights* (SAGE Publications Ltd. 2014) 1016.

[21] Ibid.

human rights. Environmental rights overlap all three generations of rights and so represent, by their nature, cross-cutting rights. It has been noted that environmental rights possibly serve the nature of all categories of rights.[22] Apart from the basic formulation of environmental rights falling in the scope of socio-economic rights; in the sense of traditional civil and political rights, these rights can be used by individuals and groups to access environmental information, judicial remedies and political processes.

Some scholars[23] observe that while there may be many approaches that human rights standards feature in environmental rights, there are three ways that these manifest. They are noticeable in: a qualitative abstract sense that points out the aspirations in relation to the interest protected; quantitative concrete sense that is specific in the quantity or computable value in relation to the interest protected; and a directive sense that creates a specific minimum obligation required for performance.[24] In their abstract form, human rights are essential to the ideal of constitutionalism and the rule of law. They fulfil a crucial role in the provision of an abstract minimum core of protection that circles on formulations of natural law, justice, democracy and legitimacy.[25] An example of this qualitative formation in international law is the human rights standard that all human beings are born free and equal in dignity and rights. In the concrete, quantitative sense rights clearly provide quantifiable standards within the right itself and an example of this would be minimum standards of types of harm that qualify as violations of a right.[26]

Other scholars[27] have -formulated three theories to show the connection between human rights and environmental rights. The three theories are the expansion theory, the environmental democracy theory and the

[22] Ibid.

[23] Louis J Kotze and Anel du Plessis 'The African Charter on Human and Peoples' Rights and Environmental Rights Standards' in Stephen J Turner (ed) *Environmental Rights the Development of Standards* (Cambridge University Press 2019) 95.

[24] Ibid.

[25] Ibid.

[26] Ibid, 95.

[27] See, Fitzmaurice (n 14 above) 71.

genesis theory.[28] The expansion theory confers an environmental interpretation on well-established human rights, for example, the right to life, the right to health and the right to privacy and this is popularly known as the greening of existing human rights.[29] In some instances these rights are usually referred to as derivative rights[30] which terminology implies that they have or could potentially be invoked in an environmental context.[31] As the term reads, the environmental democracy theory carries democratic governance into the ecological sustainability realm.[32] In light of this theory, environmental procedural rights like the right to participation, access to justice and remedies for redress in instances of violation become essential so that citizens are capacitated to participate in, influence and challenge environmental law and policy. Procedural environmental rights are an important part of international human rights law and are already entrenched in the Universal Declaration of Human Rights (UDHR).[33] Moreover, due to the interdependence between human rights and environmental rights noted in the United Nations Conference on Environment and Development (UNCED), there has been development and a wide acceptance of environmental procedural rights.[34] According to the genesis theory, the right to development[35] is requisite for the full enjoyment of basic human rights, although this approach may lead to an overemphasis on human rights for the sole purpose of environmental protection.[36] However scholars have rationalized theories on the interrelationship between human rights and environmental rights, it is pertinent to note that the relationship does exist. It might exist through the reading in of

---

[28] Ibid.

[29] See generally, Alan E. Boyle and Patricia Birnie, *International Law and the Environment* (Oxford University Press, 2002 2nd ed.).

[30] Robin Churchill, 'Environmental Rights in Existing Human Rights Treaties' in Alan E. Boyle and Michael R. Anderson (eds) *Human Rights Approaches to Environmental Protection* (Clarendon Press 1996) 90.

[31] Ibid.

[32] See, Michael Mason *Environmental Democracy: A Contextual Approach* (Earthscan 1999).

[33] See, UDHR, Articles 8, 19, 21, and 26.

[34] Report of the Joint OHCHR-UNEP Seminar on Human Rights and the Environment 16 Jan. 2002, UN Doc E/CN.4/2002/WP.7 (2002), Annex II (15).

[35] See, Roland Rich 'The Right to Development as an Emerging Human Right' (1982–1983) 23 *Virginia Journal of International Law* 320.

[36] Ibid.

environmental rights into existing rights or finding human rights standards in environmental rights. As observed in this discussion, international human rights discourse recognizes and seeks to promote environmental conservation and protection.

The Emergence of Human Rights in the Field of Regional Environmental Governance

The evolution of environmental protection norms and standards at the African regional level dates back to the colonial era when the colonial powers introduced conventions that focused on the conservation of nature and natural resources, which was driven essentially by utilitarian objectives. The colonial powers developed the Convention on the Preservation of Wild Animals, Birds, and Fish in Africa (the 1900 Convention) to prevent 'indiscriminate slaughter' and to safeguard 'the preservation throughout their possessions in Africa of the various forms of animal life existing in a wild state which are either useful to man or are harmless'.[37] They also adopted the Convention Relative to the Preservation of Fauna and Flora in their Natural State (the 1933 Convention) to preserve fauna and flora by limiting or prohibiting their collection or destruction, as they were in danger 'of extinction or permanent injury'.[38] After the end of colonialism, the OAU adopted the African Convention on the Conservation of Nature and Natural Resources (the African Convention) in 1968 to protect the natural environment by considering the specific interests of Africans.[39] The African Convention addressed a range of issues relating to the protection of the environment, and has been pronounced as a 'modern approach to the conservation of nature' as it is not limited to utilitarian objectives.[40] It introduces ground-breaking

[37] Convention on the Preservation of Wild Animals, Birds, and Fish in Africa, 19 May 1900, (not in force, pmbl.).

[38] Convention Relative to the Preservation of Fauna and Flora in Their Natural State, adopted on 8 November 1933, entered into force 14 January 1936, pmbl.

[39] African Convention on the Conservation of Nature and Natural Resources, 25 September 1968, entered into force 16 June 1969, OAU Doc. CAB/LEG/24.1.

[40] IUCN Environmental Law Programme, An Introduction to the African Convention on the Conservation of Nature and Natural Resources, IUCN Environmental Policy and Law Paper No.

approaches to nature conservation such as making environmental management as a common duty of African states (in addition to individual state action). The African Convention was, nevertheless, inadequate as it did not establish institutional structures for its effective application.[41] It was also failing in recognizing environmental protection as a human right.

The African Convention was therefore revised in 2003 due to developments in international and regional environmental law and policy. The Revised African Convention, unlike its predecessor, emphasizes the right to a satisfactory environment in the African Charter as a key principle.[42] The Revised African Convention also seeks to enhance the protection of the environment through the 'conservation and sustainable use of natural resources', as well as the harmonization and coordination of environmental protection policies 'with a view to achieving ecologically rational, economically sound and socially acceptable development policies and programmes'.[43] States Parties to the Revised African Convention are to be guided by three key principles when undertaking actions in relation to their obligations: (1) 'the right of all peoples to a satisfactory environment favourable to their development'; (2) 'the duty of States, individually and collectively to ensure the enjoyment of the right to development'; and (3) 'the duty of States to ensure that developmental and environmental needs are met in a sustainable, fair and equitable manner'.[44] The Convention further obligates States to 'adopt and implement all measures necessary' for environmental protection including taking 'preventive measures and the application of the precautionary principle, and with due regard to ethical and traditional values as well as scientific knowledge in the interest of present and future generations'.[45] The

56 (2004), p. 4.

[41] Philippe Sands and Jacqueline Peel, Principles of International Environmental Law, 3d ed. (Cambridge University Press, 2012), p. 25; Steiner, Introduction to the African Commission, p. 4.

[42] See, African Convention on the Conservation of Nature and Natural Resources (2003, Revised Version).

[43] Ibid., art. 2.

[44] Ibid., art. 3. Article 16 of the Convention requires the parties to adopt measures necessary to ensure dissemination of, and public access to, environmental information; public participation in decision-making with a potentially significant impact on the environment, and access to justice in environmental matters.

[45] Revised African Convention, art. 4.

Revised African Convention also reaffirms the responsibility of States Parties 'to ensure that activities within their jurisdiction or control do not cause damage to the environment of other states or of areas beyond the limits of national jurisdiction'.[46]

Furthermore, the need to find durable solutions to the dumping of hazardous and toxic waste in Africa led to the OAU adopting the Bamako Convention on the Ban of Import into Africa and the Control of Transboundary Movement and Management of Hazardous Waste Within Africa (Bamako Convention) in 30 January 1991.[47] The Bamako Convention recognizes that hazardous wastes and their transboundary movement are a menace to the African environment, and that 'the reduction of their generation to a minimum in terms of quantity and/or hazard potential' is an effective approach for protecting the environment.[48] The Convention has been perceived as 'raising the bar in international environmental regulation of transboundary movement of hazardous wastes'.[49] Even though the Bamako Convention does not explicitly recognize environmental protection as a right, it recalls the African Charter, which as observed earlier provides for the right to general satisfactory environment.[50]

The adoption of the African Charter on Human and Peoples' Rights (African Charter)[51] thus recognized the interconnections between human rights and the environment making it the first international treaty to recognize the right of peoples to 'a general satisfactory environment favourable to their development'.[52] The provision on the right to environment reflects the recognition that a satisfactory environment is

---

[46] Ibid., pmbl.

[47] Bamako Convention on the Ban of Import into Africa and the Control of Transboundary Movement and Management of Hazardous Waste Within Africa, 30 January 1991, entered into force 22 April 1998.

[48] Ibid., pmbl.

[49] Adebola Ogunlade, 'Can the Bamako Convention Adequately Safeguard Africa's Environment in the Context of Transboundary Movement of Hazardous Wastes?' University of Dundee (2010), p. 19.

[50] Bamako Convention, pmbl.

[51] African Charter on Human and Peoples' Rights, June 27, 1981, in force October 21, 1986, 1520 UNTS 217.

[52] African Charter, art. 24.

significant for economic, social and cultural development as well as the realization of other human rights in Africa.[53] The 1990 African Charter on the Rights and Welfare of the Child (African Children's Charter) does not have any provision on the right to a satisfactory or healthy environment. It however requires 'the development of respect for the environment and natural resources' through education.[54] The 2003 Protocol to the African Charter on Human and Peoples' Rights on the Rights of Women in Africa (the Maputo Protocol) also recognizes the right of women to 'a healthy and sustainable environment'.[55] The African Charter establishes the African Commission to promote, protect and interpret the rights in the Charter.[56] This enforcement mechanism was later reinforced with the creation of the African Court, which has jurisdiction over cases and disputes relating to the interpretation and application of the Charter and other relevant human rights treaties.[57] The provision in Article 24 of the African Charter that—'all peoples shall have the right to a general satisfactory environment favorable to their development'[58]—largely mirrors that of many contemporary regional human rights instruments around the world. Article 24 of the African Charter is further supplemented, in scope, by articles 13(3) and 16(1), which provide for access to public services (water, waste and sanitation services) and the right of every individual to 'enjoy the best attainable state of physical and mental health' respectively.[59] This right functions alongside the provision in the Constitutive Act of the AU, which lists in

[53] Ibid., p. 62.

[54] African Charter on the Rights and Welfare of the Child, 1 July 1990, entered into force 29 November 1999, OAU Doc. CAB/LEG/24.9/49 (African Children's Charter), art. 11(2)(g).

[55] Protocol to the African Charter on Human and Peoples' Rights on the Rights of Women in Africa, 13 September 2000, entered force 25 November 2005, AU Doc. CAB/LEG/66.6, art. 18.

[56] Ibid., art. 45.

[57] Protocol to the African Charter on Human and Peoples' Rights on the Establishment of an African Court on Human and Peoples' Rights, 10 July 1998, entered into force 25 January 2004, OAU Doc. OAU/LEG/EXP/AFCHPR/PROT (III) (African Court Protocol), art. 3. As at 22 January 2020, the African Court Protocol had been ratified by 30 African states.

[58] Article 24 of the ACHPR. See, also, Werner Scholtz 'Human Rights and the Environment in the African Union Context' in Werner Scholtz and Jonathan Verschuuren (eds) *Regional Environmental Law: Transregional Comparative Lessons in Pursuit of Sustainable Development* (2015) 102.

[59] See, generally, Ebenezer Durojaye 'The Approaches of the African Commission to the Right to Health under the African Charter' (2013) 7 Law, Democracy and Development 393.

its objectives that the AU seeks to 'promote sustainable development at the economic, social and cultural levels as well as the integration of African economies'[60] and to 'promote and protect human and people's rights in accordance with the African Charter on Human and People's Rights and other relevant human rights instruments'.[61]

The connection between human rights and environmental protection is also recognized by the African Commission in its 2009 Resolution 153.[62] The overarching environmental framework convention for AU member states, the 2003 Revised African Convention, also recognizes the significance of a rights-based approach in the achievement of the Convention's objectives.[63] Section 24 infers that development is an essential part of the environmental right to the extent that 'favourable development' is dependent on a 'general satisfactory environment'. Not only does this mean that the premise of the environmental right is anthropocentric but also that standards set out by the environmental right have to ensure the 'favourable development' of people, including to ensure their right to development under article 22. Therefore, the wording of the right to development recommends that the standard it sets is cast in abstract qualitative terms.[64]

In the *SERAC v Nigeria*, a Communication was brought to the African Commission as an *actio popularis* by the Social and Economic Rights Action Center (SERAC) and the Centre for Economic and Social Rights (CESR).[65] According to that Communication, the government of Nigeria had been involved in oil production through the State-owned Nigerian National Petroleum Company and the operations caused environmental contamination which led to health problems among the Ogoni people, an indigenous community in Nigeria.[66] Furthermore, the Communication

[60] Article 3(j) of the Constitutive Act of the African Union of 2000.

[61] Article 3(h) of the Constitutive Act of the African Union of 2000.

[62] The African Commission on Human Rights and Peoples' Rights (African Commission), meeting at its Forty-sixth Ordinary Session held from 11–25 November in Banjul, The Gambia.

[63] Article III of the Convention which explicitly refers to article 24 of the ACHPR.

[64] See Kotze' and Plessis (n 23 above) 101.

[65] Social and Economic Rights Action Centre (SERAC) and Another v Nigeria (2001) AHRLR 60.

[66] ACHPR 'Communication 155/96: Social and Economic Rights Action Center (SERAC) and Center for Economic and Social Rights (CESR) /Nigeria' para 10 available online at www.achpr.org/files/sessions/30th/comunications/155.96/achpr30_155_96_eng.pdf (accessed 25 November 2019).

alleged the violation of the right to health,[67] the right to free disposal of wealth and resources[68] and the environmental right[69] of the African Charter. The African Commission found that the Nigerian government had failed to uphold the duties it derives from the African Charter, including those emanating from environmental right. Another case which did not directly address article 24 of the African Charter but provided insights on the notion of environmental standards in the context of African human rights law was that of *African Commission on Human and Peoples' Rights v. the Republic of Kenya*.[70] The case concerned a notice of eviction, issued by the Kenyan Government in 2009, to the Ogiek indigenous minority community of the Mau Forest. Even though it was based on the need to protect the Mau Forest, a valuable water catchment area; the Ogiek argued that their eviction would have extensive negative consequences for the political, social and economic survival of the community.[71] Although the case involved a 2009 matter, the case was eventually decided in 2017 and the court held that the government of Kenya had violated articles 1, 2, 8, 14, 17(2) and (3), 21 and 22 of the African Charter. The government was ordered to 'take all appropriate measures within a reasonable time frame to remedy all the violations established and to inform the Court of the measures taken within six (6) months from the date of (the) judgment'.[72]

Although the violation of the article 24, environmental right was not argued, it is relevant in the context of the African Charter to point out that abstract qualitative environmental standards may arise from the collective protection of rights other than the environmental right, such as the rights of cultural minorities and the right of access to ancestral land.[73] In the *Centre for Minority Rights Development, Minority Rights Group*

[67] ACHPR, Article 16.

[68] Ibid, at Article 21.

[69] Ibid, at Article 24.

[70] Application 006/2012: African Commission on Human and Peoples' Rights v the Republic of Kenya available online at www.african-court.org/en/index.php/56-pending-cases-details/864-app--no-006–2012-african-commissionon-human-and-peoples-rights-v-republic-of-kenya-details (Accessed 26 November 2019).

[71] Ibid, para 4.

[72] Ibid.

[73] See, Kotze and Plessis (n 23 above) 106.

International and Endorois Welfare Council[74] case, the link between the natural environment and the right to development of indigenous peoples came under consideration. It claimed violation of the Endorois' rights to non-discrimination, property, access to natural resources on their land, religious and cultural life because they were excluded from consultation in the forced displacement of their community. The African Commission declared the expulsion of the community from their ancestral land illegal. Furthermore, the Commission found that the Kenyan government failed to recognize and protect the Endorois' ancestral land rights and to provide sufficient compensation.[75] These cases are a clear illustration of the fact that regionally, human rights approaches have been used in the interpretation and application of environmental rights.

3 Key Themes in the Connection Between Human Rights and the Environment: African Perspectives

African scholarship has largely been silent in this area of research. In this volume, 16 essays focus on the historical, contemporary and future aspects of human rights law and the environment in Africa. The editors and contributors are Africans based in Africa. They are situated in four out of the five main geopolitical blocks in Africa (West, Eastern, Central and Southern Africa) and include reviews of national case studies from Cameroun, Ethiopia, Kenya, Nigeria, South Africa, Tanzania, Uganda and Zimbabwe. Regional contributions are also discussed, making this edited volume a comprehensive collection of African voices on the topic. It blends the use of case studies and doctrinal legal research approaches that espouse the transformations in and of human rights law and environmental governance in Africa and its future orientation from the perspectives of indigenous African scholars, experts and practitioners.

Part I, which is devoted to the new frontiers in human rights and environmental protection in the AU, starts with Michael Addaney, Chantelle

[74] Application 276/03: *Centre for Minority Rights Development (Kenya) and Minority Rights Group.*
[75] Ibid.

Gloria Moyo and Thabang Ramakhula's forceful call for the recognition of the African Union law as a distinct legal order and thus lays legal and conceptual foundations for the interactions between human rights and the environment in Africa. The recognition and guaranteeing of an autonomous human right to general satisfactory environment in the African Charter, they argue, reflects the appreciation that a satisfactory environment is significant for economic, social and cultural development as well as the fulfilment of other human rights in Africa. But for the African human rights regime to address the violation of environmental rights in a meaningful way, Mulesa Lumina examines the scope and content of this multifaceted right and concomitant duties of African states through an examination of regional human rights instruments, treaty-body decisions, domestic laws, commentary and scholarly literature on the subject. She observes that for a considerable length of time, the exact content of Article 24 of the African Charter lacked definitive interpretation but regional case law, scholarly literature on human rights and the environment, soft law instruments of the ACHPR, especially the state reporting guidelines on article 24 and domestic law provide valuable guidance on the right to a general satisfactory environment favorable to development. Ademola Oluborode Jegede then extends the discussion of human rights to a safe, clean and healthy environment to a safe climate. Examining current developments in the scholarship on human rights and climate governance, he discerns that the broad recognition that climate change undermines the enjoyment of human rights has however not crystalized into the formulation of a right to a safe climate. In response, Jegede proposes that a human right to a safe climate should be recognized under AU human rights law. He argues that this right satisfies the six criteria required for a human right to be recognized, such as the right reflecting a fundamental social value; not repetitive of existing body of international human rights law; capable of attaining high degree of international consensus; compatible with state practice; and be precise to give rise to identifiable right and obligation.

Funmi Abioye next underscores that the protection of the environment is key to sustainable development and consequently the right to environment is guaranteed in several human rights instruments at the regional and sub-regional levels. She argues that this enhances the rule of law as the African Commission guarantees that deliberate value and emphasis should be placed on environmental protection within the African human rights system. Abioye therefore broadly examines the manner in which the African regional human rights system has sought to reinforce the international protection to a healthy environment and the environmental rule of law through the jurisprudence that have emanated from these bodies. By zeroing in on the interactions between women's human rights at the continental level, Elsabé Boshoff provides an exposition of the international framework for the protection of women's environmental rights in Africa. She focusses on the contribution of the African Commission and the far-reaching provisions of the Maputo Protocol in the development of this framework. She argues that States where the Maputo Protocol has been adopted have seen improvements, particularly in women's access to land, representation in decision-making bodies and in having their concerns reflected in development policies. Boshoff concludes that States should be encouraged to submit reports to the African Commission in compliance with the Maputo Protocol in order to promote local awareness of the Protocol to ensure its full enforcement. To conclude the discussion on the regional law, Oluwabusayo Temitope Wuraola takes on the emerging recognition of the personhood and the rights of natural entities in Africa. She observes that with the growth of eco-centric approach to protecting non-human natural entities and the increase in the number of countries granting legal personhood to non-human natural entities in different parts of the world, it is worth questioning whether this approach can be extended to Africa. The chapter juxtaposes the existing approach of granting legal personhood to non-human natural entities and argues that this will aid in determining how viable the concept of granting legal personhood to non-human natural entities will be in Africa due to the continent's inclination to protecting non-human natural entities from anthropocentric to eco-centric.

Part II, which is devoted to institutional and human rights developments in international and regional environmental governance in Africa, considers the role of judicial and non-judicial bodies in the creation and enforcement of environmental human rights in Africa. Emma Charlene Lubaale investigates the role of the judiciary in enforcing environmental rights, and ultimately environmental conservation in Africa. Using the experience of Uganda, Kenya and Tanzania, she argues the manner in which the right to environment is couched as well as the nature of a country's rules on access to court, play a critical role in ensuring that courts effectively execute their mandate in safeguarding environmental human rights in Africa. Michael Addaney and Chantelle Gloria Moyo look at advances in environmental protection and the experiences of African regional human rights courts and tribunals. After examining relevant cases at the regional and sub-regional levels, they argue that the approaches used by African human rights courts and tribunals can be a basis for development of unique African approaches to environmental protection and the realization of related human rights.

Eghosa Osa Ekhator highlights the utilization of sub-regional judiciaries in accessing justice in environmental issues by NGOs, activists, communities and individuals due to the non-justiciability of the right to environment in many African countries. Using the Economic Community of West African States (ECOWAS) Community Court of Justice (ECCJ) and the East African Court of Justice (EACJ) as case studies, he responds to the question of whether the rise of environmental governance or litigation in sub-regional judiciaries leads to better environmental protection for the victims and communities. He underscores that there is a bourgeoning environmental governance architecture under the sub-regional judiciaries in Africa. Focusing on the ECCJ and EACJ as well as considering the basis of the decisions by the African Commission (*SERAC* case), the African Court of Human and Peoples Rights (the *Ogiek* case), ECCJ (*SERAP*) and EACJ (*ANAW*), he observes that there is definitely a case for the regional (and especially sub-regional) judiciaries adjudicating on environmental issues in Africa. Ademola Oluborode Jegede shines the spotlight on the trends and significance of constitutionalizing climate protection in Africa. He notes that while most states have constitutional systems that protect the environment or recognize environmental rights, thus far, this development has not been effective in averting or stabilizing a changing climate. He therefore establishes the basis for and evidence on the incorporation of a climate change clause in constitutions and its significance for states in Africa.

Part III explores regional and national case studies on environmental protection and human rights in Africa. Desalegn Amsalu begins the discussion by analyzing the coverage of environmental rights in the Ethiopian environmental legal regime. He reveals that environmental human rights are generously recognized in the 1995 Constitution of Ethiopia and the 1997 Environmental Policy. He however observes that the realization of the rights suffers from lack of adequate framework laws and standards as well as a lack of coordination, inadequate human resource and poor technological infrastructure in the environmental protection sector. Charlotte Kabaseke extends Boshoff's discussion in the national context, drawing lessons about women's participation in environmental decision-making from examples in Uganda. She analyses the adequacy of environmental and human rights framework in ensuring the realization of women's right to participation in environmental decision-making in Uganda. She argues that in spite of the recognition and protection of women's rights at the international and regional level in Africa, women's right to participate in environmental decision-making processes is inadequately safeguarded in Uganda. Kabaseke proposes that the NEA, a recently enacted law, should be reformed in order to capture the current trends on the role of women in environment protection, especially their participation in environmental decision-making. Jean-Claude N. Ashukem moves the discussion from Uganda to Cameroon, examining the successes, limitations and prospects of using human rights norms, principles and standards in ensuring environmental protection in Cameroon. Based on the assessment, he argues that Cameroon's legal framework is characterized by normative, procedural and institutional limitations and thus, makes the framework inadequate and ineffective. Consequently, Ashukem charges the government to address these shortcomings, if truly it is committed to protecting the environment and other rights-based entitlements of the people.

In the context of constitutional environmental rights, Brewsters Caiphas Soyapi draws attention to the little consideration given to scholarship and commentary on the right to a healthy environment in Zimbabwe and on the cases in which the right has been an issue (both before and after the 2013 Constitution). Accordingly, he offers some insight into how, within the context of Zimbabwe's governance setting, the courts have adjudicated the right to a healthy environment and to review what hindrances they have faced in doing so. Collins Odote continues the discussions by exploring the extent to which a human rights lens has been adopted in protecting the environment in selected jurisdictions in Africa. Using Kenya, Nigeria and South Africa as case studies, Odote argues that applying the human rights lens will help provide an additional layer for environmental conservation as long as it is not applied as the sole or primary method for environmental management, due to the limits and dangers of a purely anthropocentric approach.

Part IV looks at business practices, human rights and the environment in Africa. Here, Lenin Tinashe Chisaira begins by legally appraising renewable energy projects such as dam construction and biofuel production for energy provisions in Southern Africa. He highlights a crooked balancing act between the neoliberal desire for maximum power generation for industry and a deliberate lack of pro-human and environmental rights policy and legal framework for rural communities. Chisaira argues that neoliberal governance systems harm human and environmental rights and thus, States in Southern Africa should adopt measures in order to cushion their populace from unchecked neoliberalism. Omphemetse S. Sibanda Snr. follows with an examination of the possible impact of the African Continental Free Trade Area (AfCFTA) on human rights and environmental sustainability by focusing on the extractive industries sector. He highlights the egregious violations of human rights and threats to environmental sustainability in the extractive industries sector. He however bemoans the lack of an AU normative and institutional framework to regulate these harmful activities. He therefore argues, among others, for the development of continental legal and policy framework to regulate the activities in the extractive sector so as to protect the environment while safeguarding rights in Africa.

References

'AU Commission Chairperson in Kigali for consultative meeting on AU institutional reforms' 08 May 2017, https://au.int/en/pressreleases/20170508/au-commission-chairperson-kigali-consultative-meeting-au-institutional (accessed on 12th January 2020).

Boyle, A.E. & Birnie, P. *International Law and the Environment* (Oxford University Press, 2002 2nd ed.).

Bulto, T.S. 'The Environment and Human Rights' in Anja Mihr (ed) *The Sage Handbook of Human Rights* (SAGE Publications Ltd 2014).

Cameron, J., & Mackenzie, R., 'Access to Environmental Justice and Procedural Rights in International Institutions' in Boyle, A.E. & Anderson, M. (eds) *Human Rights Approaches to Environmental Protection* (Clarendon Press 1996).

Churchill, R. 'Environmental Rights in Existing Human Rights Treaties' in Boyle, A.E. and Anderson, M.R. (eds) *Human Rights Approaches to Environmental Protection* (Clarendon Press 1996).

Durojaye, E. 'The Approaches of the African Commission to the Right to Health under the African Charter' (2013) 7 Law, Democracy and Development 393.

Emeka, P.A., 'Enhancing Environmental Protection and Socio-Economic Development in Africa: A Fresh Look at the Right to a General Satisfactory Environment under the African Charter on Human and Peoples' Rights,' (2009) 5 Law, Environment and Development Journal 58–72, 62.

Fitzmaurice, M., Ong, D.M. & Merkouris, P. (eds) *Human Rights and the Environment* (Martinus/Nijhof Publishers 2011).

Francioni, F. 'International Human Rights in an Environmental Horizon' (2010) 1 *European Journal of International Law* 42.

IUCN Environmental Law Programme, An Introduction to the African Convention on the Conservation of Nature and Natural Resources, IUCN Environmental Policy and Law Paper No. 56 (2004).

Kotze, L.J. & du Plessis, A. 'The African Charter on Human and Peoples' Rights and Environmental Rights Standards' in Turner, S.J. (ed) *Environmental Rights the Development of Standards* (Cambridge University Press 2019).

Mason, M. *Environmental Democracy: A Contextual Approach* (Earthscan 1999).

New Partnership for Africa's Development (NEPAD), Review of the Implementation of the Action Plan of the AU/NEPAD Environment (2012).

Ogunlade, A. 'Can the Bamako Convention Adequately Safeguard Africa's Environment in the Context of Transboundary Movement of Hazardous Wastes?' University of Dundee (2010).

Olufemi, A. *African Union Law: The Emergence of a Sui Generis Legal Order,* 1st Edition (Routledge 2020).

Redgwell, C. 'Access to Environmental Justice', in Francioni, F. (ed) *Access to Justice as a Human Right* (Oxford University Press 2007).

Sands, P. & Peel, J. *Principles of International Environmental Law*, 3rd edn. (Cambridge University Press, 2012).

Scholtz, W. 'Human Rights and the Environment in the African Union Context,' in Grear, A. & Kotzé, L.J. (eds.), *Research Handbook on Human Rights and the Environment* (Edward Elgar Publishing, 2015a).

Scholtz, W. 'Human Rights and the Environment in the African Union Context' in Scholtz, W. & Verschuuren, J. (eds) *Regional Environmental Law: Transregional Comparative Lessons in Pursuit of Sustainable Development* (Edward Elgar Publishing 2015b).

Steiner, W. An Introduction to the African Convention on the Conservation of Nature and Natural Resources (International Union for Conservation of Nature Environmental Law Centre, 2004).

Report of the Joint OHCHR-UNEP Seminar on Human Rights and the Environment 16 Jan. 2002, UN Doc E/CN.4/2002/WP.7 (2002), Annex II (15).

Rich, R. 'The Right to Development as an Emerging Human Right' (1982–1983) 23 *Virginia Journal of International Law* 320.

2

The Right to a Clean, Safe and Healthy Environment Under the African Human Rights System

Mulesa Lumina

1 Introduction

Undoubtedly, the quality of the natural environment has a substantial bearing on human health, as an unhygienic, polluted and hazardous environment has a detrimental impact on access to food, water and housing.[1] A clean and healthy environment is regarded as a precondition for the realization of various human rights including the rights to life, health and an adequate standard of living.[2] While the importance of a clean

[1] Ben Boer and Alan Boyle, 'Human rights and the environment – Background Paper for the 13th Informal ASEM Seminar on Human Rights' (21–23 October 2013) Sydney Law School Research Paper No. 14/14, http://ssrn.com/absract=2393753 accessed 8 May 2016.

[2] See, United Nations, 'Report of the Independent Expert on the issue of human rights obligations relating to the enjoyment of a safe, clean, healthy and sustainable environment, John Knox' (2012) A/HRC/22/43, para 34; see also *Case Concerning the Gabcikovo-Nagymaros Project* [1997] ICJ Rep. 7, 91–92: in his separate opinion, Judge Weeremantry recognized that the enjoyment of human rights is dependent on environmental protection; Alexandre-Charles Kiss, 'Concept and possible

M. Lumina (✉)
The Human Rights & Governance Consultancy (HURICON), East London, South Africa

M. Addaney, A. Oluborode Jegede (eds.), *Human Rights and the Environment under African Union Law*, https://doi.org/10.1007/978-3-030-46523-0_2

environment for the fulfilment of these rights is now considered trite, there remains no universal human rights treaty that expressly guarantees the right to a clean, safe and healthy environment. In fact, the existence and utility of such a right remains a contentious issue under international human rights law.[3] Under the African Charter on Human and Peoples' Rights (African Charter), however, peoples are explicitly guaranteed the right to 'a general satisfactory environment favourable to their development.'[4] While the African Commission on Human and Peoples' Rights (ACHPR) has attempted to clarify the content of this right, some confusion persists as to the exact nature, scope and content of the right and matters related to its justiciability.[5]

This chapter analyzes the normative framework governing the right to a healthy environment under the African human rights system through an examination of other regional and African Union (AU) human rights instruments, treaty-body decisions, domestic laws, commentary and scholarly literature on the subject. It also considers the relationship between human rights, health, development and the environment with the aim of coming to an understanding as to the exact content and scope of the right to a healthy environment and, concomitant duties of African states parties in this regard. Moreover, it explores the responsibilities of non-state actors. The chapter is divided into six sections. Following the introduction, Sect. 2 discusses the connection between health, development, the environment and human rights. Section 3 then explores the foundations of the right to a healthy environment in international human rights law while Sect. 4 briefly unpacks the nature and scope of a right to

---

implications of the right to environment in Kathleen Mahoney and Paul Mahoney (eds), *Human rights in the twenty-first century: a global challenge* (Martinus Nijhoff Publishers 1993) 551 and 553: Kiss contends that environmental degradation and damage has an adverse impact on normal ecological processes of nature as well as human health, both physical and mental.

[3] United Nations, 'Report of the Special Rapporteur on the issue of human rights obligations relating to the enjoyment of a safe, clean, healthy and sustainable environment' (24 January 2018) A/HRC/37/59, paras 11–13.

[4] Art 24.

[5] See Lilian Chenwi, 'The right to a satisfactory, healthy, and sustainable environment' in John Knox and Ramin Pejan (eds), *The human right to a healthy environment* (Cambridge University Press 2018) 66 and 69: Chenwi points out that although the phrasing of article 24 draws a normative link between the environment and development, it is rather vague.

a healthy environment. Section 5 looks at the normative framework governing the right to a healthy environment under the African human rights system specifically as well as its content. Section 6 is the conclusion.

2 Human Rights, Health, Development and the Environment

There is an indisputably clear link between human rights, human health and the environment. Since 1972, which saw the adoption of the Stockholm Declaration at the United Nations (UN) Conference on the Human Environment,[6] it is clear that the international community recognizes the salient connection between environmental protection and conservation on the one hand and human rights on the other, in particular, the impact of the quality of the environment on human health and dignity.[7] This recognition is captured in Principle 1 of the Stockholm Declaration, which states:

> Man has the fundamental right to freedom, equality and adequate conditions of life, in an environment of a quality that permits a life of dignity and well-being, and he bears a solemn responsibility to protect and improve the environment for present and future generations.

Evidently, this Principle emphasizes both the importance of the natural environment for supporting human life and humankind's charge in maintaining the environment for the enjoyment of present and future generations. Sohn contends that this Declaration has, to a large extent, been the foundation for a human right to a healthy environment.[8]

Since the Stockholm Conference, there has been continued discussion and examination of the relationship between the environment, human rights and sustainable development. The 1992 Rio Conference on Environment and Development, the Millennium Summit of 2000 and

---

[6] Declaration of the United Nations Conference on the Human Environment [1972] A/RES/2994.

[7] Boer and Boyle (n 1) 6.

[8] Louis B. Sohn, 'The Stockholm Declaration on the Human Environment' (1973) 14 *Harvard International Law Journal* 423,451–5.

the 2002 World Summit on Sustainable Development recognized that key to sustainable development is the realization of human rights in conjunction with the preservation of the quality of the natural environment as is enshrined in the standards adopted at these meetings. The Rio Declaration on Environment and Development,[9] for example, places human beings at the center of concerns for sustainable development and states that 'they are entitled to a healthy and productive life in harmony with nature.'[10]

Not only did this Declaration provide a new lens through which the link between human rights and the environment is considered, it highlighted the importance of access to information and citizen participation in environmental decision-making processes and for access to remedies for environmental harm:

> Environmental issues are best handled with the participation of all concerned citizens, at the relevant level. At the national level, each individual shall have appropriate access to information concerning the environment that is held by public authorities, including information on hazardous materials and activities in their communities, and the opportunity to participate in decision-making processes. States shall facilitate and encourage public awareness and participation by making information widely available. Effective access to judicial and administrative proceedings, including redress and remedy, shall be provided.[11]

As regards environmental factors that affect human health, the Malé Declaration on the Human Dimension of Global Climate Change urges the UN to treat, as a matter of urgency, the apparent and glaringly adverse effects of climate change on the full enjoyment of human rights.[12] The recognition of climate change as a threat to human health and well-being and as a grave concern for the realization of human rights is also articulated in the Paris Agreement of 2015.[13] The Agreement calls on States, when taking action on climate change, to respect, promote and consider

[9] [1992] A/CONF.151/26.

[10] Ibid, principle 1.

[11] Ibid, principle 10.

[12] Declaration on the Human Dimension of Global Climate Change [2007] Preamble.

[13] Preamble.

their human rights obligations, including those pertaining to the right to health.

The Outcome Document of the Rio+ 20 Summit reaffirmed the rights to safe drinking water, health and sanitation, the realization of which are affected by the quality of the natural environment.[14] Despite the strides the Summit made as far as sustainable development is concerned, very little was achieved with respect to progress on the recognition of the right to a healthy environment.[15] Nevertheless, one of the most important consequences of Rio+20 was the adoption of Agenda 2030 and consequent creation of the Sustainable Development Goals (SDGs) in September 2015. As part of a new sustainable development agenda, UN Member States set 17 SDGs aimed at protecting and preserving natural life on the planet, eradicating poverty and ensuring prosperity for all.[16] The SDGs also encapsulate the many ways in which the quality of human life is substantially affected by environmental factors such as improper waste management and pollution.[17]

It is also worth noting that the UN regularly conducts comprehensive studies on the relationship between human rights and the environment. A 2011 report by the UN Office of the High Commissioner for Human Rights (OHCHR), for instance, identified a number of environmental threats to human rights, such as atmospheric emissions, degradation in water quality, land degradation, deforestation, desertification, hazardous waste, loss of biodiversity and natural disasters.[18] It also examines a number of human rights and environmental instruments indicating the link between environmental protection and the enjoyment of both substantive and procedural human rights, noting that the exercise of human

---

[14] United Nations, *The Future We Want* (27 July 2012) U.N. Doc. A/CONF.216/L.1 paras 8, 108 and 138 https://sustainabledevelopment.un.org/futurewewant.html accessed 7 September 2016.

[15] Boer and Boyle (n 1) 7.

[16] See United Nations 'Sustainable development goals' http://www.un.org/sustainabledevelopment/sustainable-development-goals/ accessed 7 May 2016.

[17] See, for example, Target 3.9 of the Sustainable Development Goals: by 2030, substantially reduce the number of deaths and illnesses from hazardous chemicals and air, water and soil pollution and contamination.

[18] OHCHR, 'Analytical study on the relationship between human rights and the environment' (16 December 2011) A/HRC/19/34 paras 15–22.

rights can equally contribute to the protection and preservation of the environment.

In 2012, the UN Human Rights Council also created a mandate for a Special Rapporteur on the issue of human rights obligations relating to the enjoyment of a safe, clean, healthy and sustainable environment, which includes conducting research on human rights obligations in the context of the environment.[19] In the Rapporteur's first report to the Human Rights Council, strong emphasis was placed on the interdependent relationship between human rights and the environment.[20] Not only is a healthy environment necessary for the fulfilment of economic, social and cultural rights like the right to health and the right to property, but the exercise of civil and political rights such as the right of access to information is also an important component of environmental protection.[21] This is in conformity with the principle that all human rights are interdependent and inter-connected and must be realized with due regard to each other.

The developments mentioned above illustrate that the international community is aware and considers that the threat of environmental problems, including transboundary water and air pollution, to the enjoyment of human rights must be dealt with as a matter of priority. Notwithstanding the very necessary contributions that these declarations, resolutions and reports have made to international human rights discourse in the areas of sustainable development and the environment, they still failed to clarify the status of the right to a healthy environment in international law.[22] Indeed, these developments may have culminated in the creation of a global sustainable development agenda that also considers the relationship between socio-economic rights and the natural environment but, 48 years after the Stockholm Declaration, the existence of a universal right to a healthy environment remains a hotly contested subject.

---

[19] *Human Rights and the environment* [2012] A/HRC/RES/19/10.

[20] United Nations A/HRC/22/43 (n 2).

[21] United Nations A/HRC/37/59 (n 3) para 2.

[22] Sueli Giorgetta, 'The right to a healthy environment, human rights and sustainable development' (2002) 2 *International Environmental Agreements: Politics, Law and Economics* 172,181.

3 The Right to a Healthy Environment in International Human Rights Law

Although most international human rights treaties regard a safe, clean or healthy environment as one of the preconditions for the enjoyment of human rights, particularly socio-economic rights,[23] there is no universal human rights instrument which explicitly guarantees the right to a healthy environment. Nevertheless, it should be noted that some decisions of international human rights supervisory mechanisms as well as both non-binding and binding international human rights law point to the existence of a right to a healthy environment.[24] In addition, this right has been recognized, in varying forms, at the regional level and at the domestic level as a constitutional right or in legislation.[25]

One of the significant determinants for attaining the highest standards of health is the quality of the natural environment which is essential for the supply of safe food and water and access to adequate shelter.[26] In fact, where key human rights instruments have included specific provisions pertaining to the environment, such provisions are usually couched in narrow terms focusing on human health.[27] For instance, Article 12(2) of the International Covenant on Economic, Social and Cultural Rights (ICESCR) refers to 'the improvement of all aspects of environmental and industrial hygiene' as a step states parties are required to take to realize of the right to health. In its General Comment No. 14, the UN Committee

---

[23] Frans Viljoen, *International Human Rights Law in Africa* (2nd edn OUP 2012) 268.

[24] Ben Kiromba Twinomugisha 'Some reflections on judicial protection of the right to a clean and healthy environment in Uganda' (2007) 3(3) *Law, Environment and Development Journal* 244,247 http://www.lead-journal.org/content/07244.pdf accessed 15 August 2016.

[25] United Nations, A/HRC/37/59 (n 3), para 11.

[26] Committee on Economic, Social and Cultural Rights (CESCR), General Comment No. 14: The Right to the Highest Attainable Standard of Health (Art. 12) (2000) E/C.12/2000/4 paras 4 and 11; Ben Kiromba Twinomugisha, *Fundamentals of health law in Uganda* (Pretoria University Law Press 2015) 1 and 91; Twinomugisha, 'Some reflections on judicial protection of the right to a clean and healthy environment in Uganda' (n 24) 248–249.

[27] See, for instance, Convention on the Rights of Persons with Disabilities, art 25; Universal Declaration of Human Rights, art 25.

on Economic, Social and Cultural Rights (CESCR) notes that improving all aspects of environmental hygiene entails the following[28]:

> [T]he requirement to ensure an adequate supply of safe and potable water and basic sanitation; the prevention and reduction of the population's exposure to harmful substances such as radiation and harmful chemicals or other detrimental environmental conditions that directly or indirectly impact upon human health.

It also requires the provision of access to adequate housing, safe and hygienic working conditions and an adequate supply of food and proper nutrition, which the Committee gives prominence.[29] Moreover, the Committee re-iterates—in General Comment No. 15 on the right to water—the importance of access to adequate amounts of safe drinking water for the preservation of human life and the maintenance of good hygiene for overall good health.[30] The Committee considers the right to health, in the context of the right to water, as requiring states parties to take steps to mitigate and prevent the contamination of water (particularly natural water resources) as a result of hazardous substances and pathogenic microbes where that has a significant effect on human health. It is thus clear that the ICESCR recognizes that clean natural surroundings are a necessary condition for ensuring the highest attainable standard of human health.[31]

Similarly, Article 24(2)(c) of the Convention on the Rights of the Child (CRC)[32] enjoins states parties to take into consideration damage and risks of environmental pollution when formulating and implementing measures to fight disease and malnutrition. The Convention on the Elimination of all forms of Discrimination Against Women (CEDAW),[33] under articles 12 and 14, provides for the right to health, emphasizing the rights of rural women to participate in development planning in the

[28] CESCR (n 26) para 15.
[29] CESCR (n 26) paras 4 and 15.
[30] CESCR, General Comment No. 15: The Right to Water (2003) E/C.12/2002/11 paras 2 and 11.
[31] Ibid, para 8.
[32] [1989] UNTS 1577 3.
[33] [1979] UNTS 1249 13.

environmental context. It does not expressly provide for a right to a healthy environment, but it has been reasoned that where the right to health is threatened by environmental harm, states must take steps to maintain the quality of the natural environment.[34]

At the regional level, numerous instruments implicitly or explicitly provide for the right to a healthy environment. Article 11 of the Additional Protocol to the American Human Rights Convention in the Area of Economic, Social and Cultural Rights,[35] guarantees that 'everyone shall have the right to live in a healthy environment and to have access to basic public services' and places a duty on states parties to 'promote the protection, preservation and improvement of the environment.' Article 38 of the 2004 Revised Arab Charter on Human Rights provides for the 'right to a healthy environment' in relation to an adequate standard of living.[36]

In Europe, there is no specific provision for a human right to a healthy environment. Nevertheless, article 1 of the Convention on Access to Information, Public Participation in Decision-making and Access to Justice in Environmental Matters (Aarhus Convention)[37] provides for the right 'of every person of present and future generations to live in an environment adequate to his or her health and well-being.' While the European Social Charter[38] does not expressly provide for the right to a healthy environment, the European Committee of Social Rights has interpreted the right to health under article 11 of the Charter to include the right to a healthy environment. In *Marangopoulos Foundation for Human Rights (MFHR) v Greece*,[39] the Committee reasoned that given Greece's obligations to prevent pollution under the international environmental treaties it has ratified; the fact that environmental quality has

[34] Office of the United Nations High Commissioner for Human Rights 'Mapping Human Rights Obligations Relating to the Enjoyment of a Safe, Clean, Healthy and Sustainable Environment' (2013) para 79.

[35] Protocol of San Salvador [1988] OASTS 69.

[36] See also, ASEAN Human Rights Declaration, (18 November 2012) ASEAN Statements and Communiques, art 28: '[e]very person has the right to an adequate standard of living for himself or herself and his or her family including . . . [t]he right to a safe, clean and sustainable environment.'

[37] [1998] UNTS 2161 447.

[38] [1961] ETS 35.

[39] Complaint no. 30/2005, European Committee of Social Rights, 6 December 2006, para 195.

a significant impact on human health; the complementarity between the right to health under the Charter and the right to life under the European Convention on Human Rights; decisions of other regional human rights bodies and commentary of the UN Committee on Economic, Social and Cultural Rights, Greece had violated article 11.[40] This decision illustrates that there is some recognition that the preservation of environmental quality is seen as an important factor in realizing the right to health.[41] Moreover, it affirms the principle of the interdependence and indivisibly of all human rights.[42]

The African Charter on Human and Peoples' Rights, as will be explored in depth further on in the chapter, is the only principal treaty at the regional level that explicitly guarantees a stand-alone right to a healthy environment. The international community has been mostly reluctant to recognize a right to a healthy environment primarily because of the uncertainty surrounding the precise content of the right and the difficulties that come with attempting to define the ideal environment required for the realization of socio-economic rights.[43] Despite existing provisions recognizing the relationship between human health, standards of living, development and the environment, there remains ongoing debate regarding the scope and potential utility of a right to a healthy environment. A 2018 report of the UN Special Rapporteur on human rights and the environment sets out a framework of 16 non-binding principles that will provide the basis for the fulfilment of human rights obligations relating to a healthy environment.[44] These have been gleaned from existing human rights obligations pertaining to the environment and reflect the manner in which these obligations have been implemented in the context of the natural environment. They provide some measure of guidance as to the nature and scope of environmental human rights.

---

[40] Ibid, paras 195–196, 200, 202 and 205.

[41] Mirja Trilsch, 'European Committee of Social Rights: the right to a healthy environment' (2009) 7(3) *I.CON* 529536.

[42] Ibid.

[43] Boer and Boyle (n 1) 15.

[44] United Nations, 'Report of the Special Rapporteur on the issue of human rights obligations relating to the enjoyment of a safe, clean, healthy and sustainable environment' (2018) A/73/188 paras 8 and 9.

4 The Nature of the Right to a Healthy Environment

Traditionally, environmental rights have at times been regarded as so-called third generation or solidarity rights after civil and political rights (first generation) and socio-economic rights (second generation).[45] The right to a healthy environment, specifically, is typically formulated as a collective right.[46] However, not all human rights practitioners accept such a classification as they feel it detracts from the realization of civil, political and socio-economic rights as a matter of priority and waters down the notion of human rights.[47] There are two other approaches regarding the classification of this right. The second considers the right to a healthy environment to be a social or economic right, subject to progressive realization and the implementation of specific law and policy.[48] This approach views the environment as a good in and of itself that is necessary for socio-economic development.[49] A third, anthropocentric approach to environmental rights favours the 'greening' of existing human rights, which directs environmental protection and conservation efforts toward protecting human life, property, privacy and health.[50] For example, civil and political rights such as the rights of access to information and access to a remedy empower aggrieved individuals to seek redress for environmental damage that causes harm to their person or property and also affords them the opportunity to influence decision-making in environmental matters.[51] This approach does not see environmental rights as stand-alone rights falling into any one specific category of rights. The right to a healthy environment is a unique human right in that it

[45] Boer and Boyle (n 1) 14; Philip Alston 'Conjuring up new human rights: a proposal for quality control' (1984) 78 *American Journal of International Law* 607; Michael Kidd *Environmental law* 2ed (2nd edn Juta and Company Ltd. 2011) 21.

[46] United Nations, A/HRC/22/43 (n 2) para 12.

[47] Boer and Boyle (n 1) 14.

[48] Boer and Boyle (n 1) 14.

[49] Boer and Boyle (n 1) 14.

[50] United Nations, A/73/188 (n 44) paras 12–27; Michael Andersen, in 'Human Rights Approaches to Environmental Protection' in Alan E. Boyle and Michael R. Andersen (eds), *Human Rights Approaches to Environmental Protection* 15 (Clarendon Press 1996).

[51] Boer and Boyle (n 1) 11 and 13.

seems to have the character of the different categories of rights and none at the same time. A review of international provisions, commentary and decisions of judicial and quasi-judicial bodies indicates that the right 'involves many things including clean water, air and soil that are free from toxic wastes or hazards that threaten human health.'[52] In countless instances, the right has been formulated as a right to a 'healthy,' 'clean' or 'decent' environment or an environment conducive to well-being or an adequate standard of living.[53]

5 The Right to a Healthy Environment Under the African Human Rights System

Regional Standards

Under article 24 of the African Charter, peoples are guaranteed the right to 'a general satisfactory environment favourable to their development.' As regards the environment, other relevant related provisions include articles 16,[54] 20,[55] 21[56] and 22.[57] The wording of article 24 is unique in that the right to a healthy environment is primarily construed in the context of development as opposed to health as it is in Europe or an adequate standard of living as it is expressed in the Arab Charter and ASEAN Declaration. Nevertheless, it has proven difficult to discern exactly what is meant by the terms 'satisfactory' and 'favourable to their development.' In 2001, the African Commission on Human and Peoples' Rights (ACHPR) offered some guidance in this regard by unpacking the state's obligations under article 24 in particular, without necessarily

[52] Twinomugisha, 'Some reflections on judicial protection of the right to a clean and healthy environment in Uganda' (n 24) 248–249.

[53] Philippe Cullet 'Definition of an environmental right in a human rights context' (1995) 13 *Netherlands Quarterly of Human Rights* 29.

[54] The right to the best attainable state of physical and mental health.

[55] The right to self-determination.

[56] The right to freely dispose of wealth and natural resources.

[57] The right to development.

clarifying the normative content of the right.[58] In the ground-breaking decision of *Social and Economic Rights Action Centre (SERAC) and the Centre for Economic and Social Rights v Nigeria*,[59] the ACHPR found that Nigeria had violated articles 16 and 24 of the African Charter due to pollution of soil, water and air which caused harm to the health of the *Ogoni* people.

While the state may be entitled to conduct industrial activities for socio-economic development, it has to balance this with protection of the rights of its citizens. In respecting the right to a healthy environment, the state must refrain from employing any legal, policy and other measures that cause, endorse or tolerate violations of the right.[60] Further noting the four core obligations of states generally to respect, protect, promote and fulfil human rights,[61] the ACHPR articulated the state's obligations to realize the right to a healthy environment as encompassing duties of both a substantive and procedural nature. The substantive duties include promoting conservation, protecting the environment from damage, preventing environmental damage and pollution, mitigating these adverse impacts, ensuring ecologically sustainable use of natural resources and preventing private actors from infringing the right.[62] Furthermore, it would require that the state carry out or allow environmental impact and scientific assessments and monitoring. According to Van der Linde and Louw, these obligations are reflective of the socio-economic character of environmental rights and their related financial resource implications.[63] The procedural duties include those related to access to a remedy as well as the right of access to information on environmental impacts.[64]

---

[58] Dinah Shelton 'Decision regarding Communication 155/96 (*Social and Economic Rights Action Center/Center for Economic and Social Rights v. Nigeria*). Case No. ACHPR/COMM/A044/1' (2002) 96(4) *American Journal of International Law* 937,942.

[59] (2001) AHRLR 60 (ACHPR 2001) (*SERAC*).

[60] Ibid, para 53.

[61] Ibid, para 44.

[62] Ibid, paras 52–53; See also, Morné Van der Linde and Lirette Louw 'Considering the interpretation and implementation of article 24 of the African Charter on Human and Peoples' Rights in light of the *SERAC* Communication' (2003) 3(1) *African Human Rights Law Journal* 167,179.

[63] Van der Linde and Louw (n 62).

[64] *SERAC* (n 59) paras 52–53: this entails 'requiring and publicising environmental and social impact studies prior to any major industrial development, undertaking appropriate monitoring

In spite of the fact that the ACHPR acknowledged the development aspect of the right to a healthy environment in the *SERAC* case, it considered article 24 in relation to the right to health. This has led to the assertion that it is doubtful whether a legal claim based on this provision could ever be divorced from the right to health in order to succeed.[65] In contrast to *Marangopoulos*, for instance, the depth of analysis of the literature and commentaries of human rights supervisory mechanisms on the right to a healthy environment is not entirely clear in the *SERAC* decision. This renders it difficult to ascertain the content of the right to a healthy environment favorable to development in the African context or the basis for the ACPHR's determination of its foundations.[66] Despite some of its shortcomings, however, this decision is helpful in that it reiterates the sentiment that the right to a healthy environment is important for the achievement of other human rights, particularly the right to health, and outlines some of state parties' obligations under article 24. Moreover, while the *SERAC* decision is non-binding in that the African Commission may only give recommendations to states parties before it, it provides salient direction as to the nature of the right under the African human rights system.

Further to the *SERAC* decision, a number of soft law instruments of the AU and resolutions of the ACHPR make reference to the now-trite notion that human rights and the environment are interdependent.[67] In its General Comment No. 3, the ACHPR holds that the right to life must be accorded a broad interpretation and that as part of its obligations to protect rights-holders from threats to their lives, state action must also include 'preventive steps to preserve and protect the natural environment and humanitarian responses to natural disasters, famines, outbreaks of infectious diseases, or other emergencies.'[68]

---

and providing information to those communities exposed to hazardous materials and activities and providing meaningful opportunities for individuals to be heard and to participate in the development decisions affecting their communities.'

[65] Van der Linde and Louw (n 62) 176.

[66] Van der Linde and Louw (n 62) 179.

[67] Resolution No. 431 on the Right to Food and Nutrition in Africa, ACHPR/Res.431 (LXV) 2019.

[68] African Commission on Human and Peoples' Rights (ACHPR), General Comment No. 3 on the African Charter on Human and Peoples' Rights: The Right to Life (Article 4), paras 3 and 41.

The Principles and Guidelines on the Implementation of Economic, Social and Cultural Rights in the African Charter on Human and Peoples' Rights provide that national development plans and programmes related to land and water resource management, for instance, must be designed with the ultimate aim of creating a healthy environment that is conducive to the realization of the right to health.[69] Also, states parties are obliged to ensure that state and non-state actors alike respect the right to a satisfactory environment.[70] In 2017, the Commission took a further step toward elucidating the content and scope of article 24 of the African Charter through the introduction of its reporting guidelines on the same article in the context of extractive industries.[71] Most of the decisions of the African Commission on environmental protection and human rights have focused on the rights of indigenous persons in the context of extractive industries and forced removals, which is a common problem in Africa. However, although these guidelines and principles specifically pertain to extractive industries, they do provide useful guidance and fill numerous gaps left by *SERAC*. For example, the guidelines provide a definition for the term 'people,' and unpack, in depth, the content of the right as well as the corresponding duties of both state and non-state actors in this regard.

At the sub-regional level, two important judgments from the Court of Justice for the Economic Community of West African States (ECOWAS Court) and East African Court of Justice (EACJ) provide additional direction as to the content of the right to a healthy environment in Africa. The facts of *SERAP v Nigeria*[72] arose from oil spillage in the Niger delta which local non-governmental organization, the Socio-Economic Rights and Accountability Project (SERAP), alleged was caused by oil prospection carried out by private companies in the area. The ECOWAS Court had to decide, among other issues, whether the Federal Republic of Nigeria had in fact violated the right to a general satisfactory environment

[69] Para 67.

[70] Para 44.

[71] African Commission on Human and Peoples' Rights (ACHPR), *State Reporting Guidelines on Articles 21 and 24 of the African Charter on Human and Peoples' Rights Relating to the Operations of Extractive Industries; Resolution of the African Commission on the Niamey Declaration on Ensuring the Upholding of the African Charter in the Extractive Industries Sector*, ACHPR/Res. 367 (LX) 2017.

[72] Judgment No. ECW/CCJ/JUD/18/12.

under article 24 of the African Charter due to inadequate environmental protection measures. In finding that this was indeed the case, the Court considered both articles 1 and 24 of the Charter together.[73] The Court characterized the obligation of the state under article 24 as an 'obligation of attitude and an obligation of result' and reiterated that human health, life and dignity are dependent on the quality of the natural environment.[74] In reasoning that the provision requires states to adopt every legislative, administrative and other measure to preserve the quality of the environment, 'understood as an integrated whole,' to the satisfaction of human beings living in it and to ensure sustainable development,[75] the ECOWAS Court found that Nigeria had violated this right, read with article 1, and failed to adequately fulfil the rights of people in the Niger delta region.

The 2014 case of *African Network for Animal Welfare (ANAW) v Attorney General of The United Republic of Tanzania,*[76] though not specifically concerning the right to a healthy environment, provides important commentary regarding the duties of states with respect to maintaining the quality of the environment. In finding that articles 5(3)(c), 8(1)(c), 111(2) and 114 of the East African Community Treaty had been infringed,[77] the EACJ held that it needed to strike a balance between its interpretative jurisdiction and the needs of partner states to develop land and natural resources. Moreover, the court needed to consider issues under scrutiny in the case such as environmental protection, sustainable development, environmental rule of law and the role of the State in environmental policymaking, through the lens of the need to protect the ecosystem of the Serengeti for future generations.[78]

---

[73] Article 1 of the African Charter on Human and Peoples' Rights states: 'The member states of the Organization of African Unity parties to the present Charter shall recognize the rights, duties and freedoms enshrined in this Charter and shall undertake to adopt legislative or other measures to give effect to them.'

[74] *SERAP* (n 72) para 100.

[75] *SERAP* (n 72) para 101.

[76] Reference No. 9 of 2010 (*ANAW*).

[77] These provisions require partner states to conserve, protect and co-operate in the management of natural resources and the environment.

[78] *ANAW* (n 76) para 85.

Other AU instruments that proclaim the right to a healthy environment include the Protocol to the African Charter on Human and Peoples' Rights on the Rights of Women in Africa (Maputo Protocol),[79] the African Charter on the Rights and Welfare of the Child (ACRWC)[80] and the Convention on the Conservation of Nature and Natural Resources (Natural Resources Convention).[81] Article 18(1) of the Maputo Protocol provides that women 'shall have the right to live in a healthy and sustainable environment' and article 18(2) places a duty on states parties to implement measures to, inter alia, ensure that proper toxic waste disposal and management standards are followed and that women are empowered to participate in environmental decision-making regarding natural resources. Moreover, under article 16, the right to adequate housing includes the right of equal access to 'housing and to acceptable living conditions in a healthy environment.' The ACRWC refers to environmental sanitation in the context of the right to health under article 14(2)(h).

One of the underlying principles of the Natural Resources Convention is the right to a healthy environment under the African Charter.[82] States parties are also enjoined to prevent pollution of water sources and maintain ecological processes to protect human health.[83] This Convention further underscores the importance of procedural rights for environmental decision-making.[84]

Domestic Standards

Some 140 constitutions across the world include a state obligation to protect the environment or a right to a clean, safe, healthy, favorable or ecologically balanced environment.[85] All African states recognize the

[79] [2003].

[80] [1990] CAB/LEG/24.9/49.

[81] [2003].

[82] Art 3(2).

[83] Art 1(a)(b).

[84] Art 16.

[85] See OHCHR (n 18) para 30; see also OHCHR and United Nations Environment Programme *Human Rights and the Environment: Rio+20 Joint Report* (2012).

right to a healthy environment through their constitutions, judicial decisions, environmental legislation and/or through their ratification of the African Charter. The right to life under the constitutions of Nigeria and Tanzania has been interpreted by courts in those countries as implicitly including the right to a healthy environment.[86] Roughly 33 states in Africa of the ones recognizing the right have incorporated an explicit right to a healthy environment in their constitutions as of 2013.[87] Notable examples include C'ôte D'Ivoire,[88] South Africa[89] and Zimbabwe.[90] These provisions acknowledge the importance of maintaining a clean environment to ensure human health and in many instances make mention of considerations of sustainable development and use of natural resources. Article 27 of the Ivorian constitution, for instance, states that '[i]t is recognized that everyone throughout the national territory has the right to a healthy environment.'

Where there is an explicit individual right to a healthy environment, there is sometimes an additional provision setting out the duties of individuals to protect and conserve the environment and natural resources.[91] Other constitutional provisions have simply declared the state's obligations vis-à-vis the environment and natural resources to include the pursuance of environment-friendly and sustainable plans, programmes and projects, promoting conservation and specifically maintaining the quality of the environment for the health and safety of citizens.[92] Some provisions include both the right as well as the duties of states. For example, the South African constitution has an elaborate provision under section 24,

[86] David R. Boyd, *The Environmental Rights Revolution: A Global Study of Constitutions, Human Rights, and the Environment* (UBC Press 2011) 156.

[87] Ibid, 149. This now includes Zimbabwe where amendments to the constitution saw the inclusion of a right to a healthy environment.

[88] Law No. 2016–886 of 8 November 2016 on the Constitution of the Republic of Côte d'Ivoire, art 27.

[89] Constitution of the Republic of South Africa, 11 December 1996, s 24.

[90] Constitution of Zimbabwe Amendment (No. 20) Act, 2013, 22 May 2013, s 73.

[91] Eg. Section 15 of the Constitution of Mali, 1992: '[e]very person shall have the right to a healthy environment. The protection, defence and promotion of the environment shall be obligations for all and for the State.'

[92] Kaniye SA Ebeku 'The right to a satisfactory environment and the African Commission' (2003) 1 *African Human Rights Law Journal* 149,151.

stipulating that 'everyone has the right to an environment that is not harmful to their health or well-being.' It further provides that the state should adopt 'reasonable' measures to prevent pollution and ecological degradation, promote conservation and ensure sustainable development and use of natural resources. Section 73 of the 2013 Zimbabwean constitution is worded exactly like the South African provision however, subsection 2 stipulates that the 'state must take reasonable legislative and other measures, within the limits of the resources available to it, to achieve the progressive realization of the rights set out in this section.'

Content and Scope of the Right to a Healthy Environment Under the African Human Rights System

Foundational Values and Principles

The natural environment and its resources are an integral part of the religious, cultural and social life of many Africans,[93] but this relationship is often compromised by large-scale, land-based operations which—though aimed at development—have numerous adverse impacts on the environment and health of communities living in close proximity to these activities.[94] Moreover, with economies that are heavily dependent on extractive industries, a number of African states face the challenge of striking a balance between achieving socio-economic development and carrying out environmental conservation efforts effectively.[95] Against this backdrop, article 24 of the African Charter was drafted as a composite right, concerned with both environmental protection and the promotion of development on the continent.[96] It is rooted in the principle of solidarity in that although the state bears the primary responsibility of maintaining

[93] Chenwi (n 5) 59.

[94] ACHPR Working Group on Indigenous Populations/Communities, 'Extractive Industries, Land Rights and Indigenous Populations'/Communities' Rights 2017 East, Central and Southern Africa' (2017) 8 and 17.

[95] African Union (AU) & Economic Commission for Africa (UNECA), *Illicit Financial Flows: Report of the High Level Panel on Illicit Financial Flows from Africa* (2015) http://ww.uneca.org/sites/default/files/publications/iff_main_report_english_pdf, accessed 4 April 2016.

[96] Chenwi (n 5) 66.

environmental quality and managing natural resources in the interests of citizens, the preservation and protection of a generally satisfactory environment favorable to development is also the responsibility of individuals, communities and legal persons, especially multinational companies engaged in extractive activities.[97] The provision also has, as part of its normative basis, the principle that the use and exploitation of natural resources should be in the exclusive interest and benefit of citizens.[98] Furthermore, in accordance with the principle of effective participation, environmental decision-making processes pertaining to land-based projects, such as those involving the exploration and use of natural resources, must include those affected by such activities and such persons should have access to information on such projects. Further underlying article 24 are the principles of equality and non-discrimination, as recognized in article 2 of the Charter. The right to a healthy environment must be realized with due regard for the situation of economically and socially marginalized persons such as indigenous people and, where possible, in a manner that rectifies such inequalities in society.[99] Lastly, article 24 is premised on the principle of the interdependence and indivisibility of human rights as is underscored in the preamble of the African Charter and reiterated in the *SERAC* decision.[100] As Chenwi illustrates, this provision, in its particular phrasing, creates normative unity between environmental protection and development. In addition to the right to development, the realization of the right to a healthy environment under article 24 is both dependent on and essential for the fulfilment of the rights to life and health.[101]

[97] ACHPR *State Reporting Guidelines on Articles 21 and 24* (n 71) para 3; Resolution ACHPR/Res.224 (LI) 2012 on a human rights-based approach to the governance of natural resources.

[98] ACHPR *State Reporting Guidelines on Articles 21 and 24* (n 71) para 4.

[99] ACHPR *State Reporting Guidelines on Articles 21 and 24* (n 71) para 7.

[100] ACHPR *State Reporting Guidelines on Articles 21 and 24* (n 70) para 51.

[101] Chenwi (n 5) 67.

Content and Scope of the Right

It is also noteworthy that the right to a healthy environment is enshrined in the African Charter as a collective right afforded to peoples specifically, a distinguishing feature of the African human rights system.[102] This has raised questions as to what constitutes a 'people' under article 24, as the Charter is largely silent on this.[103] As regards article 21 of the African Charter which provides for the right of peoples to freely dispose of their wealth and natural resources, the ACHPR has remarked that the origin of the right can be traced back to the era of colonialism, noting in particular the exploitation of Africa's natural resources without any lasting benefits to her people, especially indigenous people.[104] In *SERAP v Nigeria*, the ECOWAS Court recognized the right to a healthy environment in West Africa and demonstrated support for environmental protection to ensure the health, well-being and development of the land of peoples in particular—in that case, the group of people occupying the polluted area of the Niger delta.[105] In some instances, the term 'people' has been used to refer to people of a particular African country, the people of Africa as a whole or a particular indigenous group.[106]

The right to a healthy environment is at times also formulated in such a way as to afford human rights protection in the environmental context to particular vulnerable groups including women, children, indigenous people and minorities, for example.[107] This acknowledges the disadvantaged position of these groups in society and their resultant vulnerability to harm arising out of environmental degradation. Moreover, this is in

[102] Solomon A. Dersso 'The jurisprudence of the African Commission in Human and People's Rights with respect to people's rights' (2006) 6(2) *African Human Rights Law Journal* 358,359.

[103] In *Katangese Peoples' Congress v Zaire* (2000) AHRLR 72 (ACHPR 1995) para 3, the ACHPR acknowledges this with respect to peoples' right to self-determination under article 20 of the African Charter.

[104] *SERAC* (n 59) para 56.

[105] *SERAP* (n 72) para 114; Moustapha Fall 'Valuable Precedents: The Protection of the Right to Environment by Sub-regional African Courts' (*The ACtPHR Monitor*, 13 September 2016) http://www.acthprmonitor.org/valuable-precedents-the-protection-of-the-right-to-environment-by-sub-regional-african-courts/, accessed 15 June 2019.

[106] Dersso (n 102) 361–362.

[107] United Nations, 'Issue of human rights obligations relating to the enjoyment of a safe, clean, healthy and sustainable environment' (8 January 2019) A/HRC/40/55 para 31.

accordance with the principle of equality and non-discrimination, which in the context of article 24, guarantees that the right to a healthy environment be afforded to members of society who are most economically and socially marginalized and be respected, protected and fulfilled with the aim of correcting societal inequalities. While the Maputo Protocol and the ACRWC clearly provide for the right to a healthy environment for women (especially rural women) and children, it is not entirely clear, from the wording of article 24 which specific group the provision is referring to as a 'people.'

The abovementioned state reporting guidelines on article 24 once again, give some guidance here and, where *SERAC* did not, define what constitutes a 'people' based on previous decisions of the ACHPR[108]:

> the term '**peoples**' under Articles 21 and 24 of the African Charter refers to a group of individuals having a common identity on account of objective markers of shared language, racial or ethnic makeup, historical experience, religious, cultural or ideological affinity, connection to a particular territory and the subjective manifestations of self-identification and awareness as a distinct group possessing of shared identity. Peoples are therefore not to be equated solely with 'Nation' or 'State', but include subnational groups including communities of a particular region exhibiting the above characteristics, or people whose resources are exploited under foreign occupation.

Furthermore, the guidelines state that local communities or individuals directly affected by extractive activities can be covered by this definition and thus lay claims for damages under article 24. The inquiry into the concept of a 'people' is largely determined on a case-by-case basis.[109] As was emphasized in *SERAC*, determining the content of the rights in the Charter, including article 24, requires a consideration of the entire text of Charter in its uniqueness and the various types of rights provided therein.[110]

---

[108] ACHPR *State Reporting Guidelines on Articles 21 and 24* (n 70) 11. See also *Katangese Peoples' Congress v Zaire* (n 103); *Centre for Minority Rights Development (Kenya) and Minority 223 Rights Group International on Behalf of Endorois Welfare Council v Kenya* (2009) AHRLR 75 (ACHPR 2009).

[109] Dersso (n 102) 380–381.

[110] *SERAC* (n 59) para 68.

As discussed above, there are many instances in which the right to a clean, safe and healthy environment is provided in a treaty in relation to adequate housing, food, water, health and even workplace safety, for example. The African Charter takes this a step further as the right is given a different normative lens—development. A 'general satisfactory environment' has been interpreted as enjoining states parties to adopt legislative, administrative and other measures aimed at maintaining an environment that is habitable and ensures the dignity and best attainable levels of health for the people living in it. Additionally, such environment must also be 'favourable to their development' which establishes a clear link between human rights, the environment and development. In this regard, the state can only implement development projects that will promote human rights, ensure sustainable use of land and other natural resources and benefit the people—of both present and future generations.[111] This also entitles rights-holders to be protected from pollution and environmental degradation and requires that a variety of interests—religious, cultural, economic and social—are weighed up and balanced out in the name of sustainability.[112] Essentially, the environment must be clean, safe and healthy to the extent that is conducive to the realization of not only the right to life but to the development of individuals and peoples as well.[113] This is reiterated in the Maputo Protocol, which specifically entitles African women to both participate in environmental decision-making and to live in an environment that does not threaten their health and dignity.

Obligations of State Actors

While the right to a satisfactory environment under the African human rights system is generally regarded as a socio-economic right, it also appears to have characteristics of a civil or political right. Article 24 covers both socio-economic rights such as the rights to health, water and housing, as well as civil and political rights like the rights to life, dignity

---

[111] Ebeku (n 92) 165.

[112] ACHPR *State Reporting Guidelines on Articles 21 and 24* (n 71) para 28.

[113] ACHPR *State Reporting Guidelines on Articles 21 and 24* (n 71) para 27.

and political participation. Overall, it is, evidently, a multifaceted right[114] but essentially has two main aspects: substantive and procedural.

Firstly, the substantive dimensions of the right relate to the quality of the environment in terms of which the state is obliged to conduct proper waste management, promote conservation efforts, prevent environmental pollution and contamination and remedy adverse impacts to protect the health of citizens. Given that environmental degradation can be trans-boundary, African states must also cooperate with other states on the continent to combat the effects of pollution and other forms of environmental damage.[115] Furthermore, states parties must safeguard affected persons from environmental harm caused by private persons (including multinational corporations) through the enactment of legislative, policy and other measures and must refrain from any conduct that is complicit in such harm.[116] For example, states must create legal frameworks that enable citizens to access to land, guard against arbitrary deprivations of land, ensure that expropriation of land is for a public purpose only and that natural resources are exploited for the benefit of the people.[117]

Secondly, the procedural aspects are characterized by the rights to access 'timely and accessible' information, access a remedy and participate in environmental decision-making as well as the concomitant duty of states to ensure that citizens can contribute to policymaking and the design of environment-friendly development projects.[118] This is to ensure that those most greatly affected by development programmes and projects pertaining to the environment have a say in how these projects are

---

[114] Twinomugisha, 'Some reflections on judicial protection of the right to a clean and healthy environment in Uganda' (n 24) 249.

[115] Hennie Strydom, 'Introduction to Regional Environmental Law of the African Union' in Werner Scholtz and Jonathan Verschuuren *Regional Environmental Law: Transregional Comparative Lessons in Pursuit of Sustainable Development* (Edward Elgar Publishing, 2015) 34.

[116] See also, United Nations, *Guiding Principles on Business and Human Rights: Implementing the United Nations 'Protect, Respect and Remedy' Framework:* Report of the Special Representative to of the Secretary-General on the issue of human rights and transnational corporations and other business enterprises (21 March 2011) A/HRC/17/31: Under Principle 1 states must ensure that their citizens are protected from human rights abuses by businesses by taking appropriate steps to prevent, investigate and redress such abuses including through legislative, policy, judicial and other measures.

[117] ACHPR *State Reporting Guidelines on Articles 21 and 24* (n 71) para.

[118] ACHPR *State Reporting Guidelines on Articles 21 and 24* (n 71) paras 32.

designed, are not negatively affected by their implementation and derive some benefit from said implementation. Related to this, governments are also obliged to conduct environmental, social and human rights impact assessments prior to the commencement of industrial activities and mitigate any identified risks that surface during such assessments.[119] The Maputo Protocol and Natural Resources Convention both give prominence to participation in environmental decision-making as a vital component of the right to a healthy and sustainable environment. In essence, access to remedy entails that people affected are entitled to have access to appropriate judicial and non-judicial grievance mechanisms in order to seek and obtain redress for environmental harm.[120] Moreover, where damage has occurred, affected persons should be entitled to compensation. As demonstrated above, state obligations under the right to a healthy environment are wide-ranging but these obligations encompass the overall duties of state parties to the Charter to respect, protect, promote and fulfil all human rights.[121]

The substantive content of the right to a healthy environment has arguably been more difficult to discern than its procedural components due to the vague wording of article 24.[122] The *SERAC* case and the ACHPR's state reporting guidelines on articles 21 and 24 have, to a large extent, helped elucidate this component of the right. As with other socio-economic rights in the African Charter, the right to a healthy environment under article 24 has equivocal wording and does not contain any specific internal limitations that delimit the scope of particular entitlements of rightsholders as well as the duties of the state.[123] De Vos contends that this was a premeditated move on the part of the drafters that is illustrative of their deference to the idea of the indivisibility and interde-

[119] ACHPR *State Reporting Guidelines on Articles 21 and 24* (n 71) para 30.

[120] ACHPR *State Reporting Guidelines on Articles 21 and 24* (n 71) para 36.

[121] See *SERAC* (n 59); ACHPR *State Reporting Guidelines on Articles 21 and 24* (n 71) para 40.

[122] Emeka Polycarp Amechi, 'Enhancing environmental protection and socio-economic development in Africa: a fresh look at the right to a general satisfactory environment under the African Charter on Human and Peoples' Rights' (2009) 5(1) *Law, Environment and Development Journal* 58 63.

[123] Pierre de Vos, 'A new beginning? The enforcement of social, economic and cultural rights under the African Charter on Human and Peoples' Rights' (2004) 8(1) *Law, Development & Democracy* 1 10.

pendence of all human rights and was done to ensure that this right is interpreted by relevant judicial and quasi-judicial bodies in conjunction with the other provisions of the Charter.[124]

Obligations of Non-state Actors

The activities of non-state actors such as multinational companies carrying out large-scale, land-based projects (especially extractive activities) often result in adverse environmental impacts that are detrimental to the health of communities living in close proximity to such activities. While state actors bear the primary duty of maintaining a clean, healthy and safe environment favorable to the development of citizens, it is accepted that non-state actors also have some responsibilities in this regard.[125] In general, they must respect the rights of peoples to a healthy environment and ensure that their operations do not result in harm to the environment to the extent that would infringe article 24.[126] This would imply a further obligation to ensure that such operations are in conformity with universal human rights and labour and environmental standards. In addition, in order for companies to ascertain which measures are required of them to prevent potential adverse human rights impacts, they would need to conduct human rights, social and environmental impact assessments, in conjunction and consultation with, local communities directly affected by their operations. Where business operations result in infringements or limits on the right to a healthy environment, administrative, civil and criminal liability may arise. Added to that, any resultant harm to affected people would trigger the duty of companies to pay compensation for damages suffered, including costs of rehabilitation and cleaning.[127]

[124] Ibid; see also the *SERAC* decision in which this principle is reaffirmed by the ACPHR, holding that articles 21 and 24 should be interpreted through an 'integrated and complementary reading and application of civil and political rights (Articles 2 and 4) and economic, social and cultural rights (Articles 14, 16 and 18).'

[125] ACHPR *State Reporting Guidelines on Articles 21 and 24* (n 71) para 56.

[126] ACHPR *State Reporting Guidelines on Articles 21 and 24* (n 71) para 44; United Nations, A/HRC/17/31 (n 116), Principle 11.

[127] ACHPR *State Reporting Guidelines on Articles 21 and 24* (n 71) para 59.

The Charter does not expressly recognize the duties of non-state actors with respect to the right to a healthy environment. However, Chapter II of the African Charter,[128] which provides for the duties of individuals, provides a sturdy legal and moral foundation for the obligations of non-state actors in this regard.[129] In line with articles 27 to 29 of the Charter, companies would also be obliged to contribute to the development of host communities including through the provision of support to community-based employment and economic diversification schemes and local development, health, educational or agricultural projects.[130]

6 Conclusion

Most international human rights treaties regard a hygienic environment as a precondition for the enjoyment of, for example, the rights to life, dignity, health, food, water and an adequate standard of living. In the African context, the sentiment is no different. However, the African human rights system sets itself apart from other regional systems in that the right to a healthy environment is explicitly stated as a stand-alone collective right that not only recognizes that a clean environment is necessary for the maintenance of high standards of human health but that it is also an important good for socio-economic development. Essentially, the purpose of the article 24 is to ensure that development projects, especially large-scale, land-based investments on the continent, promote sustainability and human rights and benefit citizens, particularly vulnerable groups such as indigenous people, minorities, women and children. Article 24, as it has been interpreted by the ACHPR, may provide some useful guidance for the recognition of a universal right to a healthy environment. For a considerable length of time, its exact content lacked definitive interpretation but regional case law, scholarly literature on

[128] Article 27 states that: '[e]very individual shall have duties towards his family and society, the state and other legally recognised communities and the international community. The rights and freedoms of each individual shall be exercised with due regard to the rights of others, collective security, morality and common interest.'

[129] ACHPR *State Reporting Guidelines on Articles 21 and 24* (n 71) para 58.

[130] ACHPR *State Reporting Guidelines on Articles 21 and 24* (n 71) para 65.

human rights and the environment, soft law instruments of the ACHPR (especially the state reporting guidelines on article 24) and domestic law provide valuable guidance on the right to a general satisfactory environment favorable to development. An analysis of the same provision indicates that this right evidently has the character of a socio-economic right with some features of a civil and political right. It also encompasses a range of different components and is in align with the global sustainable development agenda. In addition, the above discussion demonstrates that the right to a healthy environment under the African human rights system is a multifaceted environmental human right, comprising both substantive and procedural aspects. It also has, as a key aspect, the maintenance of an environment of the quality that ensures both the protection of human health and the goal of sustainable development and use of natural resources.

References

Alston, P 'Conjuring up new human rights: a proposal for quality control' (1984) 78 *American Journal of International Law* 607.

Amechi, EP 'Enhancing environmental protection and socio-economic development in Africa: a fresh look at the right to a general satisfactory environment under the African Charter on Human and Peoples' Rights' (2009) 5(1) *Law, Environment and Development Journal* 58.

Andersen, MR in 'Human Rights Approaches to Environmental Protection' in Boyle, AE and Andersen, MR (eds), *Human Rights Approaches to Environmental Protection* 15 (Clarendon Press 1996).

Boer, B and Boyle, A 'Human rights and the environment – Background Paper for the 13th Informal ASEM Seminar on Human Rights' (21-23 October 2013) Sydney Law School Research Paper No. 14/14, http://ssrn.com/absract=2393753 accessed 8 May 2016.

Boyd, DR *The Environmental Rights Revolution: A Global Study of Constitutions, Human Rights, and the Environment* (UBC Press 2011).

Chenwi, L 'The right to a satisfactory, healthy, and sustainable environment' in Knox, J and Pejan, R (eds), *The human right to a healthy environment* (Cambridge University Press 2018).

Cullet, P 'Definition of an environmental right in a human rights context' (1995) 13 *Netherlands Quarterly of Human Rights* 29.

Dersso, SA 'The jurisprudence of the African Commission in Human and People's Rights with respect to people's rights' (2006) 6(2) *African Human Rights Law Journal* 358.

De Vos, P 'A new beginning? The enforcement of social, economic and cultural rights under the African Charter on Human and Peoples' Rights' (2004) 8(1) *Law, Development & Democracy* 1.

Ebeku, K 'The right to a satisfactory environment and the African Commission' (2003) 1 *African Human Rights Law Journal* 149.

Fall, M 'Valuable Precedents: The Protection of the Right to Environment by Sub-regional African Courts' (*The ACtPHR Monitor*, 13 September 2016) http://www.acthprmonitor.org/valuable-precedents-the-protection-of-the-right-to-environment-by-sub-regional-african-courts/ accessed 15 June 2019.

Giorgetta, S 'The right to a healthy environment, human rights and sustainable development' (2002) 2 *International Environmental Agreements: Politics, Law and Economics* 172.

Kidd, *Environmental law* 2ed (2nd edn Juta and Company Ltd. 2011).

Kiss, A 'Concept and possible implications of the right to environment in Mahoney, K and Mahoney, P (eds) *Human rights in the twenty-first century: a global challenge* (Martinus Nijhoff Publishers 1993).

Office of the United Nations High Commissioner for Human Rights (OHCHR) 'Mapping Human Rights Obligations Relating to the Enjoyment of a Safe, Clean, Healthy and Sustainable Environment' (2013).

Shelton, D 'Decision regarding Communication 155/96 (*Social and Economic Rights Action Center/Center for Economic and Social Rights v. Nigeria*). Case No. ACHPR/COMM/A044/1' (2002) 96(4) *American Journal of International Law* 937 942.

Sohn, L 'The Stockholm Declaration on the Human Environment' (1973) 14 *Harvard International Law Journal* 423.

Strydom, H 'Introduction to Regional Environmental Law of the African Union' in Werner Scholtz and Jonathan Verschuuren *Regional Environmental Law: Transregional Comparative Lessons in Pursuit of Sustainable Development* (Edward Elgar Publishing 2015).

Trilsch, M 'European Committee of Social Rights: the right to a healthy environment' (2009) 7(3) *I.CON* 529.

Twinomugisha, BK *Fundamentals of health law in Uganda* (Pretoria University Law Press 2015).

Twinomugisha, BK 'Some reflections on judicial protection of the right to a clean and healthy environment in Uganda'(2007) 3(3) *Law, Environment and Development Journal* 244 http://www.lead-journal.org/content/07244.pdf accessed 15 August 2016.

Van der Linde, M and Louw, L 'Considering the interpretation and implementation of article 24 of the African Charter on Human and Peoples' Rights in light of the *SERAC* Communication' (2003) 3(1) *African Human Rights Law Journal* 167.

Viljoen, F *International Human Rights Law in Africa* (2nd edn OUP 2012).

United Nations, *Guiding Principles on Business and Human rights: implementing the United Nations 'Protect, Respect and Remedy' Framework:* Report of the Special Representative to of the Secretary-General on the issue of human rights and transnational corporations and other business enterprises (21 March 2011) A/HRC/17/31.

United Nations, 'Report of the Independent Expert on the issue of human rights obligations relating to the enjoyment of a safe, clean, healthy and sustainable environment, John Knox' (2012) A/HRC/22/4.

United Nations, 'Report of the Special Rapporteur on the issue of human rights obligations relating to the enjoyment of a safe, clean, healthy and sustainable environment' (24 January 2018) A/HRC/37/59.

United Nations 'Issue of human rights obligations relating to the enjoyment of a safe, clean, healthy and sustainable environment' (8 January 2019) A/HRC/40/55.

3

Should a Human Right to a Safe Climate Be Recognized Under the AU Human Rights System?

Ademola Oluborode Jegede

1 Introduction

Climate change is one of the greatest challenges of the twenty-first century.[1] The global nature of the adverse effects of climate change on human populations in every region of the world has been the subject of a range of reports of the Intergovernmental Panel on Climate Change (IPCC),[2] which articulates that climate change is a result of human

[1] UNGA, 'Protection of global climate for present and future generations of mankind' UNG.A. Res. 43/53, 70th plenary meeting 6 December 1988 (UNGA Resolution 43/53).

[2] Established by the World Meteorological Organisation and the United Nations Environment Programme in 1988, the IPCC reviews and accesses the most recent scientific, technical and socio-economical information relating to climate change, see UNGA Resolution 43/53 (n 1); Thomas Stocker and others (eds), *The Physical Science Basis. Contribution of Working Group I to the 5th Assessment Report of the Intergovernmental Panel on Climate Change* (Cambridge University Press 2013) (IPCC Summary for Policymakers).

A. Oluborode Jegede (✉)
Department of Public Law, University of Venda, Thohoyandou, South Africa
e-mail: ademola.jegede@univen.ac.za

M. Addaney, A. Oluborode Jegede (eds.), *Human Rights and the Environment under African Union Law*, https://doi.org/10.1007/978-3-030-46523-0_3

activities and will have widespread effects on human life and natural systems.[3] However, there will be more negative consequences than positive consequences from climate change for Africa, more than in other regions.[4] While the continent contributes little to its cause,[5] this is changing in the light of carbon laden economic path which has become the easy choice of countries including those in Africa.[6] For instance, the carbon profile of the two leading economies in Africa, South Africa and Nigeria, show that in the case of South Africa it is a leading CO_2 emitter in Africa that is ranked among the top 12 emitters in the world,[7] while Nigeria's economic growth is no less characterized by CO_2 emissions.[8] Yet, populations in Africa are vulnerable to climate change.[9]

Based on the vulnerability of human populations to climate change, scholarship has drawn the link between climate change to human rights.[10] The African human rights system which is based on the African Charter on Human and Peoples' Rights (African Charter),[11] and the activities of its treating monitoring body, the African Commission on Human and Peoples' Rights (African Commission),[12] for instance, has generated a

[3] Watson Roberts and others, 'Greenhouse Gases and Aerosols' in JT Houghton, BA Callander & SK Varney (eds), *The Supplementary Report to the IPCC Scientific Assessment* (Cambridge University Press 1992); Stocker and others (n 2).

[4] Paul Collier, Gordon Conway and Tony Venables, 'Climate Change and Africa' (2008) 24 *Oxford Review of Economic Policy* 337.

[5] Ibid.

[6] Gilbert Rist, *The History of Development: From Western Origins to Global Faith* (Zed Books 2009) 21–24

[7] Patrick Sekoai and Michael Daramola, 'Biohydrogen as a Potential Energy Fuel in South Africa' (2015) 6 *Biofuel Research Journal* 223.

[8] Chindo Sulaiman and Abdulsamad Abdul-Rahim, 'Population Growth and CO2 Emission in Nigeria: A Recursive ARDL Approach' (2018) *SAGE Open* 1.

[9] Collier, Conway and Venables (n 4).

[10] Graham McDowell, James Ford and Julie Jones, 'Community-level Climate Change Vulnerability Research: Trends, Progress, and Future Directions' (2016) 11 (3) *Environ Research Letters* 1; Aleksi Räsänen and others, 'Climate Change, Multiple Stressors and Human Vulnerability: A Systematic Review' (2016) 16 (8) *Regional Environmental Change* 2291.

[11] OAU, African Charter on Human and Peoples' Rights, CAB/LEG/67/3 rev. 5, 21 ILM 58, 27 June 1981 (entered into force 21 October 1986).

[12] Makau Mutua, 'The African Human Rights System: A Critical Evaluation' http://hdr.undp.org/sites/default/files/mutua.pdf accessed 21 January 2020.

number of relevant resolutions. Resolutions 153,[13] 148[14] and 271[15] of the African Commission point out the link of human rights to climate change.[16] Apart from the African Charter, which confers individual and collective rights and imposes obligations on states, there are other relevant instruments under the African Union (AU) which arguably relate to climate change. In the main, these instruments are the African Union Convention for the Protection and Assistance of Internally Displaced Persons in Africa (Kampala Convention),[17] and the African Convention on the Conservation of Nature and Natural Resources (Revised version),[18] to which different states in Africa have been parties. Additionally, states in Africa are signatories to climate-related instruments relevant to human rights such as the Paris Agreement and the United Nations Sustainable Development Goal.[19] While the foregoing developments are indicative of a system that recognizes the potential link of climate change to human rights, the right to a safe climate has not crystalized within the AU human rights system.

Alston argues that for a new right to emerge, six criteria are required: such a right should reflect a fundamental social value; relevant throughout the world value system; recognized as an interpretation of Charter obligations, a reflection of customary law or a formulation that is

[13] AU, African Commission of Human and Peoples' Rights, ACHPR/Res153 (XLVI) 09: Resolution on Climate Change and Human Rights and the Need to Study Its Impact in Africa (25 November 2009).

[14] AU, African Commission of Human and Peoples' Rights, 148: Resolution on the Establishment of a Working Group on Extractive Industries, Environment and Human Rights Violations in Africa, adopted at 46th ordinary session held in Banjul, The Gambia (11–25 November 2009).

[15] AU, African Commission on Human and Peoples' Rights '271: Resolution on Climate Change in Africa', adopted at the 55th ordinary session of the African Commission on Human and Peoples' Rights held in Luanda, Angola (28 April–12 May 2014).

[16] Ademola Oluborode Jegede, 'Climate Change in the Work of the African Commission on Human and Peoples' Rights' (2017) 31 (2) *Speculum Juris* 136.

[17] African Union Convention for the Protection and Assistance of Internally Displaced Persons in Africa (Kampala Convention), adopted by the Special Summit of the Union held in Kampala, Uganda, 23rd October 2009.

[18] The African Convention on the Conservation of Nature and Natural Resources (Revised version) (Conservation Convention 2003).

[19] Paris Agreement under the United Nations Framework Convention on Climate Change 2015, FCCC/CP/ 2015/L.9/Rev.1, adopted by Conference of the Parties, 21st Session Paris, 30 November–11 December 2015.

declaratory of general principles of law; not repetitive of existing body of international human rights law; capable of attaining high degree of international consensus; compatible with state practice; be precise to give rise to identifiable right and obligation.[20] Whether there is basis for arguing a right to safe climate should exist under the AU human right system is the focus of this contribution. A caveat is, however, necessary from the onset. In making the argument in this paper, it is acknowledged that considering the nature of climate change causation, efforts of states in Africa alone cannot halt a changing climate. However, it is an essential feature of the human rights concept that in the matter of the realization of rights, a state cannot refuse to discharge its obligations by citing the actions or inactions of other states.[21] Hence, this paper argues the vulnerability of human populations in Africa and the essential compliance with yardsticks for a new right as the bases for the right to a safe climate to emerge under the AU human rights system.

2 Defining Human Vulnerability and Climate Change in Africa

Vulnerability denotes susceptibility to harm and has been applied in different fields and disciplines.[22] In the climate change field, vulnerability research focuses on where, how and why human systems are affected by changes in the climate.[23] According to Liverman, vulnerability 'has been related or equated to concepts such as resilience, marginality, susceptibility, adaptability, fragility, and risk'.[24] In the context of climate change,

---

[20] Philip Alston, 'Conjuring up New Human Rights: A Proposal for Quality Control' (1984) 78 *American Journal of International Law* 607.

[21] Oliver De Schutter and others, 'Commentary to the Maastricht Principles on Extraterritorial Obligations of States in the Area of Economic Social and Cultural Rights' (2012) 34 *Human Rights Quarterly* 34.

[22] James Ford and others, 'Vulnerability and Its Discontents: The Past, Present, and Future of Climate Change Vulnerability Research' (2018) 151 *Climatic Change* 189.

[23] Ibid.

[24] Diana Liverman, 'Vulnerability to Global Environmental Change' in R Kasperson and others (eds) *Understanding Global Environmental Change: The Contributions of Risk Analysis and Management* (Earth Transformed Program, Clark University 1990) 27–44

vulnerability has been defined as 'the degree to which geophysical, biological and socio-economic systems are susceptible to, and unable to cope with the adverse impacts of climate change'.[25] Existing scholarship shows that human vulnerability to the adverse effects of climate change is real. Drawing attention to this fact, Aminzadeh urges that 'human beings are the icon of climate change'.[26] Equally, the concept of vulnerability has found expression in human rights discourse at the AU human rights level. In 2009, with the adoption of Resolution 153, titled 'Resolution on Climate Change and Human Rights and the Need to Study its Impact in Africa',[27] the African Commission elaborated the link between climate change and human rights. In that Resolution, the African Commission expressed the concern that human rights standards are lacking in 'various draft texts of the conventions under negotiation' and that this lack could jeopardize 'the life, physical integrity and livelihood of the most vulnerable members of society notably isolated indigenous and local communities, women, and other vulnerable social groups'.[28] According to that resolution:

> African regional standards for the protection of the environment, management of natural resources and human rights are consistent with provisions of the Convention on Biological Diversity.

Arguably, based on the above position, the Commission appears to have set the stage for the application of human rights to climate change in the region. The reference to the 'African regional standards' for the protection of human rights and the environment presupposes that there exists the prospect at the regional level to address the plight of a population or an individual facing the impact of climate change. Reinforcing the position that the reference to human rights in this Resolution is far

---

[25] Martin Perry and others (eds), *Climate Change 2007: Working Group II: Impacts, Adaptation and Vulnerability* (Cambridge University Press 2007) 976 .

[26] Ibid.

[27] African Commission of Human and Peoples' Rights, ACHPR/Res153(XLVI)09: Resolution on Climate Change and Human Rights and the Need to Study its Impact in Africa, November 25, 2009 (Resolution 153).

[28] Ibid, preamble.

from being casual, the Commission made recommendations. Among others, it urged the Assembly of Heads of State and Government of the African Union (AU) to ensure

> that human rights standard safeguards, such as the principle of free, prior and informed consent, be included into any adopted legal text on climate change as preventive measures against forced relocation, unfair dispossession of properties, loss of livelihoods and similar human rights violations… (2) that special measure of protection for vulnerable groups such as children, women, the elderly, indigenous communities and victims of natural disasters and conflicts are included in any international agreement or instruments on climate change.[29]

With the adoption of Resolution 127 which requires the Working Group on Extractive Industries to carry out an in-depth investigation into the impact of climate change on human rights,[30] there is little doubt that the African Commission takes climate change to be a serious issue in Africa.

The foregoing trend is not surprising in that the adverse effects of climate change are indeed overwhelming for Africa. As Zewdie found, after a careful review of scientific evidence on climate change effects on food production, climate change has an impact on food accessibility and utilization in sub-Saharan Africa.[31] African countries are more affected by climate change because of their reliance on agriculture as well as their lower financial, technical and institutional capacity to adapt.[32] In relation to health damages and loss associated with climate change, according to United Nations Economic Commission for Africa (UNECA), African countries will suffer health consequences due to impacts of climate change as many African countries have populations that are among the

[29] Ibid.

[30] Ibid.

[31] Alemu Zewdie, 'Impacts of Climate Change on Food Security: A Literature Review in Sub Saharan Africa' (2014) 5 (8) *Journal of Earth Science and Climatic Change* 225.

[32] Abhinav Signgh and Bharathi Purohit, 'Public Health Impacts of Global Warming and Climate Change' (2014) 26(1) *Peace Review-A. Journal of Social Justice* 112; Rose Rindap, 'The Impact of Climate Change on Human Security in the Sahel Region of Africa' 2015 1 (2) *Donnish Journal of African Studies and Development* 9.

most vulnerable to climatic changes in the world.[33] Studies have established that health impacts from climate change are expected to be particularly serious in Central and Southern Africa.[34] Wright argues that climate change will cause serious damage to health care delivery.[35] The vulnerability of the continent is exacerbated by existing problems of poverty, weak institutions and armed conflict, which limit a population's capacity to deal with the additional health challenges posed by climate change.[36]

Climate change is considered to be largely water change.[37] Globally, water insecurity is the main way through which the loss and damages associated with climate change will be more felt (OECD).[38] Water is a fundamental element of the climate system in that a change in climate will ultimately come down to changes in the availability, quality, quantity and delivery of water.[39] These changes are apt to produce risks to water security, which is generally one of the main objectives of water resources management at all levels.[40] Although Africa is likely to record a small increase in mean annual flow of water, it will experience a 200% increase in demand, exacerbating existing pressure on water availability.[41] In general terms, Southern Africa will experience reduced rainfall and hence

[33] United Nations Economic Commission for Africa African Climate Policy Centre, 'Climate Change and Health Across Africa: Issues and Options' (2011) UNECA Documents and Publishing Unit Working Paper 20, https://www.uncclearn.org/sites/default/files/inventory/uneca15.pdf accessed 21 January 2020.

[34] Jonathan Myers and others, 'A Public Health Approach to the Impact of Climate Change on Health in Southern Africa – Identifying Priority Modifiable Risks' (2011) 101 *South African Medical Journal* 101.

[35] Caradee Wright, 'Climate Change and Its Effect on Healthcare Systems' http://www.bizcommunity.com/Article/196/330/175557.html accessed 21 January 2020.

[36] n 33 above

[37] Tuula Honkonen, 'Water Security and Climate Change: The Need for Adaptive Governance' (2017) 20(1) *Potchefstroom Electronic Law Journal* 1.

[38] OECD, 'Water and Climate Change Adaptation: An OECD Perspective' (2013) 18, https://www.oecd.org/env/resources/Water%20and%20Climate%20Change%20Adaptation-%20brochure.pdf accessed 21 January 2020.

[39] Christiana Cook and Karen Bakker, 'Water Security: Debating an Emerging Paradigm' (2012) 27(1) *Global Environmental Change* 27 94

[40] Patricia Wouters, *Water Security: Global, Regional and Local Challenges* (Institute for Public Policy Research 2010).

[41] Vivideconomic, 'The Cost of Adaptation to Climate Change in Africa' Report prepared for the African Development Bank Group (2012), http://www.vivideconomics.com/wp-content/uploads/2015/03/Cost_of_Adaptation_in_Africa.pdf accessed 21 January 2020.

reduced overall water availability, with Botswana, Swaziland, Lesotho and South Africa all expected to experience around a 20% reduction in water availability.[42] In all, the potential impacts of climate change and related risks will have implications for water security in Africa.[43] In affecting different aspects that touch on quality of human welfare, it is beyond doubt that vulnerability of populations to adverse effects of climate change is a basis for crystalizing climate change as a human right issue in Africa. However, whether and how this link can form the basis for a new right to a safe climate under the AU human rights system merits articulation.

3 The Criteria for Establishing Safe Climate as a Human Right: Testing the Waters

According to Alston, for a new right to emerge, six criteria are required: such a right should reflect a fundamental social value; relevant throughout the world value system; recognized as an interpretation of Charter obligations, a reflection of customary law or a formulation that is declaratory of general principles of law; not repetitive of existing body of international human rights law; capable of attaining high degree of international consensus; compatible with state practice; be precise to give rise to identifiable rights and obligations.[44] The yardsticks set out by Alston have been tested and deployed in a number of works covering areas such as safe motherhood,[45] and the right not to be displaced.[46] Arguably, establishing a human right to safe climate under the AU meets the criteria set out by Alston.

---

[42] Ibid.

[43] Mike Muller, 'Urban Water Security in Africa: The Face of Climate and Development Challenges' (2016) 33 (1) *Development Southern Africa* 67.

[44] Alston (n 20).

[45] Hélène Julien. 'Is There a Human Right to Safe Motherhood Within the United Nations Legal System?' (2015) 2 (1) *QMHRR* 1

[46] Maria Stavropoulou, 'The Right Not to Be Displaced' (1994) 9 (3) *American University International Law Review* 689

Firstly, according to Alston, a new right should 'reflect a fundamentally important social value'.[47] Climate change has adverse effects on social life which compromise the human dignity of vulnerable populations. In affecting the security of water, food and livelihood, climate change undermines human dignity, hence, a right seeking to address the adverse effects of climate change on human wellbeing constitutes the protection of an important social value.

Secondly, Alston points out that to qualify as a human right, a right has to 'be relevant, inevitably to varying degrees, throughout a world of diverse value systems'.[48] As earlier indicated, the adverse effects of climate change are universal, and Africa is not an exception. The IPCC reports have clearly articulated the adverse effects of climate change on populations in Africa. The universal nature of the adverse effects of climate change in itself connotes that a universal normative value should not be out of contemplation while advancing appropriate legal response to the adverse effects of climate change. An important feature of human rights is their universality. The principle of universality and inalienability connotes that human rights apply to everyone everywhere in the world and that negotiations or 'trade-offs' should not result in human rights violations.[49] The notion of universality and inalienability is a core feature of the human rights approach in Africa. It is recognized in the human rights instruments at the continental level. For instance, the Preamble of the African Charter recognizes the universality of all rights, while article 46 of the Children Charter claims that the sources of inspiration for the instruments are the Universal Declaration on Human Rights,[50] the

---

[47] Alston (n 20) 615.

[48] Ibid.

[49] Vienna Declaration and Programme of Action, adopted at World Conference on Human Rights in Vienna 1993 UN doc.A/CONF.157/23 paras 1 and 5; the idea of universality of human rights is however challenged by relativists who view that human rights vary from culture to culture, see Sally Merry, *Human Rights and Gender Violence: Translating International Law into Local Justice* (University of Chicago Press 2006) 40; however human rights is at least universal in the sense that it has become a subject of attention all over the world, see Jack Donelly, *International Human Rights Law: Universal, Relative, or Relatively Universal* in Mashood Baderin and Manisuli Ssenyonjo (eds), *International human rights law: Six decades after the UDHR and beyond* (Aldershot 2010) 31–48, who argues, in my view, rightly that universality remains the core feature of human rights.

[50] Universal Declaration on Human Rights, UNGA Res 217 A(III), adopted 10 December 1948

International Convention on the Rights of the Child,[51] and African values and traditions. The Maputo Protocol affirms in its preamble that human rights are 'inalienable, interdependent and indivisible'.[52] This connotes that except the adverse effects of climate change on vulnerable populations give rise to a right to safe climate, the true meaning of universality and inalienability of rights within the African context is disputed in that it fails to address one of the greatest crisis facing mankind continuous existence. Such a development mocks, if not entirely defeats the purpose of the general scope of universality and inalienability of rights under the AU system.

According to Alston's third consideration, in order to qualify as a human right, a right has to derivate from 'an interpretation of UN Charter obligations,[53] a reflection of customary law rules or a formulation that is declaratory of general principles of law'.[54] The obligations under the United Nations (UN) relate mainly to international peace.[55] All states in Africa are members of the UN, hence, they are committed to international peace. Climate change is accepted as a global challenge to mankind,[56] and its link to conflict brings the phenomenon to the center of international peace. The link of climate change to conflict is reinforced in the 2009 report of the UN Secretary General to the United Nations General Assembly which affirms that climate change has security implications,[57] and subsequently in 2015 by the report of G7 which stresses the link of climate change to fragility of states.[58] Accordingly, formulating the right to a safe climate will be reinforcing the commit-

[51] United Nations Convention on the Rights of the Child, 1577 UNTS 3, 28 ILM 1456, 20 November 1989 (entered into force 2 September 1990).

[52] African Union Protocol to the African Charter on Human and Peoples' Rights on the Rights of Women in Africa, adopted by the 2nd Ordinary Session of the Assembly of the Union Maputo (11 July 2003).

[53] Alston (n 20) 615.

[54] Ibid.

[55] United Nations, *Charter of the United Nations*, 24 October 1945, 1 UNTS XVI.

[56] UNGA Resolution 43/53 (n1).

[57] UNGA, 'Climate Change and Its Possible Security implications' Report of the Secretary-General, A/64/350 11 September 2009.

[58] G7, 'A New Climate for Peace' https://www.newclimateforpeace.org/#report-top accessed 21 January 2020.

ment of states in Africa to international peace, an obligation that arises under the United Nations Charter.

The customary law status of a rule depends on whether the principle has been referred to, or put into operation, in a treaty, in a soft law instrument, in judicial or semi-judicial decisions, or in other expressions of state practice.[59] Arguing a right to a safe climate under the AU is not inconsistent with customary international law. As earlier mentioned the link of climate change to human rights has been made in a number of soft law instruments such as the resolutions made by the African Commission.[60] This development signifies that it is not legally impossible to assert that the right to a safe climate could soon ascend to international customary law status. While no judicial and or quasi-judicial institutions have considered a dispute in relation to climate change and human rights at the AU level, the evidence of state practice is evolving in the judiciary of states within the AU. For instance, the judiciary of at least two member states have engaged with climate change in relation to project execution and right to life. In Nigeria, *Gbemre* v. *Shell Petroleum Development Company Nigeria Limited and others*[61] arose from gas flaring activities in the Niger Delta area involving communities and other non-state actors, namely, Shell, ExxonMobil, Chevron-Texaco, and the Nigerian National Petroleum Corporation.[62] The communities contended that the practice of gas flaring and the failure by the corporations to complete an environmental impact assessment violated Nigerian gas flaring regulations. Finding that gas flaring contributes to climate change and violates human rights, the Court ordered the companies to stop gas flaring in the Niger Delta.[63]

---

[59] Pierre-Marie Dupuy, 'Formation of Customary International Law and General Principles' in Daniel Bodansky, Jutta Brunnée and Ellen He (eds), *The Oxford Handbook of International Environmental Law* (Oxford University Press, 2007) 449; James Anaya, 'Customary International Law: The Challenge of Non-State Actors' (1998) 92 (1–4) *Proceedings of the Annual Meeting (American Society of International Law)* 41

[60] Jegede (n 16)

[61] Gbemre v. Shell Petroleum Development Company Nigeria Limited and Others [2005]) AHRLR 151 (NgHC 2005).

[62] Ibid, para 51

[63] Ibid, paras 5–7.

While it does not engage with human right but environmental impact assessment, the South Africa's case of *Earthlife Africa Johannesburg* v. *Minister of Environmental Affairs and others*[64] is very instructive in terms of compatibility with a right to safe climate. It concerned a 1200 MW coal-fired power station near Lephalale in the Limpopo Province. A party seeking to construct a new coal-fired power station requires, among other things, an environmental authorization to be granted by the Minister of Environmental Affairs who granted the same in the matter. The applicant—Earthlife Africa—successfully approached the court to review the decision to grant the environmental authorization. In granting the review, the Court approved a climate impact assessment, which comprises an analysis of the extent to which a proposed coal-fired power station will contribute to climate change over its lifetime by quantifying its GHG emissions during construction, operation and decommissioning; the resilience of the coal-fired power station to climate change, taking into account how climate change will impact on its operation, through factors such as rising temperatures, diminishing water supply, and extreme weather patterns; and how these impacts may be avoided, mitigated or remedied. The reference to climate impact assessment demonstrates the willingness of the court to be pro-climate in deserving circumstances by linking projects to climate change and human rights where the need arises. Based on the foregoing, the conclusion can be reached that a right to safe climate may crystalize in Africa customary law rules.

The fourth requirement set up by Alston in order to recognize a human right is that a new right 'be consistent with, but not merely repetitive of, the existing body of international human rights law'.[65] The right to a safe climate stresses the connection between several other human rights and is not repetitive of any. There is no such specific right, that is, the right to a safe climate, known to international human rights law nor its application in Africa. Hence, the formulation of such a right is not repetitive of any existing right. At best what is evident in Africa is the recognition of the right to a general satisfactory environment favorable to the development

[64] Earthlife Africa Johannesburg v. Minister of Environmental Affairs and others [2017] JOL 37526 (GP).

[65] Alston (n 20) 615.

guaranteed under article 24 of the African Charter. Arguably, this is not the same as the right to a safe climate in that it does not speak to the specificity of climate change and more importantly, article 24 accommodates other issues in the environment such as littering, disposal of waste, air pollution, to mention a few, which may or may not be discussed without reference to climate change. Arguably, the right to a safe climate overlaps with the environment but it goes farther in that it touches on climate system which binds all mankind together and crystalizes a right which has not been so recognized under the AU system.

The fifth requirement that Alston enunciates on establishing a new right is that this right should 'be capable of achieving a very high degree of international consensus'.[66] The application of this requirement to a safe climate does not raise any difficulty. In fact, stabilizing the climate system has been the desire of larger proportion of the world system. This is more so as it has been found that it is nearly impossible for conventional climate response measures to achieve net zero carbon dioxide (CO_2) emissions and stabilize global temperatures below 2°C on their own.[67] The evidence of such consensus is found in the near 185 nations that are signatories to the Paris Agreement.[68] The Agreement does not only commit state parties to limit the global average temperature increase to 'well below' 2°C.[69] It urges parties, when taking action to address climate change, to respect, promote and consider their respective obligations on human rights.[70] With 12 more ratifications, the Paris Agreement will have the same number of parties as the UNFCCC, which, with 197 Parties, is nearly universal. While this may not necessarily translate into automatic universal endorsement of the right to a safe climate, it does indicate that the idea of recognizing a right to a safe climate is not alien to international community as it centers on an ongoing global concern.

---

[66] Ibid, 615.

[67] Royal Society & Royal Academy of Engineering, *Greenhouse Gas Removal* (Royal Society & Royal Academy of Engineering 2018) 7; Stocker and others (n 2) 25.

[68] 'Paris Agreement - Status of Ratification' https://unfccc.int/process/the-paris-agreement/status-of-ratification accessed 21 January 2020.

[69] Paris Agreement (n 2) art. 2.

[70] Paris Agreement (n 2) preamble.

The sixth requirement laid out by Alston is that a new right should 'be compatible or at least not clearly incompatible with the general practice of States'.[71] In this regard, it is evident that a right to a safe climate is not clearly incompatible with the general practice of states. The Paris Agreement aims to address the threat of climate change by requiring states to make nationally determined contributions to greenhouse gas mitigation (NDCs) and working toward meeting the commitments in their NDCs, and to transparently report on their activities toward fulfilling those commitments (the transparency framework which ensures that all countries are playing their part in addressing the challenge of climate change). The COP24 parties agreed to a 'rule-book' dubbed the Katowice Climate Package (KCP), a set of guidelines for implementing the Paris Climate Change Agreement.[72] This 'rule-book' gives NDCs more meaning or 'bite' and promises to support a hardening of NDCs into free-standing binding commitments.[73] Of the 185 members, 184 states have submitted their first NDCs.[74] Along similar line, as mentioned earlier, state practice is also evident in cases of *Gbemre*[75] and *Earthlife Africa Johannesburg v. Minister of Environmental Affairs*[76] where court respectively ruled to safeguard the climate in the context of implementing development projects.

The seventh and last of Alston's requirements is that the right 'be sufficiently precise as to give rise to identifiable rights and obligations'.[77] Human rights that are impacted by adverse effects of climate change are entrenched in a range of instruments ratified by states under the AU system. In particular, it impacts on the African Charter an instrument which has been interpreted as conferring the obligations to protect, fulfil and

[71] Alston (n 20) 615.

[72] 'UNFCCC' https://unfccc.int/news/new-era-of-global-climate-action-to-begin-under-paris-climate-change-agreement-0 accessed 21 January 2020.

[73] Frederic Sourgens, 'Paris Agreement: Regained or Lost? Initial Thoughts' *EJIL: Talk!* https://www.ejiltalk.org/paris-agreement-regained-or-lost-initial-thoughts/ accessed 21 January 2020.

[74] UNFCCC, 'NDC Staging' https://www4.unfccc.int/sites/ndcstaging/Pages/Home.aspx accessed 21 January 2020.

[75] *Gbemre case (*n 61).

[76] *Earthlife case* (n 64).

[77] Alston (n 20) 615.

promote human rights.[78] Climate change threatens a range of rights under the Africa Charter such as the right to life, the right to have access to sufficient food and water, the right to have access to health care services, the right to property, the right to adequate housing, and the right to environment. Article 4 of the African Charter urges states to make effort to avoid threat to human life. The right to life can be hindered by adverse effects of climate change such as sea level rise which may occasion flooding.[79] Climatic occurrences such as sea level rise and flooding may also breach article 14 of the African Charter which guarantees the right to property, providing that it can be limited only in the interest of public policy and in accordance with the provision of the law.

Arguably for the reasons that climate change negatively impacts on a range of rights and state obligations, there is the necessity for a right to a safe climate to emerge. Climate change implicates the right to food, which although not categorically stated under the African Charter, is evident from the jurisprudence of the Commission which considered the right to food in an environmental context. In *Ogoniland* case, the Commission interpreted articles 4 (right to life), 16 (right to health) and 22 (right to economic, social and cultural development) to ground a violation of the right to food.[80] Climate change may challenge the availability, accessibility and affordability elements of the right to water of populations in Africa and thereby constitutes a breach of article 16(1) of the African Charter. Climate change can affect the settlement of populations and therefore constitutes a threat to their right to housing. This is particularly so in Africa, where, notwithstanding a low adaptive capacity, global warming will generate problems, including heat waves, flooding, pollution and a rise in sea level.[81] All the elements in the right to adequate housing, namely, legal security of tenure, availability of services, materials, facilities and infrastructure, accessibility, location, affordability,

---

[78] Ibid.

[79] Ibid.

[80] Communication 276/03, *Centre for Minority Rights Development (Kenya) and Minority Rights Group (on behalf of Endorois Welfare Council)* (*Endorois* case) 27th Activity Report: June–November 2009,

[81] Camilla Toulmin, *Climate change in Africa* (Zed Books 2009) 87.

habitability, and cultural adequacy,[82] are disturbed when populations are displaced or forcefully evicted from their shelter in the event of climate change and implementation of adaptation and mitigation initiatives without alternative provisions that meet with their rights. In addition to highlighting that displacement can arise from 'natural or human- made disasters', the Kampala Convention specifically refers to the obligations of states to protect and assist persons displaced by climate change. In particular, article 5(4) requires states to take measures to protect and assist persons who have been internally displaced due to natural or human-made disasters, including climate change.

The impact of climate change will worsen an already calamitous situation of access to water in Africa: water stress exists in various countries including Tunisia, Algeria, Morocco, Sudan and, indeed, in most parts of sub-Saharan Africa.[83] The right to water is not expressly mentioned in the African Charter, but can be derived from article 16(1). Climate change may challenge the availability, accessibility and affordability elements of the right to water,[84] and thereby constitutes a breach of the right of populations to water as implied under article 16(1) of the African Charter. The African Convention on the Conservation of Nature and Natural Resources has a range of interesting provisions that have direct and indirect potential on the protection of vulnerable populations.[85] More specifically, it calls for states to take measures to prevent land and soil degradation, protect and conserve vegetation cover which include forest covers[86] and manage water resources.[87]

Where accessibility to natural resources such as water is hindered as a result of adverse impacts of climate change and response measures, it is capable of driving populations into clashes with settled populations and

[82] United Nations, General Comment No. 7: The Right to Adequate Housing, Art 11 (1) of the Covenant: Forced Evictions' (1997) 16th session (United Nations General Comment No. 7).

[83] Toulmin (n 81) 40.

[84] 'United Nations, General Comment No. 15: The right to water, arts 11 and 12' (2000) para 1 (United Nations General Comment No. 15).

[85] African Convention on the Conservation of Nature and Natural Resources, Algeria on 15th September 1968 OAU Doc CAB/LEG/24.1.

[86] Ibid, art. VI.

[87] Ibid, art. VIII.

will result in their displacement from their traditional territories. For instance, the incidence of conflict in Darfur has been connected with drought and famine which led to disruption in social and economic life and loss to herding and farming families.[88] Further reiterating such a possibility is the evidence in West African states, including Ghana and Burkina Faso, where climate change has been found to have security implications.[89] The failure of an appropriate legal framework to anticipate that scarcity of natural resources due to the adverse impacts of climate change can lead to conflict and will constitute a threat to the right to peace under article 23 of the African Charter which safeguards the right of 'all peoples' to national and international peace and security. In fact, it derives from obligations of states as expressed under the AU system. The obligation to comply with internationally recognized human rights requires three levels of duty from states: the duty to respect, protect and fulfil human rights. The obligation to respect signifies that states must refrain from interfering with or hindering the enjoyment of human rights. The obligation to protect demands that individual and groups should be protected from human rights abuses, especially by non-state actors. The obligation to fulfil requires states to take positive action to facilitate the enjoyment of basic human rights.[90] The conceptualization of these obligations owes its introduction and current influence on international human rights law to the pioneering work of Shue and Eide.[91] In Shue's view, the tripartite typology of duties include (1) duties to avoid the deprivation of the right concerned, (2) duties to protect rights holders from deprivation, and (3) duties to aid rights holders who have been

[88] Toulmin (n 81) 111; Oli Brown, Anne Hammill and Robert Mcleman, 'Climate Change as the 'new' Security Threat: Implications for Africa' (2007) 83 *International Affairs* 1141, 1143.

[89] Oli Brown and Alec Crawford, *Assessing the Security Implications of Climate Change for West Africa' Country Case Studies of Ghana and Burkina Faso* (IISD 2008) 9–39.

[90] Oliver De Schutter, 'Economic, Social and Cultural Rights as Human Rights: An introduction,' (2013) CRIDHO Working Paper 2, 6 http://cridho.uclouvain.be/documents/Working.Papers/CRIDHO-WP2013-2-ODeSchutterESCRights.pdf accessed 19 January 2020; OHCHR, 'International Human Rights Law." http://www.ohchr.org/en/professionalinterest/pages/internationallaw.aspx accessed 19 January 2020.

[91] Ibid, 5.

deprived.[92] The tripartite obligations have since gained international acceptance, first among scholars working on the right to food, and then in the broader area of economic, social and cultural rights.[93] Reinforcing the foregoing view, Eide argues that the tripartite obligations entail the negative obligation to abstain from acts contrary to human rights principles and a positive duty as a 'protector and provider' of rights.[94]

The African human rights system offers four layers of obligations. In *Ogoniland* case,[95] the African Commission, in the context of environmental claims over the degradation of the land of Ogoni people developed jurisprudence on a four-layer of obligation in respect of the rights, civil, political and socio-economic rights, guaranteed under the African Charter. The obligation to 'respect', 'protect', 'promote' and 'fulfil'. According to the African Commission, the obligation to respect entails that states should not interfere in the enjoyment of human rights. Also, it signifies that there should be respect on the part of the state for 'right-holders, their freedoms, autonomy, resources, and liberty of their action'.[96] In relation to the situation of a collective group, the obligation to respect entails that resources collectively belonging to this group should be respected.[97] In discussing the obligation to protect, the African Commission enjoins the state to adopt measures, including legislation, and provide effective remedies in protection of right holders 'against political, economic and social interferences'. It requires the regulation of non-state actors to ensure that their operation does not hinder the realization of rights.[98] Corresponding to the obligation to protect human

[92] Henry Shue *Basic Rights: Subsistence, Affluence, and U.S. Foreign Policy* (2nd ed. Princeton University Press 1996) 52.

[93] United Nations, General Comment No. 12: The Right to Adequate Food, U.N. ESCOR, Comm. on Econ., Soc. & Cult. Rts., 20th Sess., 14–20, U.N. Doc. E/C.12/1999/5 (1999) (United Nations General Comment No.12); General Comment No. 13:The Right to Education, U.N. ESCOR, Comm. on Econ., Soc. & Cult. Rts., 21st Sess., 46–48 (1999) (United Nations General Comment No.13).

[94] Austin Eide, 'Realisation of Social and Economic Rights and the Minimum Threshold Approach' (1989) 10 *Human Rights Law Journal* 35.

[95] Social and Economic Rights Action Centre (SERAC) and Another v Nigeria (2001) AHRLR 60 (ACHPR) (*Ogoniland* case).

[96] Ibid, para 45.

[97] Ibid.

[98] Ibid, para 46.

rights, according to the Commission, is the obligation to promote the enjoyment of all human rights,[99] which entails that the state should ensure 'that individuals are able to exercise their rights, for example, by promoting tolerance, raising awareness, and even building infrastructures'.[100] The obligation to fulfil, according to the Commission, requires the state to mobilize 'its machinery towards the actual realisation of the rights'.[101]

Arguably, failure by a state to formulate appropriate legislation for the protection of its vulnerable populations in the context of climate change at the national level is incompatible with the levels of duties imposed on states by African regional jurisprudence. It offends the obligation to respect because it signifies that states in Africa can be involved in climate change response projects without an appropriate legal basis to address its consequences. This situation flies in the face of the provision of the Kampala Convention that requires parties to ensure that communities are not displaced from their lands, except for compelling and overriding public interests.[102]

Furthermore, ignoring the need to formulate the right to a safe climate may constitute a breach of the obligation to protect because the inadequacy of the climate change regulatory framework at the national level represents a contrast to the formulation of legislation for the provision of effective remedies, and the regulation of non-state actors which the obligation to protect embodies. Failure to do this runs foul of article 3(h) of the Kampala Convention which places an obligation on the State in respect of the accountability of non-state actors, including 'multinational companies and private military or security companies, for acts of arbitrary displacement or complicity in such acts'. The obligation toward accountability extends to situations where non-state actors are involved in the 'exploration and exploitation of economic and natural resources leading to displacement'.[103] An inadequate climate change regulatory

---

[99] Ibid.

[100] Ibid.

[101] Ibid, para 47.

[102] Kampala Convention (n 17) art 4(5).

[103] Kampala Convention (n 17) art 3 (i).

framework means that the involvement of non-state actors in projects such as REDD+ which may displace indigenous peoples can remain largely unchecked. This is inconsistent with the spirit and letter of the Kampala Convention. Indeed, it is incompatible with article IV of the Conservation Convention which affirms the obligation of state parties to adopt and implement preventive measures.

Also not to formulate a right to safe climate may fail to address the inadequate attention by the national legal framework to climate initiatives, a development that is inconsistent with the obligation of states in Africa to promote the enjoyment of rights. Contrary to the promotion of a culture of tolerance and awareness-raising that the obligation entails, a weak framework indicates that there remains a lack of tolerance for the culture and lifestyle of vulnerable populations even in the face of the adverse impacts of climate change. It sends a wrong signal to non-state actors and international organizations involved in the implementation of projects, demonstrating that respect for the plight of vulnerable populations is not required in Africa.

4 Conclusion

Just as it is with the science of climate change, the vulnerability link of the adverse effects of climate change to human rights should be taken seriously. The latter is not yet the case despite the developments at the AU level, which as indicated in the discussion, are compatible with such a relationship and indeed the crystallization of a right to a safe climate. Departing from Alston model as a tool of assessing the necessity of a new human right, this contribution asserts that the right to a safe climate can emerge at the AU level. Such a right would offer specific protection or prevention of activities that result in climate change which have implications for human rights of the populations. The central attribute of this protection would be undeniably related to carbon generating activities such as fossil fuel combustion, coal-related extraction and of course climate mitigation and adaptation measures which infringe on rights of populations. According to this right, every person would be able to enjoy as a right holder the entitlement to litigate state and non-state actors as the duty bearer for remedies over their activities linked to climate change which infringe on the rights of populations.

References

Alston P, 'Conjuring up New Human Rights: A Proposal for Quality Control' (1984) 78 *American Journal of International Law* 607.

Anaya J, 'Customary International Law: The Challenge of Non-State Actors' (1998) 92 (1–4) *Proceedings of the Annual Meeting (American Society of International Law)* 41.

AU, African Commission of Human and Peoples' Rights, 148: Resolution on the Establishment of a Working Group on Extractive Industries, Environment and Human Rights Violations in Africa, adopted at 46th ordinary session held in Banjul, The Gambia (11–25 November 2009).

AU, African Commission of Human and Peoples' Rights, ACHPR/Res153 (XLVI) 09: Resolution on Climate Change and Human Rights and the Need to Study its Impact in Africa (25 November 2009).

AU, African Commission on Human and Peoples' Rights '271: Resolution on Climate Change in Africa', adopted at the 55th ordinary session of the African Commission on Human and Peoples' Rights held in Luanda, Angola (28 April–12 May 2014).

Brown O and Crawford A, *Assessing the Security Implications of Climate Change for West Africa' Country Case Studies of Ghana and Burkina Faso* (IISD 2008) 9–39.

Brown O, Hammill, A and Mcleman R, 'Climate Change as the 'new' Security Threat: Implications for Africa' (2007) 83 *International Affairs* 1141.

Collier P, Conway G and Venables T, 'Climate Change and Africa' (2008) 24 *Oxford Review of Economic Policy* 337.

Cook C and Bakker K, 'Water Security: Debating an Emerging Paradigm' (2012) 27(1) *Global Environmental Change* 27.

De Schutter O, 'Economic, Social and Cultural Rights as Human Rights: An introduction,' (2013) CRIDHO Working Paper 2, 6 http://cridho.uclouvain.be/documents/Working.Papers/CRIDHO-WP2013-2-ODeSchutterESCRights.pdf accessed 19 January 2020.

De Schutter O and others, 'Commentary to the Maastricht Principles on Extraterritorial Obligations of States in the area of Economic Social and Cultural Rights' (2012) 34 *Human Rights Quarterly* 34.

Donelly J, 'International Human Rights Law: Universal, Relative, or Relatively Universal' in Baderin M and Ssenyonjo M (eds), *International human rights law: Six decades after the UDHR and beyond* (Aldershot 2010) 31–48.
Dupuy P, 'Formation of Customary International Law and General Principles' in Bodansky D, Brunnée J and He E (eds), The *Oxford Handbook of International Environmental Law* (Oxford University Press, 2007.
Eide A, 'Realisation of Social and Economic Rights and the Minimum Threshold Approach' (1989) 10 *Human Rights Law Journal* 35.
Ford J and others, 'Vulnerability and its Discontents: The Past, Present, and Future of Climate Change Vulnerability Research' (2018) 151 *Climatic Change* 189.
G7, 'A New Climate for Peace' https://www.newclimateforpeace.org/#report-top accessed 21 January 2020.
Honkonen T, 'Water Security and Climate Change: The Need for Adaptive Governance' (2017) 20(1) *Potchefstroom Electronic Law Journal* 1.
Jegede AO, 'Climate Change in the Work of the African Commission on Human and Peoples' Rights' (2017) 31 (2) *Speculum Juris* 136.
Julien H, 'Is There a Human Right to Safe Motherhood Within the United Nations Legal System?' (2015) 2 (1) *QMHRR* 1.
Liverman D, 'Vulnerability to Global Environmental Change' in Kasperson, R and others (eds) *Understanding Global Environmental Change: The Contributions of Risk Analysis and Management* (Earth Transformed Program, Clark University 1990) 27–44 .
McDowell G, Ford J and Jones J, 'Community-level Climate Change Vulnerability Research: Trends, Progress, and Future Directions' (2016) 11 (3) *Environ Research Letters* 1.
Merry S, *Human Rights and Gender Violence: Translating International Law into Local Justice* (University of Chicago Press 2006) 40.
Muller M, 'Urban Water Security in Africa: The Face of Climate and Development Challenges' (2016) 33 (1) *Development Southern Africa* 67.
Mutua M, 'The African Human Rights System: A Critical Evaluation' http://hdr.undp.org/sites/default/files/mutua.pdf accessed 21 January 2020.
Myers J and others, 'A Public Health Approach to the Impact of Climate Change on Health in Southern Africa – Identifying Priority Modifiable Risks' (2011) 101 *South African Medical Journal* 101.
OAU, African Charter on Human and Peoples' Rights, CAB/LEG/67/3 rev. 5, 21 ILM 58, 27 June 1981 (entered into force 21 October 1986).

OECD, 'Water and Climate Change Adaptation: An OECD Perspective' (2013) 18, https://www.oecd.org/env/resources/Water%20and%20Climate%20Change%20Adaptation-%20brochure.pdf accessed 21 January 2020.

OHCHR, 'International Human Rights Law.' http://www.ohchr.org/en/professionalinterest/pages/internationallaw.aspx accessed 19 January 2020)

Paris Agreement under the United Nations Framework Convention on Climate Change 2015, FCCC/CP/2015/L.9/Rev.1, adopted by Conference of the Parties, 21st Session Paris, 30 November–11 December 2015.

'Paris Agreement – Status of Ratification' https://unfccc.int/process/the-paris-agreement/status-of-ratification accessed 21 January 2020.

Perry M and others (eds), *Climate Change 2007: Working Group II: Impacts, Adaptation and Vulnerability* (Cambridge University Press 2007) 976.

Räsänen A and others, 'Climate Change, Multiple Stressors and Human Vulnerability: A Systematic Review' (2016) 16 (8) *Regional Environmental Change* 2291.

Rindap R, 'The Impact of Climate Change on Human Security in the Sahel Region of Africa' 2015 1 (2) *Donnish Journal of African Studies and Development* 9.

Rist G, *The History of Development: From Western Origins to Global Faith* (Zed Books 2009).

Roberts W and others, 'Greenhouse Gases and Aerosols' in Houghton, JT, Callander, BA & Varney SK (eds), *The Supplementary Report to the IPCC Scientific Assessment* (Cambridge University Press 1992)

Royal Society & Royal Academy of Engineering, *Greenhouse Gas Removal* (Royal Society & Royal Academy of Engineering 2018).

Sekoai P and Daramola M, 'Biohydrogen as a Potential Energy Fuel in South Africa' (2015) 6 *Biofuel Research Journal* 223.

Shue H, *Basic Rights: Subsistence, Affluence, and U.S. Foreign Policy* (2nd ed Princeton University Press 1996).

Signgh A and Purohit B, 'Public Health Impacts of Global Warming and Climate Change' (2014) 26(1) *Peace Review-A. Journal of Social Justice* 112.

Sourgens F, 'Paris Agreement: Regained or Lost? Initial Thoughts' *EJIL: Talk!* https://www.ejiltalk.org/paris-agreement-regained-or-lost-initial-thoughts/ accessed 21 January 2020.

Stavropoulou M, 'The Right Not to be Displaced' (1994) 9 (3) *American University International Law Review* 689.

Stocker T and others (eds), *The Physical Science Basis. Contribution of Working Group I to the 5th Assessment Report of the Intergovernmental Panel on Climate Change* (Cambridge University Press 2013).
Sulaiman C and Abdul-Rahim A, 'Population Growth and CO2 Emission in Nigeria: A Recursive ARDL Approach' (2018) *SAGE Open* 1.
The African Convention on the Conservation of Nature and Natural Resources (Revised version) (Conservation Convention 2003).
Toulmin C, *Climate change in Africa* (Zed Books 2009) 87.
'UNFCCC' https://unfccc.int/news/new-era-of-global-climate-action-to-begin-under-paris-climate-change-agreement-0 accessed 21 January 2020.
UNFCCC, 'NDC Staging' https://www4.unfccc.int/sites/ndcstaging/Pages/Home.aspx accessed 21 January 2020.
UNGA, 'Climate Change and its Possible Security implications' Report of the Secretary-General, A/64/350 11 September 2009.
UNGA, 'Protection of global climate for present and future generations of mankind' UNG.A. Res. 43/53, 70th plenary *meeting 6 December 1988 (UNGA Resolution 43/53)*
United Nations Convention on the Rights of the Child, 1577 UNTS 3, 28 ILM 1456, 20 November 1989 (entered into force 2 September 1990).
United Nations Economic Commission for Africa African Climate Policy Centre, 'Climate Change and Health Across Africa: Issues and Options' (2011) UNECA Documents and Publishing Unit Working Paper 20, https://www.uncclearn.org/sites/default/files/inventory/uneca15.pdf accessed 21 January 2020.
United Nations, General Comment No. 12: The Right to Adequate Food, U.N. ESCOR, Comm. on Econ., Soc. & Cult. Rts., 20th Sess., 14–20, U.N. Doc. E/C.12/1999/5 (1999).
United Nations, General Comment No. 13: The Right to Education, U.N. ESCOR, Comm. on Econ., Soc. & Cult. Rts., 21st Sess., 46–48 (1999).
United Nations, General Comment No. 15: The right to water, arts 11 and 12' (2000) para 1 (United *Nations General Comment No. 15).*
United Nations, General Comment No. 7: The Right to Adequate Housing, Art 11 (1) of the Covenant: *Forced Evictions' (1997) 16th session (United Nations General Comment No. 7).*
United Nations, *Charter of the United Nations*, 24 October 1945, 1 UNTS XVI.
Universal Declaration on Human Rights, UNGA Res 217 A(III), adopted 10 December 1948

Vienna Declaration and Programme of Action, adopted at World Conference on Human Rights in Vienna 1993 UN doc.A/CONF.157/23
Vivideconomic, 'The Cost of Adaptation to Climate Change in Africa' Report prepared for the African Development Bank Group (2012) http://www.vivideconomics.com/wp-content/uploads/2015/03/Cost_of_Adaptation_in_Africa.pdf accessed 21 January 2020.
Wouters P, *Water Security: Global, Regional and Local Challenges* (Institute for Public Policy Research 2010).
Wright C, 'Climate Change and its Effect on Healthcare Systems' http://www.bizcommunity.com/Article/196/330/175557.html accessed 21 January 2020.
Zewdie A, 'Impacts of Climate Change on Food Security: A Literature Review in Sub Saharan Africa' (2014) 5 (8) *Journal of Earth Science and Climatic Change* 225.

4

Advancing Human Rights Through Environmental Rule of Law in Africa

Funmi Abioye

1 Introduction

The environment is important and needs to be protected, as a healthy environment is key to the survival of the human species. This is probably truer in Africa than elsewhere due to the fact that African countries have suffered extreme levels of incessant conflicts, poor policymaking, deforestation and desertification, pollution of water, air, land, environmental degradation and generally bad governance. All of these are symptomatic of development challenges in Africa, which have led to high rates of poverty and environmental degradation on the continent.[1] Environmental degradation has the ability to, and the consequence of, constraining developmental goals. This is because the full effects or manifestation of

[1] United Nations 'New and Emerging Challenges in Africa Summary Report' available at https://sustainabledevelopment.un.org/index.php?page=view&type=400&nr=502&menu=1515 accessed on 10 July 2019.

F. Abioye (✉)
University of South Africa, Pretoria, South Africa
e-mail: Abioyft@unisa.ac.za

M. Addaney, A. Oluborode Jegede (eds.), *Human Rights and the Environment under African Union Law*, https://doi.org/10.1007/978-3-030-46523-0_4

developmental efforts are stymied by prevailing situations of environmental degradation.[2] This has been the case in Africa for many years, due to the cost of development being historically high on the environment due to lax laws, and non-existent enforcement in cases where laws do exist.[3] The degradation of the environment of the Niger Delta people, the effects of the continuous gas flaring in Nigeria are just some indicators of the need to have in place better protections and governance of the environment.

The need to ensure that there is adequate protection of the environment on the continent led to the inclusion of a key mandate of environmental protection within the African human rights system. The African Charter on Human and Peoples' Rights (also referred to as the Banjul Charter) is the foundational instrument of the African human rights system, providing for the protection of human and people's rights on the continent, and is binding at all levels (regional and sub-regional), on all bodies, structures and agencies within the African Union (AU), and also on state members of the AU.[4] The African Charter leads the way in environmental protection with the inclusion of article 24 which provides that '[A]ll peoples shall have a right to a generally satisfactory environment favourable to their development'.[5] The phrasing of article 24 is particularly important and served as a pioneering development in international environmental law. Through this, and as seen all through the African Charter, the provision moves away from the individualism of international law generally propagated by the west, to include the concept of the collective 'peoples', which is in line with an African world view.[6]

[2] The Ogoni crisis stems from the environmental degradation that prevailing in the region due to years of unfettered and unmanaged exploitation of oil in the region. The multiple failures of the state to hold multinational corporations responsible and liable for the damaging effects of their exploitation activities on the environment are apparent for all to see in these regions.

[3] Ibid n 1.

[4] African Charter on Human and Peoples' Rights (adopted 27 June 1981, entered into force 21 October 1986), (African Charter), article 1. This is a mandatory provision that obliges state parties to incorporate the Charter into their municipal systems.

[5] African Charter, article 24.

[6] Olujumoke Oduwole, 'International law and the right to development: a pragmatic approach for Africa', 2014 https://s3-eu-west-1.amazonaws.com/oneworld-wp/app/uploads/2014/06/07102655/www.iss.nl_fileadmin_ASSETS_iss_Documents_Academic_publications_PCC_Inaugural_Lecture_20May2014.pdf accessed 17 November 2019. This interpretation is a contri-

The African Charter subjects the monitoring of the rights provided for therein to the African Commission on Human and Peoples' Rights (African Commission) established by article 30 of the Charter, with the mandate 'to promote human and peoples' rights and ensure their protection in Africa'.[7] The African Commission is a quasi-judicial body with powers to consider claims of violation of rights by individuals, groups or their representatives in an adversarial procedure and to issue findings and remedies.[8] This is part of the protective mandate of the Commission. This mandate is complemented by the establishment of the African Court on Human and Peoples' Rights (African Court) with a judicial mechanism to enforce any of the rights protected under the African Charter.[9]

The aim of this chapter is to examine the ways in which judicial and quasi-judicial bodies within the African human rights system have been effective in enhancing environmental protection and consequently developing an environmental rule of law.[10] This is considered through an examination of some of the relevant case law that have been decided on by these bodies, thus developing a body of environmental law principles, designed to enhance the recognition of environmental protection rules and norms in Africa, and which as argued in this chapter, equate to the emergence of an environmental rule of law in Africa.

This chapter is divided into four sections. Section 2 has introduced the topic and provided a background to the issues to be dealt with in this

bution by the Continent to the discourse on the meaning and interpretation of environmental rights. See also Funmi Abioye, 'Understanding the Rule of Law in Africa through the lens of the African Commission on Human and Peoples Rights' interpretation on the right to development in Africa' 2018 (2) VRUE 269, 271.

[7] African Charter, article 30. See also article 45 which specifically provides for the comprehensive mandate of the Commission.

[8] African Charter, articles 46–58.

[9] Protocol to the African Charter on Human and Peoples' Rights on the Establishment of an African Court on Human and Peoples' rights, OAU/LEG/EXP/AFCJPR/PROT (iii) (2004), article 2.

[10] Over the years, case law has reiterated the linkage between human rights and the environment. The ICJ in the *Gabciikovo-Nagymaros Project* case (1997) ICJ Rep 7, para 91–92, per Weeremantry J, recognized that the enjoyment of internationally recognized human rights depends upon environmental protection for it is a sine qua non for numerous human rights such as the right to health and the right to life itself. At the same time, the learned Judge added that any damage to the environment could impair and undermine all the human rights spoken of in the Universal Declaration and other human rights instruments.

chapter. Section 3 presents an analysis of environmental protection and the idea of an emerging environmental rule of law in Africa. Section 4 examines the jurisprudence of regional bodies, specifically the African Commission on Human and Peoples' Rights (ACHPR) and the African Court of Human and Peoples' Rights (ACtHPR), while section 5 examines the jurisprudence of sub-regional and national jurisdictions. These two sections further analyze the ways in which these bodies have contributed to the development of an environmental rule of law within the African human rights system.[11] Certain national jurisdictions have been able to go beyond mere rhetoric to actual implementation and enforcement of the law to ensure the protection of their environment; in the process, enhancing the notion of environmental rule of law. The conclusion highlights the ways in which environmental protection jurisprudence of the African human rights system is making use of the rule of law, to contribute to the development of a body of law referred to as the environmental rule of law, and the positive impact of such on environmental protection.

2 Background to Environmental Protection in Africa: Taking Control

The 1970s marked a turning point in the global attitude to the environment, with concrete steps taken to protect the environment. First was the convening of the United Nations Conference on the Human Environment (the Stockholm Conference) in 1972, which ended with a unanimous declaration, the Stockholm Declaration.[12] This was followed by the 1992 Rio Declaration on Environment and Development (Rio Declaration),[13] and the 2002 Johannesburg Declaration on Sustainable Development (Johannesburg Declaration),[14] and many others. In Africa, an

[11] This will eventually be known as the African Court of Justice, Human and Peoples' Rights, once the Protocol on the Statute of the African Court of Justice and Human Rights comes into force.

[12] The Declaration of the United Nations Conference on the Human Environment (Stockholm Declaration) (adopted June 16, 1972).

[13] The Declaration of the 1992 United Nations Conference on the Environment and Development (UNED).

[14] Declaration of the World Summit on Sustainable Development 2002 (Earth Summit 2002).

environmental protection consciousness began earlier with the 1968 African Convention on the Conservation of Nature and Natural Resources,[15] followed by the African Charter on Human and Peoples' Rights in 1986. This showed attention gradually focusing on environmental management and sustainable use of resources.[16]

Subsequent instruments relating to the protection of the environment at both the international and regional levels have variations of environmental protection provisions herein.[17] Though the existence of an 'environmental right' is much debated by scholars,[18] the fact that there is a relationship between human rights and the environment cannot be denied.[19] For some, the consensus appears to be that of an integration of human rights and the environment under the concept of sustainable development.[20] Wondalem recognized this when he stated that 'the issue of the environment is intrinsically related to a number of other human rights both as a precondition and an outcome of the enjoyment of many

[15] African Convention on the Conservation of Nature and Natural Resources (adopted 15 Sept 1968, entered into force June 16 1969).

[16] See IUCN, 'An introduction to the African Convention on the Conservation of Nature and Natural Resources' https://portals.iucn.org/library/sites/library/files/documents/EPLP-056.pdf accessed 24 May 2019, 1.

[17] The African Union was the first of these. Thereafter, the Organisation of Economic Cooperation and Development (OECD) advocated for a 'decent' environment to be recognized as one of the fundamental human rights; the United Nations Economic Commission for Europe (UNECE) drafted the Charter on Environmental Rights and Obligations which affirms that 'everyone has the right to an environment adequate for general health and well-being'; the Organisation of American States (OAS) in its 1988 Protocol of San Salvador included a 'right to a healthy environment for everyone', and mandated state parties to promote the protection, preservation and improvement of the environment. In addition to this, the constitutions of over 150 nations in the world have adopted this right in some form or the other.

[18] See Alan Boyle, 'Human Rights and the Environment: where next?' (2012) 23(3) The European Journal of International Law 613–642; SAHRC, 'The Right to a Healthy Environment' https://www.sahrc.org.za/home/21/files/Reports/5th_esr_environment.pdf accessed 8 July 2019.

[19] Some have tried to justify this right as a derivative of civil and political rights, thereby making it a socio-economic right; others have argued for it to be seen as an independent right on its own. See Philippe Cullet, 'Definition of an Environmental right in a human right context' (1995) 13 Netherlands Quarterly of Human Rights 27; also, Downs, 'A healthy and ecologically balanced environment: an argument for a third generation rigcfzsht' (1993) 2 Duke Journal of Comparative and International Law 351–385.

[20] Office of the High Commissioner for Human Rights (OHCHR) 'Analytical Study on the Relationship between Human Rights and the Environment', UN Doc A/HRC/19/34, https://www.ohchr.org/Documents/HRBodies/HRCouncil/RegularSession/Session19/A-HRC-19-34_en.pdf accessed 25 May 2019.

human rights'.[21] He goes further to explain that as a precondition, the preservation, conservation and protection of the environment require a number of other human rights such as the right to information, the right to participation in decision-making and the right to access to justice; while as an outcome, protection and conservation of the environment plays a vital role to the enjoyment of human right including the right to life and health,[22] and many other human rights. Therefore, many international and regional environmental instruments directly or indirectly recognize the linkage between human rights and environmental protection.

At the national level, nearly all African countries have now included some formulation of the right to environment or details of environmental obligations on the part of the state in their constitutions.[23] Apart from the provision of the right in article 24 of the African Charter, there exists other rights in the Charter that impact on the fulfilment of article 24, and are also impacted by article 24. Some of these include article 4 (right to life), article 14 (right to property), article 16 (right to enjoy the best attainable state of physical and mental health) and article 22 (right to development).[24] This reinforces the fact stressed by Wondalem above, that all human rights are interdependent, indivisible, interlinked and mutually reinforcing.[25]

The African Commission on Human and Peoples' Rights (the Commission) has described article 24 as a composite right, due to the fact that it refers to the protection of the environment in a manner favorable to the development of the people.[26] This is important in improving

---

[21] Haile Andargie Wondalem, 'The Right to Environment under African Charter on Human and Peoples Right' (2015) 2(1) International Journal of International Law 206, 210.

[22] Ibid.

[23] About 32 out of the 54 countries in Africa currently have some provision on this in their constitutions. See Dinah Shelton, 'Human rights, Environmental Rights, and the Right to Environment' (1991–1992) 28 Stanford Journal of International Law 103, 128.

[24] Werner Scholtz, 'Human rights and the environment in the African Union context' in Anna Grear and Louise Kotze, *Research Handbook on Human Rights and the Environment* (eds, Edward Elgar Publishing 2015) 405 https://0-www-elgaronline-com.oasis.unisa.ac.za/view/edcoll/9781782544425/9781782544425.xml accessed on 18 June 2019; Wondalem (n 21).

[25] Ibid; Fatsah Ouguergouz, *African Charter on Human and Peoples' Rights: A Comprehensive Agenda for Human Dignity and Sustainable Democracy in Africa* 51–52 (New York: Kluwer Law International, 2003).

[26] SERAC v Nigeria, Communication 155/96 of the African Commission, para 51.

the quality of life and safety of individuals and in promoting development. Furthermore, it reinforces the principle of interdependency (and interrelatedness) of rights and reflects the normative unity between the right to satisfactory environment and development.[27]

This need for development and environmental protection gives impetus to the principle of sustainable development, through which it is understood that any type of development must be such that can be sustained through generations, while allowing for the protection of the environment.[28] The recognition that the quest for economic development needs to consider possible consequential environmental degradation, means that measures must be implemented to ensure that the environmental cost of development is not unbearable in years to come. One of such measures is the increased use of law to bolster environmental protection and development, through the development of environmental rules, and goals, and the prescription of necessary action to realizing the goals.

3 An Environmental Rule of Law?

The discourse on the relationship between human rights and the environment has in recent times included the previously unknown concept of 'environmental rule of law'. Those advocating for this concept emphasize the need to legislate developmental efforts along with the protection of the environment as a key consideration. This is what informed the adoption of this concept in Decision 27/9 on Advancing Justice, Governance and Law for Environmental Sustainability, by the United Nations Environmental Programme (UNEP) in 2013.[29] This decision is important, as it is the first document to use the term 'environmental rule of

[27] Lillian Chenwi,'The right to a satisfactory, healthy and sustainable environment in the African Regional Human Rights System' in John Knox & Ramin Pejan, *The Human Right to a Healthy Environment* (Edward Elgar Publishing, 2018) 66.

[28] OHCHR (n 20). In 2015, the global community met and adopted the Sustainable Development Goals (SDGs) to foster the realization of sustainable development across the world. See 'Transforming our world: the 2030 Agenda for sustainable Development' https://sustainabledevelopment.un.org/?menu=1300 accessed 8 July 2019.

[29] 'Decision 27/9 on Advancing Justice, Governance and Law for Environmental Sustainability' Decisions adopted by the Governing Council/Global Ministerial Environment Forum at its first

law'. As a possibly emerging concept, it seeks to recognize the importance of the rule of law in the field of environment protection, and thus serves as a means of reducing violations of environmental law and achieving sustainable development at the same time.[30] Decision 27/9

> recognises that the violation of environmental law has the potential to undermine sustainable development and the implementation of agreed environmental goals and objectives at all levels and that *the rule of law and effective governance play an essential role in reducing such violations* and invites Governments and relevant organizations to further strengthen mechanisms and … measures to increase the effectiveness of administrative, civil and criminal enforcement mechanisms, institutions and laws in the field of environmental as well as applicable education and training.[31]

This makes it clear that law is necessary to create an enabling environment for the protection of the environment and translates into development and growth for any society.[32] In essence, legal provisions (and judicial mechanisms to enforce these provisions) help to govern our interaction with our environment in order to achieve the goal of protecting the environment for the benefit of the present and future generations. Only by legislating and enforcing such legal protection mechanisms can we begin to ensure that the environment is sufficiently protected. This is what the notion of an environmental rule of law seeks to address. It has been explained as providing 'an essential platform underpinning the four pillars of sustainable development - economic, social, environmental and peace'.[33] It is said to be the means through which the protection of the

universal session, https://wedocs.unep.org/bitstream/handle/20.500.11822/12221/Governing%20Council%20Decision%2027-2.pdf?sequence=1&isAllowed=y accessed on 3 June 2019.

[30] UNEP 'Environmental Rule of Law: critical to sustainable development' https://wedocs.unep.org/bitstream/handle/20.500.11822/10664/issue-brief-erol.pdf?sequence=1&isAllowed=y accessed on 2 June 2019.

[31] Decision 27/9 (n 29) no 4 (emphasis mine).

[32] The introduction of environmental protection provisions in the 2010 Kenya Constitution is a good example of such legal protection. Articles 42, 69 and 70 of the makes provisions for environmental protection, and actually elevates the requirement for Environmental Impact Assessments (EIA) to be carried out before development constitutional requirement.

[33] UNEP 'Environmental Rule of Law: First Global Report' available at http://wedocs.unep.org/bitstream/handle/20.500.11822/27279/Environmental_rule_of_law.pdf?sequence=

environment is provided for in legal mechanisms and rules, with concomitant sanctions in the event of breaches.[34] The concept therefore encapsulates the depth and the role of law, regulation and enforcement in environmental protection arena.[35] This is a result of the work of UNEP over the years, and also is an extension of the generally accepted rule of law approach to environmental management, with the UNEP affirming that 'without environmental rule of law, development cannot be sustainable'.[36]

Environmental rule of law offers a conceptual and policy framework for strengthening the implementation of environmental law in a systematic and holistic manner, while holding all entities accountable to promulgated laws that govern the environment.[37] It leads to the creation of normative standards and modes of behavior dealing with the environment. In societies where there are environmental protection mechanisms in place, it provides the state with an opportunity to enforce laws that increase environmental protections, and ensure sustainable development of the environment.[38] One of the means of such enforcement is through the purview of judicial or quasi-judicial bodies, which are mandated by law to uphold the protection of the environment. The judiciary plays a key role of adjudicating disputes, and interpreting the law, thus building up a body of rules for governance (and in this case, for environmental protection). This much was recognized by the Global Judges Symposium in 2002, when they reiterated the importance of the judiciary in ensuring environmental protection. They held that

---

1&isAllowed=y accessed 17 June 2019, 1. These four pillars are enshrined in the UNGA '2030 Agenda for Sustainable Development' 2015.

[34] Ibid.

[35] Alexandra Dapolito Dunn and Sarah Stillman, 'Advancing the Environmental Rule of Law: a call for measurement' 2015 (21) South Western Journal of International Law 283, 284; see also Jessica Scott 'From Environmental Rights to Environmental Rule of Law: A proposal for better environmental outcomes' 2016 (6)1 Michigan Journal of Environmental & Administrative Law 203 at 220–238, wherein the author highlights the seven precepts of effective national environmental governance.

[36] Ibid.

[37] UNEP document (n 33). See the full discussion of the UNEP on this concept.

[38] Dunn and Stillman (n 35) 289.

> the fragile state of the global environment requires the judiciary as the guardian of the Rule of Law, to boldly and fearlessly implement and enforce applicable international and national laws, which in the field of environment and sustainable development will assist in alleviating poverty and sustaining an enduring civilization, and ensuring that the present generation will enjoy and improve the quality of life of all peoples, while also ensuring that the inherent rights and interests of succeeding generations are not compromised.[39]

This led to the realization that effective judiciaries and quasi-judicial bodies are an integral component of the institutional architecture for sustainable development,[40] and in the same vein, for the protection of the environment. The manner in which the jurisprudence of these bodies has contributed to the emergence of the concept of an environmental rule of law will be examined next.

4 African Regional Bodies

African Commission on Human and Peoples' Rights

As indicated above, the African Commission was established by the African Charter with both promotional and protective mandates.[41] It is made up of 11 members from different African states, who must be people of high reputation, morality, integrity and competence in matters dealing with human and people's rights.[42] In addition to its promotional and protective mandate, the African Commission has an interpretative mandate, which tasks it with the interpretation of all the provisions of the

[39] See 'The Johannesburg Principles on the Role of Law and Sustainable Development' (adopted at the Global Judges Symposium, 2002) https://www.eufje.org/images/DocDivers/Johannesburg%20 Principles.pdf accessed 8 July 2019.

[40] Collins Odote, 'The role of the Environment and Land Court in governing natural resources in Kenya' in Patricia Kameri-Mbote and others, *Law, Environment, Africa* (Nomos, Germany 2019) 337.

[41] African Charter, article 45. For an in-depth analysis of the mandates of the African Commission, see Funmi Abioye 'The African Commission on Human and Peoples' Rights: achieving its purpose?' (2014) 39 Supp SAYIL 80 at 84–95.

[42] African Charter, article 31.

Charter at the request of a state party.[43] In carrying out its protective mandate, the African Commission is granted power to examine inter-state communications and communications from non-state parties, and has the mandate to enforce the provisions of the Charter.[44] This mandate is seen in the work of the Commission in dealing with communications filed before it. These communications and the manner in which the Commission deals with them form the jurisprudence of the African Commission.[45]

As a quasi-judicial body, the non-binding status of the decisions of the African Commission has generated much debate.[46] The Commission has noted that it can 'urge' and 'recommend' actions to be taken, but lacks the power to 'order' or 'command' such action.[47] Even though, it can make recommendations which are understood to be non-binding on state parties,[48] the decisions of the African Commission carry normative weight, and are critical to the African regional human rights system.

In relation to article 24 of the Charter, the Commission has not had the privilege of considering many communications dealing directly with this provision, in comparison with other rights guaranteed by the Charter. Of the few communications dealing with article 24, only one has been found to be admissible and was decided on the merits,[49] another was ruled inadmissible,[50] while a third was referred to the African Court of Human and Peoples' Rights for final adjudication.[51] The Commission

[43] African Charter, article 45(3).

[44] African Charter, articles 46–58.

[45] Abioye (n 41) 88.

[46] Evelyn Ankumah *The African Commission on Human and Peoples Rights: Practise and procedures* (Martinus Nijhoff Publishers, 1996) 23, 74–75; Gino Naldi 'The African Union and the Regional Human Rights System' in Evans and Murray *The African Charter on Human and Peoples' Rights: the system in practise 1986–2006* (2nd edn, Cambridge University Press 2008) 20, 35, Kofi Oteng Kufor 'Safeguarding human rights: A critique of the African Commission on Human and Peoples Rights' (1993) 18 *Africa Development* 65.

[47] Communication 275/03, Article 19 v Eritrea.

[48] Abioye (n 41) 93; Chairman Okoloise 'Circumventing Obstacles to the Implementation of Recommendations by the African Commission on Human and Peoples' Rights' (2018) 18 African Human Rights Law Journal 29.

[49] Communication 155/96 SERAC v Federal Government of Nigeria.

[50] Communication 524/15 Peter Ngoge v Republic of Kenya.

[51] African Commission on Human and Peoples Rights v Republic of Kenya.

has also decided other matters that have an indirect impact on the protection of the environment; and through all of these, the Commission has been able to develop its jurisprudence on environmental protection within the African human rights system.[52]

The first opportunity for the African Commission to deal with article 24 of the Charter came in the form of the *Social and Economic Rights Action Centre (SERAC) and the Center for Economic and Social Rights (CESR)* v *Federal Republic of Nigeria* communication[53] (the SERAC case). In this case, the complaints alleged a series of violations of human rights[54] against the Government of Nigeria. They further alleged that by virtue of the actions and/or inactions of the Government of Nigeria in regulating the multinational corporations engaged in oil exploitation in Ogoniland, the rights of the people of Ogoni, including the right to a healthy environment, to housing and food, had been significantly violated.[55] This, they alleged the government of Nigeria had done by directly participating in the contamination of air, water and soil thereby harming the health of the Ogoni population; by failing to provide or require studies of potential/actual environmental and health risks caused by the oil operations,[56] and by failing to enforce existing laws and legislative safeguards to control the way and manner in which the operations of oil exploitation were being carried out.

In its findings, the Commission reiterated the importance of a clean and safe environment and that such is closely linked to economic and social rights 'in so far as the environment affects the quality of life and safety of the individual'.[57] It affirmed the justiciability of the broad range

[52] Its decisions in communications dealing with the right to development have had an indirect impact on the protection of the environment. See for example, DRC v Burundi, Rwanda and Uganda; Centre for Minority Rights Development (Kenya) and Minority Rights Group (on behalf of Endorois Welfare Council) v Kenya; Sudan Human Rights Organisation & Centre on Housing Rights and Evictions (COHRE) v Sudan, and Open Society Justice Initiative v Cote d'Ivoire.

[53] *Communication 155/96 of the African Commission.*

[54] These were articles 2 (non-discriminatory enjoyment of rights), 4 (right to life), 14 (right to property), 16 (right to health), 18 (family rights), 21 (right of peoples to freely dispose of their wealth and natural resources) and 24 (right of peoples to a satisfactory environment) of the African Charter.

[55] SERAC v Nigeria, paras 1 to 10.

[56] SERAC v Nigeria, para 50.

[57] SERAC v Nigeria, para 51.

of rights included in the Charter, especially environmental rights. It further affirmed the duties and responsibilities of states to 'take reasonable and other measures to prevent pollution and ecological degradation, to promote conservation, and to secure an ecologically sustainable development and use of natural resources'.[58] The Commission held that the Nigerian Government had violated its positive obligations imposed under the Charter by its failure to take positive measures to control the activities of oil companies that have caused enormous violation to the rights of the Ogoni people, and thus violated the provisions of the African Charter.[59] In reaching such a decision, the Commission held that

> [G]overnments have a duty to protect their citizens, not only through appropriate legislation and effective enforcement, but also by protecting them from damaging acts that may be perpetrated by private parties. This duty calls for positive action on the part of governments in fulfilling their obligations under human rights instruments.[60]

This decision by the Commission re-emphasizes the indivisibility and interconnectedness of rights in the African Charter, and the fact that these rights cannot be dealt with in isolation. It further places a positive obligation on state governments to put measures in place to ensure the implementation of the Charter. The Commission has been commended for its creativity in identifying the violation of human rights resulting from the pollution of environment. By this, the Commission showed the role that the arms of justice and quasi-judiciary bodies in Africa could play to enhance the environmental rule of law, and to ensure protection of the environment,[61] and thereby solidifying environmental rule of law on the continent.

The next case of significance that came before the Commission reinforced the wholesome view of human rights protection in Africa. Though

[58] SERAC v Nigeria, paras 68 and 52.

[59] SERAC v Nigeria, para 58.

[60] SERAC v Nigeria, para 57.

[61] Abdulkadir Abdulkadir, 'The right to a healthful environment in Nigeria: a review of alternative pathways to environmental justice in Nigeria' (2014) 3(1) Journal of Sustainable Development Law and Policy 127.

the case of *Centre for Minority Rights Development and Minority Rights Group* (on behalf of the Endorois Welfare Council) v *Kenya*[62] (Endorois case) did not deal directly with a violation of article 24 on the right to a satisfactory environment, but rather with the violation of article 22 on the right to development, the decision has a direct effect on the interpretation of article 24, and further buttresses the interrelatedness and connectivity of environmental rights. The applicants in this communication were forcefully evicted from their natural environment and their ancestral lands around the lake Bogoria area, and purportedly in violation of their rights.[63] As a result of this eviction, the applicants alleged, among other things, the violation of their right to development under the African Charter.[64] In one of its defenses, the respondent state indicated that its actions were taken in a move to protect and conserve the natural environment around the said area, and to ensure proper wildlife management and conservation.[65]

In its decision that found the state breached the rights of the applicants, the Commission found that the violation of the right to development consists of a two-pronged test: constitutive (substantive) and instrumental (procedural), and that a violation of either would result in a violation of the right to development.[66] In relation to the instrumental test, the Commission noted in respect of the need for consultations that 'governments must consult with indigenous peoples especially when dealing with sensitive issues as land';[67] and that such consultation must be 'in good faith, through culturally appropriate procedures, and with the objective of reaching an agreement'.[68] Thus, in line with the consultation requirement, even though the state had a legitimate aim (development of the area) in interfering with the right of the applicants to property,

[62] Communication 276/03 Centre for Minority Rights Development and Minority Rights Group (on behalf of the Endorois Welfare Council) v Kenya.

[63] *Endorois*, para 22.

[64] *Endorois*, para 22. These other rights including violation of the right to practice religion (art 8), the right to culture (art 17(2) & (3), and the right to property (art 14), the right to dispose of their natural resources freely (art 21).

[65] *Endorois*, para 178.

[66] *Endorois*, para 277.

[67] *Endorois*, para 281.

[68] *Endorois*, para 289.

however, such actions should have been carried out only after effective engagement/participation of the community in the decision-making itself.[69]

In relation to the constitutive element of the test, the Commission held that any development of the land (environment) of the Endorois in this case must be aimed at the empowerment of the people and toward improving their capabilities and choices.[70] The Commission further decided that the community has a 'right to reasonably share in the benefits made as a result of a restriction or deprivation of their right to the use and enjoyment of their traditional lands and of those natural resources necessary for their survival'.[71] This decision of the African Commission, while focusing on the infringement of the right to land (traditional land at that), extends to environmental right by providing that the 'the community' has the right to the use and enjoyment of their lands, and also of the natural resources necessary for their survival. By this, the African Commission developed the environmental rule of law in Africa at the same time, by laying down the requirement of proper consultation with an affected community before any action or development can commence. Furthermore, the Commission stressed the need for the type of development to be such that will have a positive impact on the community affected, 'enhancing their capabilities and choices'. Thus, development must not only be directed solely for the benefit of the state, to the detriment of the community, but the community must be a beneficiary of the said development.[72]

African Court on Human and Peoples' Rights

The African Court of Human and Peoples' Rights (the African Court, also known as the ACtHPRs) was established by protocol to the African

---

[69] *Endorois*, paras 281–282.

[70] *Endorois*, paras 282–283.

[71] *Endorois*, para 295.

[72] This is instructive for a situation like that of the Ogoni community in Nigeria, where the communities in the Nigeria Delta region have not benefited commensurately from the exploitation of oil, compared to the benefit that the Nigerian state has derived from this activity.

Charter on Human and Peoples' Rights. It was established with the purpose of enhancing the efficiency of the African Commission and thereby safeguarding the African Charter.[73] The role of the African Court is complementary to that of the Commission, and bolsters the efforts of the Commission to safeguard the African Charter and its provisions.[74] This is necessary to address the deficiencies arising from the quasi-judicial status of the Commission,[75] since the decisions of the African Court are binding on state parties and also carry remedial powers, and thus carry more weight than the communications of the Commission, which are non-binding, but merely recommendatory.[76] The African Court has broad jurisdiction to hear cases and disputes concerning the interpretation and application of the rights in the African Charter, African Women's Protocol and other relevant human rights instrument ratified by the states concerned.[77] However, the jurisdiction of the Court is optional and states may also choose whether to accept direct access to the court in line with article 34(6) of the Protocol. This restricts direct access to the Court, requiring most cases to proceed to the Court via the African Commission.[78]

---

[73] Protocol to the African Charter, preamble, article 2. See also Werner Scholtz (n 24) 417.

[74] The ACtHPRs has broad material jurisdiction which extends to 'all cases and disputes concerning the interpretation and application of the Charter, the protocol (itself), and any other relevant Human Right instrument ratified by the States concerned' see article 3. In terms of jurisdiction over persons, article 5 of the protocol indicates that the Commission, state parties, and any African intergovernmental organizations could submit cases to the Court. However, in the case of individuals instituting cases directly before it, there is a caveat by the protocol that such would only be allowed if the concerned state part has made a declaration accepting the competence of the Court to receive such cases.

[75] Werner Scholtz (n 24) 417.

[76] In 2008, the Protocol on the Statute of the African Court of Justice and Human Rights was adopted by the AU, and provides for the merger of the African Court of Justice and the African Court of Human and Peoples' Rights in order to create the African Court of Justice and Human Rights. However, this protocol is not yet in force as the required number of ratifications have not been reached at the time of writing.

[77] See Protocol to the African Charter, article 3; see also Chenwi (n 22) 77. This jurisdiction however restricts the rights of individuals to approach the court directly. Such would only be accorded in situations where the relevant state party had already deposited a declaration to that effect.

[78] Protocol to the African Charter, article 5, see Evadne Grant 'International Courts and environmental human rights: re-imagining adjudicative paradigms' in Anna Grear and Louis Kotze Research Handbook on Human Rights and the Environment (Edward Elgar, 2015) 379.

The African Court became operational in 2009 when it heard its first case.[79]

In 2017, the African Court became seized with the case of *African Commission on Human and Peoples Rights* v *Republic of Kenya*.[80] In this case, the Court commented and made indirect reference to the protection of environmental rights. The matter was transferred to the African Court from the African Commission, after the Commission failed in its attempt to resolve the matter.[81] Though, the case principally concerned the right of indigenous people, it however also touched on the need for environmental conservation. The Ogiek Community, an indigenous minority ethnic group in Kenya, faced eviction from their ancestral lands in the Mau forest.[82] The government of Kenya argued that the eviction was necessary in order for it to conserve the environment against the cultural practices of the Ogiek. The Court found inter alia that the eviction of the Ogiek violated a number of their rights guaranteed under the African Charter,[83] and held that the continued denial of access to and eviction of the Ogiek population from the Mau Forest cannot be necessary or proportionate to achieve the justification of preserving the natural ecosystem of the Mau Forest.[84]

With respect to the violation of the Ogiek's right to culture protected under article 17(2) & (3) of the Charter, the respondent state had argued before the court that it had a responsibility to ensure a balance between cultural rights and environmental conservation. It justified its position that the fact that the exercise by the Ogiek of their cultural rights (which included fishing, hunting, among others) could negatively impact on the

[79] Judgement of the African Court in Michelot Yogogombaye v Senegal Appl No 1/2008, ACtHPR (15 Dec 2009).

[80] Application No 006-2012 – African Commission on Human and Peoples Rights v the Republic of Kenya.

[81] The relationship between the African Court and the African Commission is of a complementary nature and allows for this. In this case, the African Commission had issued provisional measures to Kenya, but this had failed and there were ongoing serious or massive violations of human rights at the time.

[82] African Commission v Kenya, para 6.

[83] The rights to non-discrimination under article 2 of the charter, culture (article 17(2)), religion (article 8), property (article 14), natural resources (article 21) and development (article 22).

[84] African Commission v Kenya, para 130.

natural resources in the environment meant that such activities needed to be balanced against other public interests.[85] Here the respondent state took the position that it could infringe other rights guaranteed under the African Charter, in its responsibility to conserve and protect the natural environment of the Mau Forest complex. This argument did not carry weight before the Court, which examined whether the interference with the cultural rights of the Ogieks could be justified by the need to attain a legitimate aim under the Charter, in this case the conservation and protection of the environment.[86] In finding for the applicants, the Court reiterated its position that the respondent state had not provided sufficient evidence of its claim that the interference with the rights of the Ogiek was justified, as it had failed to provide evidence of the particular cultural activities that are inimical to the environment.[87] Thus, the purported reason to preserve the natural environment cannot constitute a legitimate justification for the exercise of the cultural rights of the Ogiek. The African Court in its approach in this case upheld the approach of the Commission on the inter-relatedness of rights, and the violation of one would necessarily impact on others. Thus, the finding of the Court that the attempt by the state to 'conserve and protect' the environment could not justify the violation of the cultural rights of the Ogieks.

5 Sub-regional and National Jurisdictions

Community Court of Justice of the Economic Community of West African States (ECOWAS Court)

The ECOWAS Court operates at sub-regional level and exercises jurisdiction over human rights matters. The Court was originally founded to resolve disputes between state parties and community institutions and was later amended to include jurisdiction over human rights violations in

[85] African Commission v Kenya, para 174.
[86] African Commission v Kenya, para 184.
[87] African Commission v Kenya, para 189.

member states; and to allow direct access to private litigants.[88] This makes a first among the bodies comprising the African human rights system. The Court can apply a range of international human rights instruments, along with African Union instruments, in this case, instruments dealing with environmental rights. In *Socio-Economic Rights and Accountability Project* v *Nigeria* (SERAP case),[89] the court was of the view that the central disputes relate to whether Nigeria had breached the right provided in article 24 of the ACHPR. The Court noted that while Nigeria had passed many laws and had effected a range of regulatory measures in response to the problems in the Niger Delta,[90] the state had failed to implement those measures and had failed to act to prevent or curtail environmental damage, and thus had violated article 24 of the charter. The Court here introduced a new dimension to the award of compensation for environmental damage. The Court deemed monetary compensation to be inappropriate for claims affecting a collective or community, and held that 'a collective benefit adequate to repair, as completely as possible, the collective harm that a violation of a collective right causes' is necessary.[91] This acknowledgment by the Court is an important step in environmental rights protection, as it recognizes the fact that in the context of African society, and the collective nature thereof, it is necessary for the judicial arm to award compensation and damages that are appropriate and relevant to the circumstances of the people. This further develops the environmental rule of law in Africa on the issue of the nature of compensation to be awarded to a group, when their environmental rights are breached. Thus, as noted by the court, an award of monetary compensation would not have repaired the collective harm to the community.

Environment and Land Court of Kenya (ELC)

The jurisprudence of the African regional and sub-regional bodies discussed above highlights the way African human rights system is

[88] Protocol on the Community Court of Justice, A/P.1/7/91.
[89] ECW/CCJ/APP/08/09.
[90] ECW/CCJ/APP/08/09, para 111.
[91] ECW/CCJ/APP/08/09, para 116.

developing its own principles and rules in relation to the protection of the environment, and thereby enhancing the concept of an environmental rule of law on the continent. This especially as the rules and principles developed by these judicial and quasi-judicial bodies serve as precedent for national jurisdictions and possibly national courts to follow.

One of such discussed here is the case of Kenya, in the creation of a dedicated court to try cases relating to the environment. Prior to 2010, the situation of environmental protection in Kenya was dire, and the National Environmental Management Authority of Kenya (NEMA)[92] had acknowledged that Kenya's economic progress till then, had been at the cost of environmental degradation, manifested in the loss of forests, pollution, declining wildlife and increased use of toxic substances.[93] This degradation of the environment led to the inclusion of provisions on environmental protection in the 2010 Constitution, thus elevating the protection of the environment to a constitutional protection in order to give more weight and more impetus to the issue of environmental protection.[94] Article 42 of the 2010 Constitution thus provides

> 42. Every person has the right to a clean and healthy environment, which includes the right;
> (a) to have the environment protected for the benefit of present and future generations through legislative and other measures, particularly those contemplated in Article 69; and
> (b) to have obligations relating to the environment fulfilled under Article 70

The Kenya Constitution goes further in creating a special court to deal with all environmental issues, in the name of the Environmental and Land Court of Kenya (ELC) in article 162.[95] Soyapi opines of the ELC

[92] Established by the Environmental Management and Coordination Act (EMCA) No. 8 of 1999 Laws of Kenya. For a more comprehensive history of what led to the Environmental and Land Court of Kenya, see Odote (n 40) 339–341.

[93] See Caiphas B Soyapi, 'Environmental Protection in Kenya's Environment and Land Court' (2019) 31 Journal of Environmental Law 151–161, 154.

[94] 2010 Constitution of Kenya, article 42, grants every person the right to a clean and healthy environment.

[95] Article 162, Constitution of Kenya 2010.

that it is probably the only constitutionally mandated specialist court dealing with environmental protection in the English-speaking parts of Africa.[96] The Court has jurisdiction to decide cases relating to the right to a clean and healthy environment, and also acts as an appellate court in matters relating to land and environmental rights.[97]

Though faced with jurisdictional and administrative challenges,[98] the Court hit the ground running and already adjudicated on matters relating to the protection of the environment, even passing judgment against an arm of government as it is empowered to do so under its enabling act.[99]

In *Joseph Leboo & 2 others v Director Kenya Forest Services & another*, the ELC had to decide on the granting of a permanent injunction to stop the logging of trees from the Lembus forest without an Environmental Impact Assessment (EIA) being carried out.[100] The court held that before any kind of economic development (logging activities in this case) can be carried out on a parcel of land, a management plan and an Environmental Impact Assessment (EIA) are compulsory steps that must be taken before such.[101] The Court was concerned with the fact that irreparable damage would be meted out on the forest if the illegal activities were allowed to continue.[102] While the Court in this case did not analyze the scope of article 42 of the constitution, it however used the right to locate a legal basis for the claim.[103]

The ELC in *Moffat Kamau & 9 Others v Aelous Kenya Limited & 9 others* held that where a project that had a prior EIA had been expanded in size and scope, then a new EIA process had to be initiated, rather than a

[96] Soyapi (n 93) 152.

[97] See Environmental and Land Court Act 2011, section 13(1).

[98] Odote (n 40) 342–346.

[99] In ACRAG case, the court held that a municipal council had failed to maintain and manage a dumping site adequately, and also that an EIA had not been concluded before operating the dump-site. In view of these and other problems emanating from the dumping site, the court held that the municipality had to close the dumping site, find another location, conduct an EIA before opening an alternative.

[100] (2013) eKLR.

[101] (2013) eKLR para 46.

[102] (2013) eKLR para 50.

[103] Soyapi (n 93) 157.

variation of the original one.[104] The court held that in balancing environmental protection and economic development, the scale must always tilt in favor of environmental protection, regardless of how 'green' the development projects are. The court here insisted in upholding the environmental rule of law to ensure that environmental laws are not bypassed.[105] The Court in this case halted the development that 'may' cause damage to the environment. By so doing, the court stressed the importance of upholding the rules and regulations, and that failure abide by the EIA led to a violation of the right under article 42 of the Kenyan Constitution.[106]

These cases and others decided by the ELC in Kenya have contributed to the environmental rule of law in Kenya and thereby stand as persuasive foreign authority for other African countries. The Court in *Kamau* applied precaution and went against the trend that we see on the continent, where commercial goals tend to trump environmental concerns in both political and legal spheres. The jurisprudence of the court has further highlighted a 'higher mandate' for everyone to protect the environment, and to ensure that sustainable development practices are in place.

6 Conclusion

This chapter has sought to establish if the newly emerging concept of environmental rule of law can actually be justified through the jurisprudence of the African human rights system. In order to do so, an analysis of the decisions of the African Commission and the other structures and bodies within the system that pertain to the protection of the environment has been carried out. It must be noted that the African Commission along with other regional and national bodies have through their jurisprudence highlighted and laid emphasis on the body of legal rules that ensure such environmental protection, taking into cognizance, the peculiarities of African human rights systems, and the cultural and social factors that inform these systems. This is what is identified as an environmental

[104] (2016) eKLR, para 84.
[105] Kamau, para 104.
[106] Ibid.

rule of law for Africa. The jurisprudence of these bodies though not necessarily binding on the entire continent, however in such cases, stands as persuasive authority for the rest of the continent. The principles and rules laid down by the judicial and quasi-judicial bodies discussed in this chapter must be extrapolated and applied to other cases that will still arise in the realm of environmental protection in the African human rights system.

References

Books

Evelyn Ankumah, *The African Commission on Human and Peoples Rights: Practise and procedures* (Martinus Nijhoff Publishers, 1996).

Collins Odote, 'The role of the Environment and Land Court in governing natural resources in Kenya' in Patricia Kameri-Mbote and others, *Law, Environment, Africa* (Nomos, Germany 2019).

Fatsah Ouguergouz, *African Charter on Human and Peoples' Rights: A Comprehensive Agenda for Human Dignity and Sustainable Democracy in Africa* 51–52 (New York: Kluwer Law International, 2003).

Gino Naldi, 'The African Union and the Regional Human Rights System' in Evans and Murray *The African Charter on Human and Peoples' Rights: the system in practise 1986–2006* (2nd edn, Cambridge University Press 2008).

Lillian Chenwi, 'The right to a satisfactory, healthy and sustainable environment in the African Regional Human Rights System' in John Knox & Ramin Pejan, *The Human Right to a Healthy Environment* (Edward Elgar Publishing, 2018).

Werner Scholtz, 'Human rights and the environment in the African Union context' in Anna Grear and Louise Kotze, *Research Handbook on Human Rights and the Environment* (eds, Edward Elgar Publishing 2015) 405.

Journal Articles

Abdulkadir Abdulkadir, 'The right to a healthful environment in Nigeria: a review of alternative pathways to environmental justice in Nigeria' (2014) Journal of Sustainable Development Law and Policy 3(1) 127.

Alan Boyle, 'Human Rights and the Environment: where next?' (2012) 23(3) The European Journal of International Law 613–642.
Alexandra Dapolito Dunn & Sarah Stillman, 'Advancing the Environmental Rule of Law: a call for measurement' 2015 (21) South Western Journal of International Law 283, 284.
Caiphas B Soyapi, 'Environmental Protection in Kenya's Environment and Land Court' (2019) 31 Journal of Environmental Law 151–161, 154.
Chairman Okoloise, 'Circumventing Obstacles to the Implementation of Recommendations by the African Commission on Human and Peoples' Rights' (2018) 18 African Human Rights Law Journal 29.
Dinah Shelton, 'Human rights, Environmental Rights, and the Right to Environment 28 Stanford' (1991–1992) 28 Journal of International Law 103, 128.
Downs, 'A healthy and ecologically balanced environment: an argument for a third generation rigcfzsht' (1993) 2 Duke Journal of Comparative and International Law 351–385.
Funmi Abioye, 'African Commission on Human and Peoples Rights: Achieving purpose?' 2014 South African Yearbook of International Law 39 Suppl 80–92.
Funmi Abioye, 'Understanding the Rule of Law in Africa through the lens of the African Commission on Human and Peoples Rights' interpretation on the right to development in Africa' 2018 (2) VRUE 269, 271.
Haile Andargie Wondalem, 'The Right to Environment under African Charter on Human and Peoples Right' (2015) 2(1) International Journal of International Law 206, 210.
Kofi Oteng Kufor, 'Safeguarding human rights: A critique of the African Commission on Human and Peoples Rights' (1993) 18 *Africa Development* 65.
Philippe Cullet, 'Definition of an Environmental right in a human right context' (1995) 13 Netherlands Quarterly of Human Rights 27.

Internet Sources

Decision 27/9 on Advancing Justice, Governance and Law for Environmental Sustainability of the UNEP https://wedocs.unep.org/bitstream/handle/20.500.11822/12221/Governing%20Council%20Decision%2027-2.pdf?sequence=1&isAllowed=y
Environmental Management and Coordination Act (EMCA) No. 8 of 1999 Laws of Kenya.

Infrastructure' available at https://www.investopedia.com/terms/i/infrastructure.asp

Infrastructure development' at African Development Bank Group website, available at https://www.afdb.org/en/knowledge/publications/tracking-africa%E2%80%99s-progress-in-figures/infrastructure-development/

IUCN, 'An introduction to the African Convention on the Conservation of Nature and Natural Resources' https://portals.iucn.org/library/sites/library/files/documents/EPLP-056.pdf accessed 24 May 2019, 1.

Office of the High Commissioner for Human Rights (OHCHR) 'Analytical Study on the Relationship between Human Rights and the Environment', UN Doc A/HRC/19/34, https://www.ohchr.org/Documents/HRBodies/HRCouncil/RegularSession/Session19/A-HRC-19-34_en.pdf

SAHRC, 'The Right to a Healthy Environment' https://www.sahrc.org.za/home/21/files/Reports/5th_esr_environment.pdf

The Johannesburg Principles on the Role of Law and Sustainable Development' adopted at the Global Judges Symposium, 2002, available at https://www.eufje.org/images/DocDivers/Johannesburg%20Principles.pdf accessed 8 July 2019.

'Transforming our world: the 2030 Agenda for sustainable Development' https://sustainabledevelopment.un.org/?menu=1300 accessed 8 July 2019.

United Nations 'New and Emerging Challenges in Africa Summary Report' available at https://sustainabledevelopment.un.org/index.php?page=view&type=400&nr=502&menu=1515

United Nations Environmental Program (UNEP) 'Environmental Rule of Law: critical to sustainable development' https://wedocs.unep.org/bitstream/handle/20.500.11822/10664/issue-brief-erol.pdf?sequence=1&isAllowed=y

UNEP 'Environmental Rule of Law: First Global Report' available at http://wedocs.unep.org/bitstream/handle/20.500.11822/27279/Environmental_rule_of_law.pdf?sequence=1&isAllowed=y

5

Women's Environmental Human Rights in Africa with Reflections on Key Provisions of the Maputo Protocol

Elsabé Boshoff

1 Introduction

> *Saving our planet, lifting people out of poverty, advancing economic growth... these are one and the same fight. We must connect the dots between climate change, water scarcity, energy shortages, global health, food security and women's empowerment. Solutions to one problem must be solutions for all.* (Ban Ki-moon, United Nations Secretary General 2007–2016)

Women are often left behind in the realization of basic human rights. The debate around women's rights as human rights gained momentum in feminist circles particularly during the 1980s and 1990s.[1] The proponents of the recognition of women's rights argue that in order to attain equality of all people, the human rights of women must be protected on

[1] C Medina "Women's Rights as Human Rights: Latin American Countries and the Organisation of American States (OAS)" in M Díaz-Diocaretz & IM Zavala (eds.) *Women, Feminist Identity, and Society in the 1980s: Selected Papers* John Benjamins Publishing.

E. Boshoff (✉)
African Commission on Human and Peoples' Rights, Banjul, Gambia

M. Addaney, A. Oluborode Jegede (eds.), *Human Rights and the Environment under African Union Law*, https://doi.org/10.1007/978-3-030-46523-0_5

an equal basis with that of men. However, as this is an argument for substantive equality as opposed to formal equality, it means that there may be instances where there is a need for additional protections for women, particularly in areas where they continue to be marginalized. In Africa, such marginalization often arises "out of the intersection of indigenous customs and colonial experiences",[2] which have resulted in women being excluded from decision-making fora, not provided with equal access to resources as their male counterparts and being subjected to less beneficial socio-economic conditions, all while continuing to bear the brunt of child-care.

This marginalization of women also impacts on women's environmental human rights.[3] For example, where women do not have equal rights to own property, in particular land, this impacts negatively on their ability to have a say in how such land is used. Similarly, in patriarchal systems where women are excluded from decision-making positions, this also limits their ability to choose the kind of development models which are followed. Such marginalization impacts on the ability of women to raise environmental concerns, particularly in issues where women are specifically impacted. Diamond and Orenstein summarize the vulnerability of female bodies to environmental destruction as follows:

> [B]ecause of women's unique roles in the biological regeneration of the species our bodies are important markers, the sites upon which local, regional or even planetary stress is played out. Miscarriage is frequently an early sign of the presence of lethal toxins in the biosphere.[4]

This stress on women's bodies, manifested in health and reproductive challenges, which does not similarly impact on men together with the violations of subsidiary rights to land and participation, provide justification for the adoption of specific protections of women's environmental

[2] F Coleman 'Pan-African Strategies for Environmental Preservation: Why Women's Rights Are the Missing Link' (2008) *Berkeley Journal of Gender, Law & Justice* 188.

[3] Coleman 188. Environmental human rights in this regard include the right to a clean and healthy environment, and developmental rights, to the extent that this is impacted by the environment.

[4] I Diamond and G Orenstein (eds) (1990) *Reweaving the World: The Emergence of Ecofeminism*, (San Francisco: Sierra Club) x.

rights. Furthermore, in many African societies, women continue to fulfil traditional roles of water collection, food preparation and child care. Women are more affected by environmental challenges such as climate change, loss of biodiversity and desertification, as they may have to walk further to reach potable water, may have more challenges in securing nutritious food for their children and in some societies are the last to eat. In addition to their own welbeing, in their continued role as primary carers for children, women require a clean and healthy environment for child rearing. Children are still developing, and they have a long life ahead of them during which the consequences of such exposure can cumulate and manifest. Women who are displaced, for example by changing climate patterns, are also more likely to suffer than men, including as a result of displacement as well as the added risk of sexual violence or sex for food or aid, which has become prevalent.[5] Because of the low levels of development in Africa, as well as the lack of sufficient adaptations to withstand the worst consequences of climate change, the consequences of climate change for African women as land cultivators and primary care givers are dire.

Acknowledging the multifaceted relationship between women and the environment and the challenges to the realization of women's environmental rights, various international human rights instruments deal with different aspects of this challenge. The next section provides a more in-depth discussion of the specific interrelationship between women in Africa and the environment and also the negative consequences suffered by women when the environment is harmed. Section 3 considers the protection provided to women's environmental rights at the international level. It starts by considering the relationship between the right to environment and developmental rights, which are often considered together at the international level. It considers the provisions under the Convention on the Elimination of All Forms of Discrimination Against Women (CEDAW) and the African Charter on Human and Peoples' Rights (African Charter) and then proceeds to critically unpack the rights under

[5] African Commission on Human and Peoples' Rights "Resolution 417 on the human rights impacts of extreme weather in Eastern and Southern Africa due to climate change – ACHPR/Res. 417 (LXIV) 2019" 14 May 2019 https://www.achpr.org/sessions/resolutions?id=441

the Protocol to the African Charter on Human and Peoples' Rights on the Rights of Women in Africa (Maputo Protocol). The chapter discusses the practical and far-reaching implications of these rights for women in Africa if it were to be realized. It also considers the duties on states to protect and promote these rights, along with the right to remedy, which is also specifically protected under Article 25 of the Maputo Protocol. In conclusion the chapter makes recommendations to States and other actors on steps to realize these rights.

2 Interrelationship Between African Women and the Environment

Apart from the challenges discussed above which may apply to all African women, a distinction should be drawn between the experiences and negative impacts suffered by women in rural and women in urban areas. In rural areas people are "basically reliant on resources obtainable from their environment for sustenance"[6], be it through farming or gathering of food and herbal medicines directly from the environment. In Sub-Saharan Africa, on average, about 50 percent of the agricultural labor force is women.[7] The higher levels of urbanization among men has also resulted in an increase in rural women headed households,[8] despite the fact that women have low levels of land ownership. This is the case mainly because of patrilineal inheritance of land, with women's access to land largely dependent on men, usually through marriage.[9] As a result, widowed or divorced women or women who decide not to marry often do not have access to land in rural areas, and do not have a say in what is done with the land.

[6] Emeka Polycarp Amechi, 'Linking Environmental Protection and Poverty Reduction in Africa: An Analysis of the Regional Legal Responses to Environmental Protection' (2010) 6/2 *Law, Environment and Development Journal* 112, 114; Coleman (2008) 195.

[7] The Food and Agriculture Organization of the United Nations "The role of women in agriculture" (2011) ESA Working Paper No. 11-02 available at http://www.fao.org/3/am307e/am307e00.pdf, 4.

[8] According to the FAO, the percentage of women agriculture increased from about 45 percent in 1980 to around 50 percent at present, Ibid.

[9] Jackson 406.

Traditional structures of authority also often exclude women from positions of power and decision-making. Women in rural areas are moreover responsible for other household activities such as water sourcing and gathering firewood—activities which are impacted by environmental degradation—resulting in women having to spend more time accessing these resources.[10] Because of the division of labor in many rural African contexts, "[t]he result is that women often bear disproportionately any negative costs associated with environmental degradation without receiving the reciprocal benefits".[11]

Other environmental impacts in rural areas are also rife. In Mozambique, severe drought resulting from climate change led to an increase in child marriage, as parents did not have sufficient money to feed their children or pay school fees.[12] Lake Tanganyika in East Africa, the second largest freshwater lake in the world and a source of livelihood for over ten million people, faces threats from "deforestation, unsustainable agricultural and fishing practices, mining activities and pollution", particularly through waste water entering the lake.[13] Women have little say in decisions related to Lake Tanganyika, despite their extensive role in "drying, smoking, and selling fish", and because their role is little valued, they are the first to lose their jobs if there is reduced resources, and they also have much lower access to credit than men.[14] The Niger delta in Nigeria is an example of where extensive oil pollution has made rural life all but impossible, which won it the title of "one of the most polluted

[10] Amechi 124.

[11] Coleman (2008) 195; Cecile Jackson "Women/Nature or Gender/History? A critique of Ecofeminist 'Development'" (1993) 20 *The Journal of Peasant Studies* 388, 405.

[12] United Nations Women "Rural women tackle drought-affected Mozambique's rise in child marriage" (2017) https://www.unwomen.org/en/news/stories/2017/11/feature-mozambique-rural-women-tackle-drought-related-rise-in-child-marriage (accessed on 1 December 2019).

[13] United Nations Development Programme "Lake Tanganyika, what the future holds" 22 March 2017 https://stories.undp.org/lake-tanganyika-what-the-future-holds (accessed on 22 July 2019).

[14] The World Bank "Lake Tanganyika Environmental Management Project" (2018) p. 7 http://documents.worldbank.org/curated/en/668911539594000200/pdf/Concept-Project-Information-Document-Integrated-Safeguards-Data-Sheet-Lake-Tanganyika-Environmental-Management-Project-P165749.pdf; Pétur Waldorff "Left Out to Dry? Gender and Fisheries on Lake Tanganyika" (2017) UNU-GEST https://unu.edu/publications/articles/gender-and-fisheries-on-lake-tanganyika.html

places on earth".[15] Women, traditionally responsible for gathering fish and shellfish, growing crops and gathering drinking water, are left destitute.[16] Artisanal miners across the continent, of which about a third are women, suffer from coming in contact with various harmful chemicals and the dust that they breathe in. Furthermore, "to meet their domestic demands women often carry out potentially hazardous mineral processing in their homes alongside regular domestic activities, putting themselves and their families at risk".[17] While some countries have taken steps to regulate artisanal mining, in many African countries it remain unregulated, without sufficient labor, health and environmental protections.

Urban dwellers do not have this same close dependence on the environment, but suffer from environmental degradation and in particular high levels of pollution, with serious repercussions for their health.[18] Women living in urban areas in cities and towns in Africa are often affected by a lack of infrastructure to support the rapid expansion in these areas, resulting in rubbish being burnt in the streets, various fuels burned for energy, exhaust fumes from millions of vehicles and lack of sewage infrastructure. Exposure is often highest among women and children "who spend the most time in houses or near emission sources".[19] Mining towns suffer their own range of consequences. The Mpumalanga region of South Africa has the highest levels of nitrogen dioxide pollution, because of more than a dozen coal fire plants located there.[20] The town of

---

[15] Amnesty International "Niger Delta Negligence" March 2018 https://www.amnesty.org/en/latest/news/2018/03/niger-delta-oil-spills-decoders/ (accessed on 22 July 2019).

[16] Ruth Maclean "Toxic mud swamps fortunes of Niger Delta women years after oil spill" (2018) *The Guardian* https://www.theguardian.com/global-development/2018/dec/19/toxic-mud-swamps-fortunes-of-niger-delta-women-years-after-oil-spill

[17] Sheila Keetharuth "Women in the Extractive Industries in Africa: Gender Perspectives" *Newsletter of the Working Group on Extractive Industries, Environment and Human Rights in Africa* of the African Commission on Human and Peoples' Rights (2nd ed) (October 2019) https://www.achpr.org/newsletter/viewdetail?id=28

[18] Patricia Kameri-Mbote 'Access, control and ownership: Women and sustainable environmental management in Africa' (2007) 72 *Agenda* 36.

[19] Business Tech 'Government is worried about alarmingly high levels of air pollution in these South African towns' (2 June 2019) *Business Tech* available at https://businesstech.co.za/news/government/317836/government-is-worried-about-the-high-air-pollution-in-these-south-african-towns/ (accessed on 22 July 2019).

[20] Lynsey Chutel "This picturesque corner of South Africa has the world's deadliest air pollution" (29 October 2018) Quartz Africa https://qz.com/africa/1441504/highest-concentration-of-deadly-air-pollution-found-in-south-africa/ (accessed on 21 July 2019).

Arlit in Niger, where the French-owned company Areva mines uranium, is exposed to extraordinary high levels of radiation, including in the drinking water which has impacted on women's general and reproductive health. Women human rights defenders (HRDs) also suffer negative consequences because of their involvement in activism. Environmental HRDs are among the most persecuted the world over. In addition to the threats to their lives and jobs which are similar to those faced by men, women environmental HRDs in addition receive threats about the safety of their families and their children.

Apart from suffering extensive impacts from environmental degradation, women also have a role to play in environmental conservation. Possibly the most pertinent example in this regard is that of Nobel Peace winner Wangari Maathai, whose Green Belt Movement has by July 2019 planted over 51 million trees in Kenya and mobilized hundreds of women to take sustainable livelihoods into their own hands.[21] Across the continent, particularly because of their ability to mobilize locally, women have played an important role in environmental conservation. This was recognized by the African Ministerial Conference on the Environment (AMCEN), which during its eighth session in 2000 stressed the "importance of integrating 'women's groups,' among other sectors of society, into the development process in order to take advantage of their particular skills and environmental knowledge".[22]

It is clear from the above that "[e]nvironmental laws, policies and interventions that fail to factor in the gender dynamic contribute to environmental degradation while accentuating women's poverty and increasing their workload".[23] The next section looks at the existing international legal framework for ensuring that laws and policies at national level provide sufficient protection to women's environmental rights.

---

[21] http://www.greenbeltmovement.org/, accessed on 22 July 2019. Wangari Maathai passed away in 2011, but her work has been continued through the Green Belt movement.

[22] Coleman (2008) 186.

[23] Kameri-Mbote (2007) 38.

3 The International Legal Framework for the Protection of Women's Environmental Rights

The so-called International Bill of Rights—the Universal Declaration of Human Rights (UDHR),[24] the International Covenant on Economic, Social and Cultural Rights (ICESCR)[25] and the International Covenant on Civil and Political Rights (ICCPR)[26] do not make any mention of environmental rights, except to the extent that it can be inferred from the right to health and clean water.[27] While there has been a lot of normative development at the international level related to the environment, this has been (a) within the field of international environmental law, which largely relates to agreements between States, and thus does not create direct rights to citizens,[28] or (b) in relation to environment and sustainable development, where developments have been in the form of non-binding declarations and goals set by the UN.[29] Similarly, the Revised African Convention on the Conservation of Nature and Natural Resources does not relate specifically to human rights.[30] Some scholars have made the case that the human right to a healthy

[24] Universal Declaration of Human Rights 1948 (GA Res 217A (III) UN Doc A/810 at 71 (1948)).

[25] International Covenant on Economic, Social and Cultural Rights 1966 (UN Doc A/6316 (1966) 993 UNTS 3).

[26] International Covenant on Civil and Political Rights 1966.

[27] Committee on Economic, Social and Cultural Rights "General Comment No. 14" (2000) The right to the highest attainable standard of health (article 12 of the International Covenant on Economic, Social and Cultural Rights) 2. See also United Nations General Assembly resolution 45/94 on the need to ensure a healthy environment for the well-being of individuals.

[28] See for example the Basel Convention on the Control of Transboundary Movements of Hazardous Waste and Their Disposal, Convention on Biological Diversity, United Nations Framework Convention on Climate Change (UNFCCC), among many others, as well as bilateral environmental agreements.

[29] For example the Sustainable Development Goals.

[30] Despite this, the Principles to guide States in the implementation of the Convention are purely rights based, and include the right to a satisfactory environment, the State's duty to ensure the right to development and the duty on States to ensure that the environmental and development needs are met.

environment has become a principle of customary international law.[31] A strong proponent of this view is John Knox, the UN special rapporteur on human rights and the environment until 2018.[32] However, this view is not universally accepted, and the UN has not yet taken the step to formalize the right in a binding treaty. This means that a comprehensive, binding human rights framework for environmental rights at the UN level is still lacking.

The Link Between Environment and Development

Within the human rights field, environmental rights are often closely related to development or sustainable development. "Development" in its classical form has come at the cost of extensive destruction of the environment through the exploitation of non-renewable resources and pollution at a global scale. However, gradually an awareness developed of the negative consequences of environmental destruction, the limitations of the natural world to absorb the costs of human activity, and the finite nature of natural resources. It also became clear that loss of biodiversity, climate change and other environmental challenges negatively impact on human health and well-being.

The international discourse on the protection of environmental rights arguably starts with the 1972 UN Conference on the Human Environment, which for the first time "brought the industrialized and developing nations together to delineate the 'rights' of the human family to a healthy and productive environment".[33] In an aim to strike a balance between the seeking for better quality of life and more wealth on the one hand and environmental consequences on the other hand, the concept of "sustainable development" was introduced through the Brundtland Report on Our Common Future of 1978. Sustainable development was defined as the use of resources to meet our present development needs

[31] John Lee, "The Underlying Legal Theory to Support a Well-Defined Human Right to a Healthy Environment as a Principle of Customary International Law" (2000) 25 COLUM. J. ENVTL. L.

[32] See for example John H. Knox & Ramin Pejan (eds) *The Human Right to a Healthy Environment* (Cambridge University Press, 2018).

[33] Gro Harlem Brundtland "Chairman's Foreword" *Our Common Future* (1978).

while allowing future generations to do the same. The Brundtland report thus situated the environment as a core element of sustainable development. Since the Brundtland report the discourses around environment and development have become inseparable. The Rio Declaration on Environment and Development (1992) sought to build on the Declaration of the United Nations Conference on the Human Environment and was followed by the World Summit on Sustainable Development in Johannesburg, South Africa, in 2002, and the United Nations Conference on Sustainable Development in 2012. The United Nations Development Programme (UNDP) went on to confirm that "integrating sustainable environmental policies into development objectives is a human rights issue".[34] The most recent development in the development/environment context was the 2015 adoption of the Sustainable Development Goals which incorporate a core environmental aspect in Goal 6: Clean, accessible water for all; Goal 7: Affordable and Clean Energy; Goal 12: Responsible Production and Consumption; Goal 13: Climate Action; and Goal 15: Sustainably manage forests, combat desertification, halt and reverse land degradation, halt biodiversity loss. It in addition puts in place Goal 5: Achieve Gender equality and empower all women and girls.

This clear recognition that considerations of environment and development cannot be separated is a crucial mind-shift which, if implemented, is to the benefit of women in Africa. This is because women in Africa suffer equally from poverty and other consequences of low levels of development and from the negative consequences of environmental destruction resulting from classical development models. Very few African States have done the necessary research to have the statistics to report accurately on their progress towards meeting the SDGs. However, a quick review of the above goals show that in terms of clean accessible water, only 24 percent of people in Sub-Saharan Africa have access to safe drinking water,[35] there is still a high reliance on energy generated through the burning of wood

[34] Flynn Coleman, "Pan-African Strategies for Environmental Preservation: Why Women's Rights Are the Missing Link" (2008) 23 *Berkeley J. Gender L. & Just.* 181, 184.

[35] United Nations Water "World Water Development Report 2019: 'Leaving no one behind'" 19 March 2019 https://www.unwater.org/world-water-development-report-2019-leaving-no-one-behind/ accessed on 22 July 2019.

or coal, access to land remains a contentious issue across the continent and while most states have adopted some form of climate change strategy, these are largely focused on adaptation rather than mitigation of climate change. As discussed in the previous section, all of these factors impact negatively on women's environmental rights, as it impacts on their water sourcing ability. Further they suffer the most from pollution from coal or wood stoves. Given the general focus of the SDGs and national provisions on attaining gender equality, across all spheres equality in environmental rights perhaps has not received the attention that it warrants.

Women's Environmental Rights Under CEDAW

The Convention on the Elimination of All Forms of Discrimination Against Women (CEDAW), is the international treaty adopted in 1979 to protect the rights of women. While not providing explicitly for environmental rights, CEDAW has quite a bit to say about the role of women in development. In the preamble, it provides that inequality between men and women not only hampers the political, social, economic and cultural life of their countries, but also that "the full and complete development of a country, the welfare of the world and the cause of peace require the maximum participation of women on equal terms with men in all fields". In relation to the substantive rights, an equality clause in Article 3 requires States to take "all appropriate measures, including legislation, to ensure the full development and advancement of women". CEDAW in Article 14 acknowledges the role of women in development and provides for a number of duties on States in relation to rural women's right to development. These duties include, to ensure the full participation of *rural* women in development, to have access to land and to enjoy adequate living conditions, including in relation to housing, sanitation, water, among others.

As there is near-universal ratification of CEDAW in Africa, these provisions would be applicable to rural African women as well. In particular CEDAW makes clear the link between development and women's access to land. However, as the CEDAW provision on environment is limited to rural women, its application in the African context is also limited, and becoming less relevant, given the speed at which Africa is urbanizing.[36] It is also silent on the various human rights challenges faced by urbanized women in Africa. In particular, given the lack of infrastructure in many African cities for waste disposal and sewerage, cities in Africa have high levels of water, air and ground pollution, with serious repercussions for women's health and well-being. Mining is also one of the main culprits of environmental degradation on the continent, and women residing in mining towns are also not considered rural dwellers, and would thus be excluded under the provisions of CEDAW on environment.

In interpreting the rights under Article 14, the Committee on the Elimination of Discrimination Against Women provides for an obligation on states to

> [a]ddress specific threats posed to rural women by climate change, natural disasters, land and soil degradation, water pollution, droughts, floods, desertification, pesticides and agrochemicals, extractive industries, monocultures, biopiracy and the loss of biodiversity, in particular agrobiodiversity. They should alleviate and mitigate those threats and ensure *that rural women enjoy a safe, clean and healthy environment*. They should effectively address the impact of such risks on rural women in the planning and implementation of all policies concerning the environment, climate change, disaster risk reduction, preparedness and management and ensure the full participation of rural women in designing, planning and implementing such policies [emphasis added].[37]

This explicit recognition of environmental degradation and climate change as culprits that hamper the achievement to the right to development in an important way enforces the view that development and environment cannot be separated. This exposition however again highlights how regrettable the limitation of the rights in Article 14 to rural women alone is, given that all women should be guaranteed participation in design and implementation of policies which impact on climate change

[36] World Bank "Urbanization in Africa: Trends, Promises, and Challenges" (2015) https://www.worldbank.org/en/events/2015/06/01/urbanization-in-africa-trends-promises-and-challenges [accessed on 17 July 2019].

[37] Committee on the Elimination of Discrimination Against Women "General recommendation No. 34 (2016) on the rights of rural women" 5.

and the environment, and given that water pollution and other environmental degradation impacts equally on urban women as on rural women.

Women's Environmental Rights Under the African Charter

At the regional level, the African Charter on Human and Peoples' Rights, which entered into force in 1986, and has been ratified by 54 out of the 55 Africa States, in Article 24 provides for the right of all peoples to a *generally satisfactory* environment, favorable to their development. It also provides that all peoples have the right to their economic, social and cultural development in the equal enjoyment of the common heritage of mankind.[38] States have a duty in this regard, to ensure the exercise of the right to development.[39] Given that the African Charter has an equality clause which prohibits discrimination on the basis of sex, these rights equally apply to African women and men.

The African Commission interpreted the provisions of the African Charter through a few cases, as well as in its recently adopted State Reporting Guidelines and Principles on Articles 21 and 24 of the African Charter (Guidelines and Principles). The most important jurisprudence of the Commission to date related to environmental rights is set out in the case of Social and Economic Rights Action Center (SERAC) and Center for Economic and Social Rights (CESR) v. Nigeria (the SERAC case).[40] This case related to "environmental degradation and health problems resulting from the contamination of the environment" in the Ogoniland of Nigeria as a result of oil spills and pollution of water, soil and air.[41] The Complainants argued that the State had failed to provide the Victims with the necessary information or ensure their participation in processes affecting them. The State further did not require of the oil

[38] Article 22(1) of the African Charter.

[39] Article 22 (2) of the African Charter.

[40] Communication 155/96 – Social and Economic Rights Action Center (SERAC) and Center for Economic and Social Rights (CESR) v. Nigeria, adopted at its 30th Ordinary session held in Banjul, The Gambia from 13th to 27th October 2001.

[41] Para 1–2.

companies to undertake impact assessments and in fact *facilitated* the violations through "placing the legal and military powers of the state at the disposal of the oil companies".[42]

The Commission in its decision reiterates the duties on State Parties to the African Charter to respect, protect and promote the human rights protected therein. The Commission finds that the right to health "requires the state to take reasonable and other measures to prevent pollution and ecological degradation, to promote conservation, and to secure an ecologically sustainable development and use of natural resources" and thus finds that Article 24 has been violated.[43] Subsequently, following non-implementation of the Commission's decision, the ECOWAS Court in the case of SERAP v Nigeria found the Respondent State in violation of the African Charter and ordered it to "take all effective measures, within the shortest possible time, to ensure restoration of the environment of the Niger Delta", "prevent occurrence of damage to the environment" and "hold perpetrators of the environmental damage accountable".[44]

In Front for the Liberation of the State of Cabinda v Republic of Angola (Cabinda case) the Commission held that there was no violation of Article 24.[45] This case dealt with "economic self-determination of the people of Cabinda and in particular with the disposition and exploitation of onshore oil, mineral and natural resources".[46] The Complainant submitted that the operation of certain companies has resulted in harm to human health and the environment, as the "activities of oil companies are not monitored and no compensation is paid in the event of damage to the environment".[47] The Respondent State denied a violation of this right, making reference to the right to environment as enshrined in the Angolan Constitution, the legislation dealing specifically with environmental protection and the steps which it had taken to ensure that persons

---

[42] Para 3, 5–6.

[43] Para 52.

[44] SERAP v Federal Republic of Nigeria, N° ECW/CCJ/JUD/18/12, para 121.

[45] Communication 328/06 – Front for the Liberation of the State of Cabinda v Republic of Angola, adopted at its 54th Ordinary Session, held from 22 October to 5 November 2013.

[46] Para 54.

[47] Para 66.

in Cabinda affected by, for example oil spills, are compensated.[48] The Commission found that there was not enough evidence to find a violation of Article 24, as the Complainants had not submitted any evidence to prove their claims and also did not dispute the submissions by the State of the various steps which it had taken to ensure environmental protection and compensation, where necessary.

While these Communications did not relate to the rights of women specifically, it provided the first interpretation of the Commission of the right to environment. In particular it recognized the indivisibility of environmental rights from other rights, such as the right to housing, health, food, right to access to information and freedom of expression, among others. Additionally, while the right to environment is formulated as a generally satisfactory environment, the Commission has interpreted this expansively, to make it clear that where there are detrimental effects on the health of people, that would constitute a violation of this right. This is particularly important for women, as their concern for the environment is often closely related to their concern to provide food and safe housing for their families, and safeguard their health and welbeing. Moreover, by emphasizing the right to freedom of expression and linking it to environmental rights, these Communications could be used by women activists to ensure that women are able to raise environmental issues affecting them specifically. Amechi notes that apart from providing for a right to a generally satisfactory environment, Article 24 also links the right to environment to development.[49] He submits that this envisages "that African citizens should not only be able to live in a undegraded and pollution-free environment, but also, be able to access the resources provided by their environment in order to develop to their full potential". This would equally apply to women.[50]

The most recent and most comprehensive interpretation by the Commission of the right to a healthy environment is found in the Guidelines and Principles on Articles 21 and 24 of the African Charter, launched in November 2018. While these Guidelines and Principles were

---

48 Para 84–86.

49 Amechi (2010) 123.

50 Amechi (2010) 123.

adopted in the context of extractive industries specifically, it is also applicable to the right to a healthy environment more generally. It clarifies that the right to a healthy environment "is not the right to an ideal environment that is completely unaffected or completely clean. It is rather an environment that is clean enough for a safe and secure life and development of individuals and people".[51] It states that the right to environment encompasses the right to have the environment used in a sustainable manner, and to be protected from environmental degradation and pollution. It further requires that appropriate environmental risk assessments be undertaken prior to industrial development, adequate waste disposal systems be put in place and that the right to timely and accessible information and to be consulted about activities that might affect them be respected. The right to environment also provides for full, adequate and effective compensation and environmental restoration for despoliation; and the right to justice through appropriate judicial and non-judicial mechanisms.[52]

Importantly, efforts were made to streamline women's rights and concerns throughout the Guidelines and Principles. The Explanatory note to the State Reporting Guidelines notes the importance of the interrelation of rights, the need for participation of all affected persons in decisions affecting them, and the principles of non-discrimination and equality, providing explicitly that the provisions in Article 24 "should be applied and operationalized with particular regard to the most marginalized and socio-economically excluded members of society, *including most notably women*" (emphasis added), on the basis of the equality clause in Article 2 of the African Charter.[53]

The rights of women are further streamlined in the Guidelines and Principles, as follows:

i. In discussing the compensation to persons removed from their land for the purpose of extractive industries, it provides that "Compensation should also take account of the additional vulnerability of certain groups, such as female-headed households"[54];

[51] State Reporting Guidelines and Principles (2018) para 27.

[52] Para 28–36.

[53] State Reporting Guidelines and Principles (2018), pp. 20–22.

[54] Para 15.

ii. The Guidelines state that there is a duty on States to ensure that there is no impunity for threats, attacks and acts of intimidation of human rights defenders, including women human rights defenders[55];
iii. In relation to access, use and ownership of land, the Guidelines state that measures to protect land rights should "also take account of the often more precarious nature of land ownership by women in Africa"[56]; and
iv. In relation to artisanal mining, the Guidelines provide for special protection for vulnerable groups, including women, through ensuring in particular environmental and safety standards.[57]

These references to women throughout the Guidelines highlight some of the particular environmental challenges affecting women, and aim to provide guidance to States in ensuring that the rights of women are protected and promoted in these contexts. It addresses some of the challenges affecting specifically vulnerable groups of women, including women human rights defenders and women working in artisanal mining, who are more prone to human rights violations than women in general. Furthermore, it takes cognizance of the African context in relation to women's limited land rights, which is also a welcome development. Ayeni identifies the "normative content with respect to the protection of women's rights" as inadequate, and one of the main shortcomings of the African Charter, in that it only discusses women in relation to the family context.[58] However, in the specific context of environmental rights, it can be argued that the normative development through the State Reporting Guidelines and Principles has been a significant positive development which has taken forward the interpretation of the right to environment and of women specifically, and the duties that this imposes on States. That being said, the Guidelines is a soft law instrument, and mainly aimed at the specific context of State Reporting and the thematic area of extractive industries. Therefore, while it has interpretative value, its application is limited.

[55] Para 24.

[56] Para 42.

[57] Para 45.

[58] Victor O Ayeni "Introduction and preliminary overview of findings" in Victor O Ayeni (ed.) (2016) *The impact of the African Charter and the Maputo Protocol in selected African states* 7.

Women's Environmental Rights Under the Maputo Protocol

The Maputo Protocol, adopted on 11 July 2003 and entering into force on 25 November 2005, has to date been ratified by 40 of the 55 African States. It is aimed at providing further protection to the rights of women that are not protected specifically under the African Charter on Human and Peoples' Rights (African Charter) or under the CEDAW.[59] Viljoen summarizes this well, as follows:

> [T]he Protocol speaks in a clearer voice about issues of particular concern to African women, locates CEDAW in African reality, and returns into its fold some casualties of quests for global consensus [...]. More specifically, the Protocol expands the scope of protection rights beyond those provided for under CEDAW and it deals with rights already covered in CEDAW with greater specificity.[60]

When it comes to the right to environment, while the African Charter protects the right of all peoples to a *generally satisfactory* environment, and the CEDAW protects the rights of *rural* women, the Maputo Protocol goes further than both these instruments. Articles 18 and 19 of the Maputo Protocol guarantees to all women the right to live in a healthy and sustainable environment and also the right to sustainable development. Given the profound and far-reaching implications of these two sections, it is necessary to consider their content in detail.

Article 18 provides thus:

1. Women shall have the right to live in a healthy and sustainable environment.
2. States Parties shall take all appropriate measures to

[59] Frans Viljoen "An Introduction to the Protocol to the African Charter on Human and Peoples' Rights on the Rights of Women in Africa" (2009) 11 *Wash. & Lee JCR & SOC. JUST.*

[60] As above.

a) ensure greater participation of women in the planning, management and preservation of the environment and the sustainable use of natural resources at all levels;
b) promote research and investment in new and renewable energy sources and appropriate technologies, including information technologies and facilitate women's access to, and participation in their control;
c) protect and enable the development of women's indigenous knowledge systems;
d) regulate the management, processing, storage and disposal of domestic waste;
e) ensure that proper standards are followed for the storage, transportation and disposal of toxic waste.

The first sub-article does not place any internal limitations on the right to a healthy and sustainable environment. It is thus a very strong right, and as noted above, it goes beyond any previous formulation of the right to healthy environment. It also immediately creates a link with Article 19, through its reference to a sustainable environment. Nsibirwa further argues that "[t]he [...] Protocol ought not to be viewed as merely advancing the rights of women, but rather as advancing the interests of society in general. This is because the whole society gains if a healthy and sustainable environment is attained".[61]

The duty on State Parties as stipulated in Article 18(2)(a) to ensure "greater participation of women in the planning, management and preservation of the environment and the sustainable use of natural resources at all levels" is an important recognition of the interest that women have in ensuring the preservation of the environment, the need for equality between men and women and enhancing the crucial role that women have been playing in ensuring environmental protection on the continent.

Sub-article 2 also places extensive obligations on States to respect, protect and promote the rights of women in this regard, some of which are very specific, for example promoting women's access to credit, training

[61] Martin S Nsibirwa "A brief analysis of the Draft Protocol to the African Charter on Human and Peoples Rights on the Rights of Women" (2001) 1 *African Human Rights Law Journal* 49.

and skills development which is easily measurable, and others which are much broader and vaguer, such as ensuring that the negative effects of globalization are reduced to the minimum for women. Contrary to CEDAW, in relation to the provision of credit, training and so forth, it provides explicitly for promotion at both the rural and urban levels. It also speaks specifically to the challenges affecting women in African States, for example in addressing the general lack of waste removal infrastructure in African cities, as well as acknowledging the role of African women as the custodians of indigenous knowledge.

Article 19 provides:

Women shall have the right to fully enjoy their right to sustainable development. In this connection, the States Parties shall take all appropriate measures to

a) introduce the gender perspective in the national development planning procedures;
b) ensure participation of women at all levels in the conceptualization, decision-making, implementation and evaluation of development policies and programs;
c) promote women's access to and control over productive resources such as land and guarantee their right to property;
d) promote women's access to credit, training, skills development and extension services at rural and urban levels in order to provide women with a higher quality of life and reduce the level of poverty among women;
e) take into account indicators of human development specifically relating to women in the elaboration of development policies and programs; and
f) ensure that the negative effects of globalization and any adverse effects of the implementation of trade and economic policies and programs are reduced to the minimum for women.

While sustainable development remains an elusive concept and is sometimes used as a "catch-all" phrase to reach agreement between divergent views placing emphasis either on the economic, social or the environmental aspects of sustainable development,[62] the Maputo Protocol, through the duties that it places on States in relation to sustainable development gives some content to this concept in the African context. It links sustainable development to access to resources, including financing and land, it links it to empowerment of people to have a say in decisions that impact them, and it is also linked to addressing poverty and the attainment of human development indicators. Interestingly, it also notes that the State should "ensure that the negative effects of globalisation and any adverse effects of the implementation of trade and economic policies and programmes are reduced to the minimum for women".[63] This is a provision which could have far-reaching implications, and should be further developed to determine what this would mean in practice.

These two provisions, when read together, provide for a very strong right to environment and development for women, which has not been seen in any of the previous instruments. Furthermore, Articles 18 and 19 contain provisions which relate to a range of other rights, such as the right to equality, right to participation and freedom of expression, right to education (19(d)), to decent employment, right to access and control of land (Article 19(c)), and a right to health, which are central to fulfilling the rights to a healthy environment and sustainable development—thereby highlighting the indivisibility of human rights.

A further right in the Maputo Protocol which is of relevance in this context, is the provisions of Article 25 in terms of which "State parties shall undertake to provide for appropriate remedies to any woman whose rights or freedoms, as herein recognised, have been violated".[64] While the right to remedy is an established principle of customary international law, its explicit inclusion in the Protocol as a substantive right emphasizes it central importance in ensuring that justice is done in relation to women whose rights have been violated. The types of remedies which should be provided for in relation to environmental damage are set out in the

[62] John McCormick, 'The Origins of the World Conservation Strategy' (1986) Environmental Review 10(3) 177, 178.

[63] Article 19(f) of the Maputo Protocol.

[64] Article 25(a) of the Maputo Protocol.

Explanatory note to the State Reporting Guidelines on Articles 21 and 24 and include (a) full, adequate and effective compensation including for "material loss, livelihood disruptions including opportunity costs and illness as well as emotional distress resulting from such environmental damage"; (b) affordable health care and support for irreparable harm suffered to their health; (c) support for the rehabilitation of their livelihoods; (d) cleaning up and restoration of the affected environment.[65]

Very little normative development has been done around these rights at the level at the African human rights system. The African Commission has to date not decided any cases on the Maputo Protocol and the cases decided by the other two international tribunals with the power to interpret the Maputo Protocol, the African Court on Human and Peoples' Rights (African Court) and the Economic Community of West African States Community Court of Justice (ECOWAS Court) deal only with first generation rights.[66] There is thus no relevant jurisprudence to rely on to assist in interpreting these provisions. Another guideline for interpretation is the African Commission's State Reporting Procedure and Concluding Observations on State Reports.[67] The Commission in its two most recent Concluding Observations, on Mauritania and Burkina Faso, for the first time clearly distinguished its recommendations on the Maputo Protocol from that on the African Charter. However, in relation to Articles 18 and 19 the Commission makes very few or no recommendations to the States for improvement. The Commission commended Burkina Faso for the adoption of Agrarian Land Reform legislation which incorporates principles of gender and equity and the law on Rural Land Tenure, in terms of which men and women have equal rights to own and access land. Both of these developments relate to the State's duty under Article 19 (c) to ensure control by women of land and productive resources. However, no further details were provided by the State on the

[65] Para 34 and 35 of the Explanatory note to the State Reporting Guidelines on Articles 21 and 24.

[66] The ECOWAS Court made its first ruling on the Maputo Protocol in 2017 and the Court made its first and only ruling on the Protocol in May 2018, both cases decided in favor of the Applicants. However, both these cases, and subsequent cases at the ECOWAS court dealt with the right to bodily integrity, cruel, degrading and inhuman treatment in the case of the ECOWAS court, and discriminatory treatment in family law, in the case before the African Court.

[67] Between 2015 and July 2019, 12 States have made submissions on their compliance with the Maputo Protocol.

implementation of these legislative developments and how they have benefitted women in practice.

In the Concluding Observations on the State Report of Mauritania, the Commission commended the encouragement by the State of the use of solar and gas ovens, adherence to international environmental agreements and enactment of national legislation on the right to environment. While the first relates to women specifically, as it is stated that it relieves "women from the drudgery associated with the supply of domestic energy", the second and third commendations are for general developments, which do not relate to women specifically. Unfortunately, the Commission lost the opportunity to recommend to the State to ensure that these developments support women's environmental rights. In addition, the Commission missed the opportunity of linking the provision of solar ovens with reduced pollution in the household. One of the reasons for this superficial engagement by the Commission with these rights in the State Reporting Procedure is that so little work has been done on the interpretation and elaboration of what these rights can and should entail, and thus the Commission has no set standards against which to measure government conduct.

Domestication of Women's International Environmental Rights

Notwithstanding the limited jurisprudence and guidance from the African human rights system on how these rights should be interpreted, there are some trends which can be established in States' compliance with the provisions under Articles 18 and 19. While it is not in most cases possible to draw a direct correlation between the ratification of international instruments and developments in State Parties, there have been a number of positive developments in State Parties to the Maputo Protocol following ratification, including the adoption of national level legislation to give effect to these international obligations.

Efforts have been made to increase women's participation in decision-making processes and particularly the introduction of quotas for representation in government. Unfortunately, there are no clear cases in which there are specific provisions for women to have a central role in management of *environmental* processes.[68] However this is countered to some extent in that many African States have adopted national (sustainable) development strategies or policies which provide specifically for the role of women.[69] Burkina Faso has further established Regional and Communal Gender Promotion Councils, which promote gender at the community level, with the mandate to, inter alia integrate gender into local development plans.

A crucial development in recent years has been in terms of access of women to land—in most countries law now provides that women can own land in their own name, as well as following divorce or death of a spouse.[70] As noted above, ownership of land is an important criterion for ensuring that women have a say in the use of the land and empowering them to decide on the development path that they want to follow. Control over natural resources by women, who are aware of the implications of pollution and degradation on their families, would also result in more responsible resource use and environmental conservation, as was seen in the case of the Green Belt Movement in Kenya. There has also been an increase in women's access to credit as required under Article 19(2)(d).[71]

Many States have also established Ministries dealing with gender matters as well as gender focal points in various ministries and budgets for gender mainstreaming through different ministries.[72] For example, the Ethiopian Ministry of Mines and Petroleum has a Directorate dedicated to Gender Mainstreaming. Such developments are positive, in that they ensure that the rights and considerations of women involved in the mining industry, including their environmental rights is given due consider-

[68] Concluding Observations on Burkina Faso, Mauritania.

[69] This is particularly in relation to Burkina Faso, Cameroon, Nigeria, Lesotho. See Some in Ayeni, pp. 21–22; Forkum in Ayeni p. 36; Nyarko in Ayeni 103.

[70] Concluding Observations on Namibia, Burkina Faso.

[71] Concluding Observations on Burkina Faso, Mauritania. For example according to the latest State Report of Burkina Faso, measures have been taken to promote women's access to credit through the establishment of micro-credit institutions such as the Women's Income-Generating Activities Assistance Fund (FAARF) and the Assistance Fund for Income-Generating Activities for Women Farmers (FAAGRA), and the establishment of the Special Assistance Fund for Women Entrepreneurship which has a budget of five billion CFAF.

[72] Gambia, Ghana, South Africa, Burkina Faso.

ation. Kenya's 2010 Constitution in Article 58 establishes an Equality Commission with the duty, inter alia to promote "gender equality and equity generally and to coordinate and facilitate gender mainstreaming in national development".[73] While not related specifically to environmental rights, it is related to development and would thus require that environmental considerations are taken into account. One of the most promising developments, the adoption of a Gender and Equal opportunities Act that would domesticate the provisions of Articles 1 to 24 of the Maputo Protocol in Nigeria, has unfortunately been stalling since 2011. This would make the provisions of articles 18 and 19 directly enforceable in the national courts and would ensure greater protection of women's environmental rights at the national level in Nigeria.

One innovative development at the sub-regional level is the Protocol on Gender and Development adopted by the Southern Africa Development Community (SADC) in 2008, which harmonizes all international, regional, sub-regional and national commitments in relation to gender equality, including the Maputo Protocol.[74] The adoption of the Protocol was followed by the establishment of a Barometer, for measuring the extent of implementation of their obligations by Member States. Unfortunately, the 2019 Barometer does not provide any specifics about state action towards women's environmental rights.[75] An Oxfam report on "Implementing the Protocol on the Rights of Women in Africa: Analysing the Compliance of Kenya's Legal Framework" similarly does not provide any information on adherence with environmental rights. Clearly, there is much that remains to be done to elaborate the content of women's environmental rights in Africa, this chapter being only an initial step, as well as to develop frameworks for tracking compliance by States.

---

[73] Oxfam "Implementing the Protocol on the Rights of Women in Africa: Analysing the Compliance of Kenya's Legal Framework" (2014) http://www.soawr.org/sites/default/files/ml-implementing-protocol-womens-rights-africa-010314-en.pdf

[74] South African Development *Community Protocol on Gender and Development* (2016) https://genderlinks.org.za/wp-content/uploads/2016/01/ADOPTED-REVISED-PROTOCOL-ON-GAD.pdf

[75] Tarisai Nyamweda and Colleen Lowe Morna "State of Women in SADC 2019 Report" (2019) https://genderlinks.org.za/wp-content/uploads/2019/08/StateOfWomenSADC2019rev2.pdf

4 Conclusion

This chapter provides an overview of the relationship between women and the environment in the African context. It identifies the various ways in which environmental degradation impacts on women's rights as well as the negative consequences for women and their bodies resulting from environmental pollution. It also draws a link between the right to a healthy environment and the right to sustainable development, and in particular notes that in order to be to the benefit of women, development in Africa must take account of and reduce environmental impacts. The chapter further considers the extent to which women's environmental rights are protected within the human rights framework applicable in Africa. While the Maputo Protocol, which provides the strongest protection for these rights, has to date been ratified by only 40 African States, the African Charter and CEDAW enjoy near-universal ratification on the African continent, and already imposes a significant number of obligations on States in the protection of women's environmental rights. However, full ratification of the Maputo Protocol would widen the net of protection to ensure that no women are left behind. African States should thus make the necessary provision for ratification and domestication of the Maputo Protocol. This chapter recognises the unique provisions of the environmental rights in the Maputo Protocol, noting that it has no internal limitations and goes beyond a satisfactory environment.

As can be seen in this chapter, States where the Maputo Protocol has been adopted have seen improvements, particularly in women's access to land, representation in decision-making bodies and in having women's concerns reflected in development policies. States should be encouraged to submit to the African Commission on compliance with the Maputo Protocol. This may also increase local awareness of the Maputo Protocol and further benefit its full realization. Sub-regional initiatives such as the SADC Barometer on Gender and Development are highly commendable and should be encouraged. This is also a way in which States can undertake peer review and share best practices for implementation of the rights in the Maputo Protocol. However, there is a need to ensure that in all reporting on the Maputo Protocol all rights receive sufficient consideration, including women's environmental rights. Further unpacking should also be done of the specific content of each of the obligations under Articles 18 and 19, in order that States and other stakeholders will know what the duties are that are imposed in protecting and fulfilling women's environmental and developmental rights.

Strategic litigation on the African continent should include bringing cases on Articles 18 and 19 of the Maputo Protocol before the African Commission, the African Court and other bodies mandated to interpret its provisions. This would ensure greater clarity from the judicial bodies on how these rights should be interpreted and applied in practice. The African Commission may, through its Special Rapporteur on the Rights of Women in Africa, together with the Working Group on Economic, Social and Cultural Rights, and the Working Group on Extractive Industries, Environment and Human Rights, prepare General Comments on the content of Articles 18 and 19, as well as on Article 24 of the African Charter, building on its existing jurisprudence and normative developments.

As noted by Viljoen, these various human rights instruments together create "overlapping and connecting [...] layers of protection"[76] and also provide a patchwork of mechanisms for victims to vindicate their rights, which together can provide for the full realization of women's environmental rights in Africa.

References

Books

Diamond, I and G Orenstein (eds) (1990) Reweaving the World: The Emergence of Ecofeminism, (San Francisco: Sierra Club).

Knox, John H. & Ramin Pejan (eds) The Human Right to a Healthy Environment (Cambridge University Press, 2018).

[76] Frans Viljoen "An Introduction to the Protocol to the African Charter on Human and Peoples' Rights on the Rights of Women in Africa" (2009) 11 *Wash. & Lee JCR & SOC. JUST* 29.

Book Chapters

Ayeni, Victor O "Introduction and preliminary overview of findings" in Victor O Ayeni (ed.) (2016) The impact of the African Charter and the Maputo Protocol in selected African states.

Medina, C "Women's Rights as Human Rights: Latin American Countries and the Organisation of American States (OAS)" in M Díaz-Diocaretz & IM Zavala (eds.) Women, Feminist Identity, and Society in the 1980s: Selected Papers John Benjamins Publishing.

Journal Articles

Amechi, Emeka Polycarp., 'Linking Environmental Protection and Poverty Reduction in Africa: An Analysis of the Regional Legal Responses to Environmental Protection' (2010) 6/2 Law, Environment and Development Journal.

Coleman, F 'Pan-African Strategies for Environmental Preservation: Why Women's Rights Are the Missing Link' (2008) Berkeley Journal of Gender, Law & Justice.

Jackson, Cecile., "Women/Nature or Gender/History? A critique of Ecofeminist 'Development'" (1993) 20 The Journal of Peasant Studies.

Kameri-Mbote, Patricia., 'Access, control and ownership: Women and sustainable environmental management in Africa' (2007) 72 Agenda.

Lee, John., "The Underlying Legal Theory to Support a Well-Defined Human Right to a Healthy Environment as a Principle of Customary International Law" (2000) 25 COLUM. J. ENVTL. L.

McCormick, John., 'The Origins of the World Conservation Strategy' (1986) Environmental Review 10(3) 177, 178.

Nsibirwa, Martin S., "A brief analysis of the Draft Protocol to the African Charter on Human and Peoples Rights on the Rights of Women" (2001) 1 African Human Rights Law Journal.

Viljoen, Frans., "An Introduction to the Protocol to the African Charter on Human and Peoples' Rights on the Rights of Women in Africa" (2009) 11 Wash. & Lee JCR & SOC. JUST.

Online Sources

African Commission on Human and Peoples' Rights "Resolution 417 on the human rights impacts of extreme weather in Eastern and Southern Africa due to climate change – ACHPR/Res. 417 (LXIV) 2019" 14 May 2019 https://www.achpr.org/sessions/resolutions?id=441

Amnesty International "Niger Delta Negligence" March 2018 https://www.amnesty.org/en/latest/news/2018/03/niger-delta-oil-spills-decoders/ (accessed on 22 July 2019).

Business Tech 'Government is worried about alarmingly high levels of air pollution in these South African towns' (2 June 2019) Business Tech available at https://businesstech.co.za/news/government/317836/government-is-worried-about-the-high-air-pollution-in-these-south-african-towns/ (accessed on 22 July 2019).

Chutel, Lynsey., "This picturesque corner of South Africa has the world's deadliest air pollution" (29 October 2018) Quartz Africa https://qz.com/africa/1441504/highest-concentration-of-deadly-air-pollution-found-in-south-africa/ (accessed on 21 July 2019).

Keetharuth, Sheila "Women in the Extractive Industries in Africa: Gender Perspectives" Newsletter of the Working Group on Extractive Industries, Environment and Human Rights in Africa of the African Commission on Human and Peoples' Rights (2nd ed) (October 2019) https://www.achpr.org/newsletter/viewdetail?id=28

Maclean, Ruth "Toxic mud swamps fortunes of Niger Delta women years after oil spill" (2018) The Guardian https://www.theguardian.com/global-development/2018/dec/19/toxic-mud-swamps-fortunes-of-niger-delta-women-years-after-oil-spill

Nyamweda, Tarisai & Colleen Lowe Morna "State of Women in SADC 2019 Report" (2019) https://genderlinks.org.za/wp-content/uploads/2019/08/StateOfWomenSADC2019rev2.pdf

Oxfam "Implementing the Protocol on the Rights of Women in Africa: Analysing the Compliance of Kenya's Legal Framework" (2014) http://www.soawr.org/sites/default/files/ml-implementing-protocol-womens-rights-africa-010314-en.pdf

The Food and Agriculture Organization of the United Nations "The role of women in agriculture" (2011) ESA Working Paper No. 11-02 available at http://www.fao.org/3/am307e/am307e00.pdf

The Green Belt Movement http://www.greenbeltmovement.org/

United Nations Development Programme "Lake Tanganyika, what the future holds" 22 March 2017 https://stories.undp.org/lake-tanganyika-what-the-future-holds (accessed on 22 July 2019).
United Nations Water "World Water Development Report 2019: 'Leaving no one behind'" 19 March 2019 https://www.unwater.org/world-water-development-report-2019-leaving-no-one-behind/
United Nations Women "Rural women tackle drought-affected Mozambique's rise in child marriage" (2017) https://www.unwomen.org/en/news/stories/2017/11/feature-mozambique-rural-women-tackle-drought-related-rise-in-child-marriage (accessed on 1 December 2019).
Waldorff, Pétur "Left Out to Dry? Gender and Fisheries on Lake Tanganyika" (2017) UNU-GEST https://unu.edu/publications/articles/gender-and-fisheries-on-lake-tanganyika.html
World Bank "Lake Tanganyika Environmental Management Project" (2018) p. 7 http://documents.worldbank.org/curated/en/668911539594000200/pdf/Concept-Project-Information-Document-Integrated-Safeguards-Data-Sheet-Lake-Tanganyika-Environmental-Management-Project-P165749.pdf
World Bank "Urbanization in Africa: Trends, Promises, and Challenges" (2015) https://www.worldbank.org/en/events/2015/06/01/urbanization-in-africa-trends-promises-and-challenges

6

The Legal Rights of Natural Entities: African Approaches to the Recognition of Rights of Nature

Oluwabusayo Temitope Wuraola

1 Introduction

African legal systems pay more attention to indigenous laws and culture than any other laws; be it common law, civil law or otherwise. Despite colonialism and several other attempts to put indigenous laws (customary laws) of Africans to various repugnancy tests, there has always been resilience to practice these "introduced" laws. Africans have always resorted more to their customary laws which depicts their way of life than any other received laws. This has resulted into pluralism of laws within the African legal system. It has brought about misunderstandings regarding which laws are applicable in some situations within jurisdictions and which laws are not. However, the focus of this research is not pluralism of laws but the approaches taken by the African legal system in tackling environmental degradation and addressing the issues arising from the protection of environmental rights. All laws within the African legal system are anthropocentric. Anthropocentrism is the theory that ascribes

O. T. Wuraola (✉)
School of Law, University of Leicester, Leicester, UK

M. Addaney, A. Oluborode Jegede (eds.), *Human Rights and the Environment under African Union Law*, https://doi.org/10.1007/978-3-030-46523-0_6

value to humans in such a way that all other existing creatures are to satisfy human values than any other entity within the ecosystem. This theory has been responsible for a whole lot of controversies within environmental law philosophy in recent times. While some argue that anthropocentrism is the root of all eco-destructions and there is a need to change this worldview, to change the world,[1] it is presumed that some other environmentalists believe it is wrong to critic anthropocentrism.[2] Ecocentrism, an environmental theory considered against anthropocentrism, covers all of the rights of nature and Earth jurisprudence tenets. Ecocentrism stands a chance with more benefits than anthropocentrism because reverence is had of all members of the ecosystem and not only of humans. However, with the above, this chapter seeks to analyze the African approach to environmental rights, which is anthropocentric, and consider the possibilities of a shift from the anthropocentric to ecocentric. This chapter will also assess the jurisdictions that have legalized this shift by granting legal personhood to non-human natural entities. Therefore, this chapter proposes a shift from anthropocentrism to ecocentrism in Africa and advocates for an extension of legal considerations toward non-human natural entities or the ecosystem as entities that deserve legal rights.

2 The Ecocentric Approach

Environmental law narratives usually attribute three eras to the evolution of international environmental law[3]: the traditional era from 1945 to 1972[4]; the modern era from 1972 to 1992[5]; and the post-modern era from 1992 till date.[6] These environmental law eras have been referred to

[1] Thomas Berry 'Great Work' (2000) 4 National Catholic Reporter 38. http://go.galegroup.com.ezproxy3.lib.le.ac.uk/ps/i.do?id=GALE%7CA59450120&v=2.1&u=leicester&it=r&p=EAIM&sw=w Accessed 23 May 2019.

[2] Helen Kopnina "Anthropocentrism, more than just a misunderstood Problem (2018) 31(1) Journal of Agricultural and Environmental Ethics. 109, 109–127.

[3] Peter H Sand "The History and Evolution of International Environmental Law" (2015) Edward Elgar Publishing Limited xii, 1–889.

[4] United Nations Stockholm Conference on the Human Environment (UNCHE).

[5] United Nations Conference on Environment and Development (UNCED) in Rio de Janeiro.

[6] Ibid 3.

as traditionally anthropocentric eras. This is because the traditional era had conservation-minded "green" policies in treaties.[7] These conservation-minded policies are more beneficial to humans than other non-human entities. For instance, the 1856 Bayonne Boundary Treaty between France and Spain aimed at preventing the destruction of fishery in the Bidassoa River which is beneficial to humans, but not for the protection of the river.[8] This approach is anthropocentric.

Romano acknowledges that there are two conflicting approaches over the role of the environment: the anthropocentric and the ecocentric approaches.[9] Ecocentrism is described as the broadest term for the world-view that recognizes the intrinsic value of all life-forms and ecosystems including their abiotic components.[10] This is in contrast with the world-view of anthropocentrism, which values other life-forms and ecosystems only in so far as they are valuable to humans, their well-being and their interests. Burdon contends that anthropocentrism no longer has any credibility in modern science because human beings exist as one equal part of a broader Earth Community.[11]. All members of a broader Earth Community with intrinsic value should be valued equally without special values accorded to only humans. There has to be a shift from subjecting other non-human entities' values to only being instrumental to humanism. Their intrinsic value should be recognized.

According to Burden, a shift from anthropocentric to the Earth-centered or ecocentric theory of law, which recognizes both human beings and nature, is relevant to the operation of law.[12] Kopnina and others also contend that ecocentrism is central to solving unprecedented

---

[7] Ibid 3.

[8] Ibid 3.

[9] Rebecca Romano 'Can Nature be Entitled to Constitutional Rights? A Historical Overview and the Innovative Approach of Bolivia and Ecuador' (2017/2018) Comparative Constitutionalism.

[10] Helen Kopina, Hadyn Washington, Paul Cryer, Bron Raymond Taylor, John J Piccolo, "Why Ecocentrism is the Key Pathway to Sustainability' (2017) The Ecological Citizen 35–41.

[11] Peter Burdon 'The Great Jurisprudence' (2011) 14 Southern Cross University Law Review http://dequinceyco.net/wp-content/uploads/2013/04/Burdon_SCU-article_2011.pdf Accessed on 17th May 2019.

[12] Ibid.

environmental crisis.[13] They gave three justifications for the essence of ecocentrism in four terms[14]: ethical terms, spiritual terms, ecological terms and evolutionary terms. The ethical terms bring about the expansion of respect beyond humanity toward care to all life. The care to all life includes care to terrestrial and aquatic ecosystems.[15] This is because there is no sound justification for not extending moral concern to all of the ecosphere and its components.[16] The evolutionary terms suggest that all species are interrelated and, thus, biological kinship should confer moral responsibilities toward all species. This brings about suppressing the notion of human supremacy and encourages the recognition of all parts of nature. The spiritual terms suggest that ecocentrism has a contention with the predominant religions in the world which have offered escape from mortality and relief from suffering that human life naturally involves.[17] Lastly, the ecological terms suggest that ecocentrism reminds us about the ecosphere and the interdependency and interconnectedness of humans and non-humans.[18] The ecocentric approach predominantly promises to ensure a more effective and a fair protection of environmental rights than the anthropocentric approach.

3 Environmental Rights Protection in Africa

The anthropocentric approach of protecting environmental rights throughout Africa has been criticized by environmental philosophers. This is because moral considerations of values of other non-human animals or non-human natural entities in Africa tilt toward the individual human person and the advancement of her welfare than any other entity

---

[13] Helen Kopnina, Hadyn Washington, Paul Cryer, Bron Raymond Taylor, John J Piccolo, "Why Ecocentrism is the Key Pathway to Sustainability' (2017) The Ecological Citizen 35, 39.

[14] Helen Kopnina, Hadyn Washington, Paul Cryer, Bron Raymond Taylor, John J Piccolo, "Why Ecocentrism is the Key Pathway to Sustainability' (2017) The Ecological Citizen 35, 39

[15] Helen Kopnina, Hadyn Washington, Paul Cryer, Bron Raymond Taylor, John J Piccolo, "Why Ecocentrism is the Key Pathway to Sustainability' (2017) The Ecological Citizen 35, 39.

[16] Ibid 39.

[17] Ibid.

[18] Ibid.

within the ecosystem. This has been "blamed" for the ineffectiveness of environmental laws and the implementation thereof.

It is pertinent to note that Uganda recently recognised the rights of nature nationally in The National Environment Act 2019 and is the first country in Africa to recognise the legal rights of nature.However, several African countries have been striving to advance the protection of environmental rights through different environmental rights protection approaches including constitutionalism of environmental rights in the various African constitutions. From 2001, at least 32 countries in Africa (approximately two-thirds) have some constitutional provisions ensuring the right to a healthy environment.[19] About two-thirds of the African countries practicing civil law have constitutional environmental provisions while about half of the common law countries in Africa have constitutional environmental provisions for the entrenchment of environmental rights protection.[20]

Constitutionalism is also considered a step in the right direction for the protection of environmental rights, but it is still an anthropocentric approach. For instance, South Africa, like several other African countries has incorporated the protection of the environment in its constitution. This significantly brings about the elevation of the rights of non-human natural entities to the "constitutional status". The constitution, being the grundnorm, will have such effect as to the implementation and effectiveness of such rights. It is envisaged that it will command reverence to the rights of nature, being one of the fundamental human rights, and command obedience from citizens in such a way that they detach from all forms of degradation and pollution of the environment. However, this elevation of the environment to a fundamental justiciable human right is seen to be a road which will lead to the goal of attaining a protected environment.[21]

---

[19] Carl Bruch, Wole Coker, and Chris VanArsdale 'Breathing Life into Fundamental Principles: Implementing Constitutional Environmental Protections in Africa" (2001) in Environmental Governance in Africa Working Paper Series. Edited by Jesse C, Ribot Peter and G Veit.13, 1–49.

[20] Ibid 14.

[21] Louis J. Kotze "The South African Environment and the 1996 Constitution: Some Reflections on a Decade of Democracy and Constitutional Protection of the Environment" (2007) Direitos Fundamentais & Justica N 1- Out. 37, 36–57.

In addition, constitutionalism still portrays anthropocentrism rather than ecocentrism. By the provisions of the South African Constitution, there are elements of human-centered provisions for the protection of environmental rights. The first part of the provision of the constitution provides for the protection of the health and well-being of citizens. The second part seeks to protect present and future generations by preventing pollution and ecological degradation. It also promotes conservation and securing ecologically sustainable development and the use of natural resources while promoting justifiable economic and social development.[22]

Most of the rights contained in constitutional environmental provisions in Africa are generalized.[23] They are not specific enough to address specific environmental issues like the prevention of hazardous wastes, toxic wastes, radioactive wastes among others.[24] Overall, it is assumed that these constitutional environmental provisions come under particular categories of the constitutions. These categories include fundamental rights and duties, or general constitutional rights and duties, or vague rights and duties.[25] Prior to the entrenchment of environmental rights in the constitution, there was no respect for and protection of human rights in some African countries. There were more of legislations and policies concerned with the facilitation of resource allocations and resource exploitation and conflicts over land and access to natural resources.[26] However, this is one of the reasons constitutional environmental rights is considered as one of the alternatives or solutions to eco-destruction but these alternatives are still anthropocentric.

---

[22] Louis J. Kotze and Anel du Plessis 'Some Brief Observations on Fifteen Years of Environmental Rights Jurisprudence in South Africa' (2010) 3(1) Journal of Court Innovation, 157–159, 157–176. See Section 24, Constitution of the Republic of South Africa, 1996.

[23] Louis J. Kotze and Anel du Plessis 'Some Brief Observations on Fifteen Years of Environmental Rights Jurisprudence in South Africa' (2010) 3(1) Journal of Court Innovation, 157–159, 157–176.

[24] Ibid 14.

[25] Carl Bruch, Wole Coker, and Chris VanArsdale 'Breathing Life into Fundamental Principles: Implementing Constitutional Environmental Protections in Africa" (2001) in Environmental Governance in Africa Working Paper Series. Edited by Jesse C, Ribot Peter and G Veit 15, 1–49.

[26] Louis J. Kotze "The South African Environment and the 1996 Constitution: Some Reflections on a Decade of Democracy and Constitutional Protection of the Environment" (2007) Direitos Fundamentais & Justica N 1- Out. 38, 36–57.

Asides legislative or constitutional protection of environmental rights in Africa, there is the religious side to its protection. This depends on the duty the religion being practiced by the citizens of those countries imposes.[27] Africans are very religious people such that they sometimes take their religious laws more serious than the codified laws in their countries. Therefore, their religions go a long way in influencing their perspectives toward the protection of environmental rights, particularly African countries that have not constitutionalized their environmental rights. African countries that do not have constitutional environmental provisions have the religious duty to protect the environment in their religious laws.[28] For instance, it is observed that Islamic African Countries like Algeria, Egypt, Mauritania, Morocco, Libya and Tunisia do not have environmental provisions in their constitutions.[29] It has been observed that the Quran is silent on human rights to clean environment but the large body of environmental ethics stress the duty of the individual Muslim to care for the natural environment. This is in addition to the belief that the Earth is the creation of Allah and therefore both the individual and the state have to be responsible to it as a religious and ethical obligation.[30] Muslims are specifically asked to treat non-human natural entities with dignity.[31] However, this is more ecocentric than anthropocentric because it emphasizes on the intrinsic values of non-human natural entities. Christianity, on the other hand, lays more emphasis on the power of humans over non-human natural entities and, as a result, the African Christian perspective toward non-human natural entities is that of dominion of humans over other entities. This is anthropocentric.[32]

Therefore, those African countries without either the constitutional environmental provisions or the religious duty or obligations to protect the environment make do with other legal mechanisms. This includes

[27] Carl Bruch, Wole Coker, and Chris VanArsdale 'Breathing Life into Fundamental Principles: Implementing Constitutional Environmental Protections in Africa" (2001) in Environmental Governance in Africa Working Paper Series. Edited by Jesse C, Ribot Peter and G Veit.13, 1–49

[28] Ibid 14.

[29] Ibid 14.

[30] Ibid 14.

[31] Quran 6: 38.

[32] Genesis 1: 26–28.

African legal principles, African legal theories and African judicial mechanisms for assured environmental rights protection. Some of these worked better in Africa than the Received Common Law principles like torts and contracts. It is noteworthy that some African countries actually also resort to redress environmental rights through private laws like the laws of contract and tort in Kenya.[33] The Environmental Management and Co-ordination Act (EMCA) 1999 was presumed too weak to address environmental rights issues. The constitution of Kenya was also presumed to be weak; hence there was a continued reliance on common law principles in court in order to tackle environmental rights issues. However, this weak constitutional and legal framework was a setback in the realization of environmental rights in Kenya.[34] The African principle of Ubuntu, on the other hand, lays more emphasis on humanity—I am because we are—than any other entity in Africa. This is also anthropocentric because it focuses more on humans and their humanity; therefore the protection or preservation of these non-human natural entities seems probable only for the benefits of humans.

4 Rights of Nature and Earth Jurisprudence

The whole idea of rights of nature was broached by Stone who proposed not only legal extensions of rights toward nature but also a legal psychic and socio-psychic trajectory for the viability of the legal construct.[35]According to him, there has always been an extension of legal rights toward persons like children and other objects and things like trusts, corporations, joint ventures and others. Therefore it would not be new to extend legal rights to nature because until the rightless thing has rights, it cannot be anything other than being instrumental for those who are holding rights at the time.[36] Stone made proposals for the extension

[33] Kirui Kiprono Calvin 'Environmental Rights in Kenya' Unpublished work found in academia.

[34] Ibid.

[35] Christopher D Stone "Should Trees have Standing? - Towards Legal Rights for Natural Objects" (1972) 45 South California Law Review, 450–501.

[36] Ibid.

of legal rights to non-human natural entities and made different proposals. Three of these will be considered.

Firstly, Stone sought for a review of the actions of existing rights holders inconsistent with the right of the potential rights holders to be done by some authoritative public body. This could be a judge, legislator or official in order to establish the legal rights of the natural object. This authoritative body would institute legal actions on behalf of non-human natural entities with all of the advantages and liability of the legal action ascribed to the natural entity.[37] Secondly, Stone called for a reconsideration of not only humans' consciousness about the environment but reconsideration of humans' consciousness in order to bring about a tangible institutional reform. He reiterates that the consciousness, which is of the view that nature is the dominion of man—an idea which might have sprung from religious beliefs—will only be effective if they are for a tangible institutional reform. All of these, according to him, depend on effecting a radical shift in our feelings about our place in the rest of nature.[38] Thirdly, Stone asked for a giving-up of the psychic sense of separateness and specialness in the universe by humans and embracing the belief that the Earth is only one organized "field" of activities and so is the human person.[39]

In continuance of the rights of nature legal construct, Thomas Berry brought about Earth Jurisprudence, a theory which postulates that all laws should be Earth-centered laws and have all of Earth's components into consideration without considering only species like humans. These approaches in Africa are new. How much these theories influence laws in Africa has to be considered for a shift in Africa. This is because African values are anthropocentric and human-centered through the dispensation of values through their cultural heritage and land rights. Africans see non-human natural entities as properties to be inherited and passed on to generations. However, this shift from anthropocentrism, as much as it is needed in Africa, could be difficult. The perspectives of Africans will be

---

[37] Ibid.

[38] Ibid.

[39] Ibid.

difficult to change except the justifications for the grant of rights of nature in other jurisdictions are persuasive enough.

In 2006, Pennsylvania, the Tamaqua Borough Sewage Sludge Ordinance[40] considered the natural communities and ecosystems as "persons" for the purposes of the enforcement of the civil rights of those residents, natural communities and ecosystems. Some other local laws in the United States were enacted to recognize and protect the rights of nature. However, some of them do not specifically label non-human natural entities or ecosystems as persons like it was done in other places like Halifax,[41] Mahanoy[42] and Nottingham.[43] These municipalities enacted entirely different ordinances that recognized the existence of the inalienable and fundamental rights of natural communities and ecosystems. Ecuador became the first country to recognize rights of nature in its constitution.[44] In 2008, the Ecuador constitution had rights for nature in order to reverse the damage done to the Amazon rainforest and its indigenous people by oil companies. In 2010, Bolivia developed her Law on the Rights of Mother Earth which gives nature rights and mandates a radical ecological transition of Bolivia's economy and society.[45]

Romano sees the grant of the right of nature as a step forward in the protection of nature exemplified by the Ecuadorian and the Bolivian constitutions.[46] Other communities and municipalities also recognized the

---

[40] Tamaqua Borough was the first US municipality that recognized the environment's legal rights in order to protect Pennsylvania's coal region from coal sludge pollution in 2006. As a result, Pittsburgh and several other communities have implemented same decrees. See SECTION 7.6. Tamaqua Borough Ordinance 2006 612.

[41] See Section 30–156.7 Ordinance to Amend the Town Code of Halifax, Virginia adding "Article VII. Corporate Mining and Chemical Radioactive Trespass" under "Chapter 50 Environment".

[42] Mahanoy Township, Schuylkill County, Pennsylvania Ordinance 2008, S7 (14).

[43] Nottingham Water Rights and Self Government Ordinance, 2018, S. 5(1).

[44] Constitution of the Republic of Ecuador, 2008, Articles 71–74. See also Conway M. 'The Colorado River Has Its Own Lawyer Now' http://www.nonprofitquarterly.org Accessed on 19 October 2019.

[45] "Bolivia's Leadership: Global Alliance for the Rights of Nature" http://therightsofnature.org Accessed on 17 February 2018.

[46] Rebecca Romano 'Can Nature be Entitled to Constitutional Rights? A Historical Overview and the Innovative Approach of Bolivia and Ecuador' (2017/2018) Comparative Constitutionalism, 1, 5.

rights of nature in this light. In 2010, Licking,[47] Packer,[48] Wales[49] and Pittsburgh[50] recognized rights of non-human natural entities by enacting ordinances to that effect. In 2011, Baldwin,[51] Forest Hills,[52] W. Homestead,[53] State College[54] and Mountain Lake Park[55] followed suit with ordinances recognizing the rights of nature. In 2012, Yellow Spring[56] and Broadview Heights[57] and, in 2013, Mora County[58] and Santa Monica[59] all recognized rights of non-human natural entities in their various local legislations.

Other jurisdictions that have granted environmental personhood rights include New Zealand,[60] Columbia and Ohio.[61] A resolution made in July 2018 by the Board of Trustees of the Town of Crestone, Colorado, officially recognized that "nature, natural ecosystems, communities, and all species possess intrinsic and inalienable rights which must be effectuated to protect life on Earth.[62] However, the purpose of the grant of rights of nature in New Zealand seems different from other grants of rights of nature. In New Zealand, non-human natural entities becoming "persons" is presumed to have begun some 140 years ago. By 2012, an agreement was

---

[47] Licking Township Ordinance; Section 7, Packer Carbon County, Pennsylvania Ordinance, S 3(6).

[48] Packer Carbon County, Pennsylvania Ordinance; S 7.

[49] Town of Wales, New York, Local Law 201, S(4).

[50] Chapter 618.0 2 3 Ordinance supplementing the Pittsburgh Code Legislation Text, City of Pittsburgh.

[51] Borough of Baldwin, Ordinance 2011, S 3(6).

[52] Borough of Forest Hills Ordinance 2011, S 3.

[53] Borough of Homestead, Pennsylvania Ordinance 2011, S 3.

[54] State College Borough Bill of Rights Ordinance 2011, S 4.

[55] Town of Mountain Lake Park, Ordinance 2011, S 2.

[56] Village of Yellow Sprigs, Ohio Ordinance 2012, S3.

[57] City of Broadview Heights, Ohio Ordinance 2012, S 1.

[58] State of New Mexico, County Ordinance, 2013, S 4(3).

[59] Chapter 4.74, Santa Monica, California Ordinance 2921.

[60] Whanganui River was also granted personhood rights in 2017 by the Te Awa Tupua Whanganui River Claims Settlement Bill. The river is recognized as one which has all the duties, liabilities and rights of a legal person.

[61] Colorado river is also about to join the league of nature with legal personhood: an environmental group in the United States of America sued on behalf of Colorado river for personhood. See Borg J., 'Colorado River 'Sues' for Personhood' http://www.mbcpathway.com accessed on 19 October 2017.

[62] Town of Crestone Colorado Resolution No. 006, 2018.

made between Whanganui Iwi and the Crown[63] settling historical treaty of Waitangi Claims of Whanganui Iwi in respect of the Whanganui River with the Maori people.[64] New Zealand's Resource Management Act recognized the rights of non-human natural entities. In December 2017, The Crown and Taranaki tribe signed a Record of Understanding to vest Egmont National Park (Taranaki Maunga) with legal personality, same as that granted to Te Urewera and Whanganui River. Section 11 of the Te Urewera Act, 2014, declared Te Urewera a legal entity with "all the rights, powers, duties and liabilities of a legal person".[65] Following this was the grant of the same personhood to Te Awa Tupua (Whanganui River)[66] in 2017.

With the rise in the number of legislations and ordinances being granted rights of nature globally, there is a need to assess the tendency of the extension of this grant of rights of nature to African jurisdictions. Earth Jurisprudence changes the entire idea of human-centered laws which is dominant in most of the legislations for the protection and preservation of non-human natural entities in Africa and beyond. This approach, which takes Earth governance from anthropocentric to ecocentric could be seen as expedient as a result of the fact that the anthropocentric approach adopted for years have not been so effective in the protection of non-human natural entities. Therefore, there is the need to try a different approach which is presumed to be more equitable than other approaches.

The prevailing feature of Earth Jurisprudence is the acknowledgment of all of the Earth's components as members of the single integral community. This will bring about each having a role to play in their own

[63] The agreement "represents the commitment of Whanganui Iwi and the Crown to progress the development of agreed Te Awa Tupua arrangements as part of the settlement of the historical Treaty of Waitangui Claims of Whanganui Iwi in respect of the Whanganui River". Whanganui refers to the river as a whole, its spiritual and physical dimensions and their unity.

[64] It is believed that the Maori people equal themselves to mountains, rivers and seas, and therefore believe they have equal rights with the river.

[65] Te Urewera Act, 2014, S,11(1) http://legislation.govt.nz Accessed 9th November, 2018. A Te Urewera Board was actually established for the governance and management of Te Urewera, and to act on behalf of, and in the name of, Te Urewera.

[66] Te Awa Tupua (Whanganui River Claims Settlement) Act 2017, S 14.

dignity.[67] The recognition of nature has always come through the preservation of nature as a role to be played in human dignity and not the dignity of nature itself. Constitutional laws which provide for the protection of environmental rights only make that provision in line with the right to life and other fundamental "human" rights.

Another basic justification of Earth Jurisprudence is the recognition of the rights of each member of the Earth community. According to Berry:

> Trees have tree rights, insects have insects' rights, rivers have their river rights. Mountains have mountain rights. So too with the entire range of beings throughout the universe. All rights are limited and relative. So too with humans. We have human rights. We have rights to the nourishment and shelter we need. We have rights to habitat.[68]

Recognition of rights by legislation would involve the enactment of laws for the recognition of these rights as done in different jurisdictions in the world.

As mentioned above, the whole theory of anthropocentrism is that it seeks to protect the environment through humans—"by regulating human interaction".[69] This is said to have resulted in degrading actions from human activities not being challenged except by proof of direct harm being caused to a human or representative body.[70] Hence, it is difficult to prove direct interest or harm caused to an individual human or representative body in order to protect the environment by preventing environmental degradation. It presumed to limit those seeking to protect the environment and thus the effectiveness of environmental protection[71] It is also contended that the environment is "indeed, by every measurable

---

[67] Thomas Berry 'Great Work' (2000)4 National Catholic Reporter 38. http://go.galegroup.com.ezproxy3.lib.le.ac.uk/ps/i.do?id=GALE%7CA59450120&v=2.1&u=leicester&it=r&p=EAIM&sw=w Accessed 23 May 2019.

[68] Ibid.

[69] Peter Burdon 'The Rights of Nature: Reconsidered' (2010) University of Adelaide Law School Research Paper 4, 69–89 http://ssrn.com/abstract=1709015 Accessed on 17th May 2019.

[70] Ibid.

[71] Ibid.

statistic in a worse condition than thirty years ago when the first environmental protection law was passed".[72]

5 The Potential Viability of Rights of Nature in Africa

All the approaches toward the protection of environmental rights will be unsuccessful if environmental laws are not effectively implemented. According to Kotze, the adequate enforcement of the constitutional protection of the environment on paper is being questioned enormously because of the more pressing priorities such as the alleviation of poverty and health issues which may shift the attention of the legal order in South Africa.[73] As against the effective implementation of environmental laws, there could be a reconsideration of human consciousness toward nature or the environment for a tangible institutional reform in Africa. This, as explained further by Stone, will bring about a radical shift in place in the rest of nature from being an entity valued only for instrumental purposes to be an entity valued intrinsically. The rights of nature would probably not work or be effective without this reconsideration of our perceptions of things. For other non-living things like corporations to be effective as legal persons in Africa, there is perhaps a chance that granting rights to nature would work in Africa as a result of the persuasiveness of the existence of these grants globally.

On the whole, by the proposition of Earth jurisprudence proponents, and partly Stone's proposal, there is a need to give up the psychic sense of separateness and specialness of humans in Africa. This will encourage the interconnectedness of all of the ecosystems. It will bring about respect for all of the Earth's components and moral responsibilities toward all species. This will be a guarantee of the practicability of a shift from anthropocentrism to ecocentrism. The spiritual system could as well contribute

[72] Ibid.

[73] Louis J. Kotze "The South African Environment and the 1996 Constitution. Some Reflections on a Decade of Democracy and Constitutional Protection of the Environment" (2007) Direitos Fundamentais & Justica N 1- Out. 36, 57.

to the viability of rights of nature in Africa. The teachers of religions or the ministers could adopt religious ethics that would recognize the rights of nature as a necessity for the protection of environmental rights and the prevention of environmental degradation. Other legislations, private laws and public laws could as well ensure the viability of rights of nature in Africa, but this will amount to a thorough understanding of its precepts by the court for there to be equity and fairness in the dispensation of justice.

There is a need to analyze the purpose and significance of having rights of nature in Africa to avoid the problem of pluralism of laws which may arise if not well structured and defined. The grant of rights of nature in Africa could be practicable if all of the tenets of African laws and traditions are juxtaposed with that of rights of nature. The viability of rights of nature may be difficult if the African legal system does not approve of it traditionally. There has to be a different worldview about non-human natural entities in Africa. They have to be seen as an entity with a good of its own as well as being good for other purposes. The intrinsic and instrumental values of non-human natural entities need to be respected and recognized culturally.

6 Conclusion

This chapter has been able to put forward a proposal for the personhood of non-human natural entities in Africa as an approach to ensuring the recognition and more effective protection of rights of nature. This is in contrast with the anthropocentric means of protection. Importantly, this chapter has been able to assess environmental rights protection in Africa and some of the mechanisms used in ensuring effective protection and preservation. However, this chapter considered some anthropocentric ways of protection. This includes constitutionalism of environmental rights, religious ways of protection through Islam and Christianity, as well as some African legal principles of Ubuntu and common law principles. Therefore, this chapter concludes that all of these approaches are anthropocentric. This chapter posits that the rights of nature and Earth jurisprudence concepts are ecocentric and are more viable in protecting the rights of nature in Africa. The shift from anthropocentrism to ecocentrism in Africa is probable with a change in the worldview of ascribing value to humans to ascribing value to nature. Other African countries may adopt the Ugandan approach of recognition of rights of nature.

References

Berry Thomas 'Great Work' (2000) 4 National Catholic Reporter 38. http://go.galegroup.com.ezproxy3.lib.le.ac.uk/ps/i.do?id=GALE%7CA59450120&v=2.1&u=leicester&it=r&p=EAIM&sw=w Accessed 23 May 2019.

Bruch Carl & others, 'Breathing Life into Fundamental Principles: Implementing Constitutional Environmental Protections in Africa" (2001) in Environmental Governance in Africa Working Paper Series. Edited by Jesse C, Ribot Peter and G Veit 1–49

Burdon Peter 'The Great Jurisprudence' (2011) 14 Southern Cross University Law Review http://dequinceyco.net/wp-content/uploads/2013/04/Burdon_SCU-article_2011.pdf Accessed on 17th May 2019

Burdon Peter, 'The Rights of Nature: Reconsidered' (2010) University of Adelaide Law School Research Paper 4, 69–89 http://ssrn.com/abstract=1709015 Accessed on 17th May 2019

Kopnina Helen "Anthropocentrism, more than just a misunderstood Problem (2018) 31(1) Journal of Agricultural and Environmental Ethics, 109–127

Kopnina Helen & others "Why Ecocentrism is the Key Pathway to Sustainability' (2017) The Ecological Citizen 35–41

Kotze J. Louis "The South African Environment and the 1996 Constitution: Some Reflections on a Decade of Democracy and Constitutional Protection of the Environment" (2007) Direitos Fundamentais & Justica N 1-Out, 36–57

Kotze Louis and Du Plessis Anel 'Some Brief Observations on Fifteen Years of Environmental Rights Jurisprudence in South Africa' (2010) 3(1) Journal of Court Innovation, 157–176.

Romano Rebecca 'Can Nature be Entitled to Constitutional Rights? A Historical Overview and the Innovative Approach of Bolivia and Ecuador' (2017/2018) Comparative Constitutionalism

Sand H Peter "The History and Evolution of International Environmental Law" (2015) Edward Elgar Publishing Limited xii, 1–889

Stone D Christopher "Should Trees have Standing? – Towards Legal Rights for Natural Objects" (1972) 45 South California Law Review, 450–501

Part II

Enforcement of Environmental Rights by Judicial and Quasi-Judicial Institutions

7

Judicial Enforcement of Environmental Human Rights in Africa

Emma Charlene Lubaale

1 Introduction

From forests, water resources to wildlife, the environment across the globe continues to be threatened. In the context of Sub-Saharan Africa, various environmental challenges including soil erosion, deforestation, wetland degradation biodiversity loss, extreme vulnerability to climate change and desertification remain prevalent.[1] For countries in Africa, these risks are heightened by multiple factors including the fact that they remain on the receiving end of problematic technology and prohibited waste and chemicals from their more developed counterparts.[2] In addition to the foregoing vulnerability, environmental crimes have

[1] Akin L. Mabogunje, *The Environmental Challenges in Sub Saharan Africa*, African technology Forum. 8 (1998), http://web.mit.edu/africantech/www/articles/EnvChall.htm; United Nations Environment Program "UN Environment's Africa Office is supporting the continent's journey towards sustainable development" (2019), https://www.unenvironment.org/regions/africa

[2] Avitus Agbor, *The Ineffectiveness and Inadequacies of International Instruments in Combatting and Ending the Transboundary Movement of Hazardous Wastes and Environmental Degradation in Africa*, 9 African Journal of Legal Studies. 236 (2016).

E. C. Lubaale (✉)
Rhodes University, Grahamstown, South Africa

M. Addaney, A. Oluborode Jegede (eds.), *Human Rights and the Environment under African Union Law*, https://doi.org/10.1007/978-3-030-46523-0_7

become the order of the day, with commentators submitting that this category of crimes presently constitutes a large criminal enterprise as to fall within the ranks of renowned heinous crimes such as drug and human trafficking.[3] Threats to the environment have a negative impact not only on human life, but also on animal life and the general interaction that goes on within the environment. Worthy to note, most people in Africa still heavily rely on land, forest and water resources for survival.[4] This dependence entails recourse to crops that grow on land's soil and animals that inhabit such land and water resources. These resources, however, continue to become scarce, with the situation getting worse by the day.[5] Critical to note also, in many African countries, wood and residues of agriculture remain the major source of energy.[6] The implication of such dependence for the environment has been increased deforestation.[7] This has had adverse effects including the compromising of fauna, flora, animal habitats and, more generally, biodiversity.[8] This, along with other pitfalls including soil erosion, has left the environment in disarray.

Threats to the environment such as these can be traced back to conduct or omissions by individuals, companies and even state organs. Such conduct or omission may include smuggling of animal parts such as ivory, deforestation, dumping of rubbish and hazardous waste, illegal property development, pollution and corruption across state departments in so far as adherence to environmental procedures is concerned. Like never before, the state of the environment across many countries continues to progress in a downward spiral. The need to address the various threats to the environment have seen various states mandate courts to entertain matters pertaining to actual and likely threats, infringements and violations to the environment.[9] The courts in various countries are

[3] Christian Nellemann, Rune Henriksen, Arnold Kreilhuber, Davyth Stewart, Maria Kotsovou, Patricia Raxter, Elizabeth Mrema and Sam Barrat, *The rise of environmental crimes* § 7 (2016).

[4] John Diamond (ed), *Land, Water, and Forests: Assets for Climate resilient Development in Africa* § 7–15 (2011).

[5] Ibid.

[6] Ibid., 8.

[7] Ibid.

[8] Ibid.

[9] See, for example, article 50(1) of the Constitution of the Republic of Uganda, 1995; article 70(1) of the Constitution of the Republic of Kenya 2010.

therefore able to exercise both criminal and civil jurisdiction over matters pertaining to the environment.[10] This step is in the right direction and is one to be praised. One judicial commentator has in fact so aptly put it that "an independent judiciary in a political and legal system that values integrity and transparency is vital in addressing environmental degradation and upholding the environmental rule of law worldwide."[11] Indeed, the effective implementation of environmental laws has the potential to advance environmental conservation. It is also indisputable that effective enabling environmental laws and rules of procedures constitute prerequisites for the effective judicial enforcement of environmental rights. In as far as the role of the judiciary in the enforcement of environmental rights is concerned, mention can be made of the Johannesburg Principles on the role of law and Sustainable Development adopted at the Global Judges Symposium held in Johannesburg, South Africa on 18–20 August.[12] These Principles affirm that "the fragile state of the global environment requires the judiciary as a guardian of the rule of law, to boldly and fearlessly implement and enforce applicable international and national laws." At this Symposium, it was also underscored that the judicial mandate to apply and interpret law is critical to the realization of the goals on sustainable development which includes, amongst others environmental conservation. Commenting on these principles, Mushkat opines that in implementing the Principles enunciated at this Symposium, the judiciary becomes involved in concretizing and molding of emerging principles on subjects of international environmental law.[13] What, how-

[10] The criminal jurisdiction of these courts is derived from multiple legislation in these countries. For example, in terms of section 96 of the National Environment Act of Uganda Chapter 153, the failure to prepare or submit or prepare an Environment Impact Assessment amounts to a criminal offence the penalty of which is imprisonment for 18 months or a fine worth 180,000 Ugandan Shillings and not more than 18 million Ugandan shillings. In certain cases, the guilty party can be imprisoned and also be required to pay a fine.

[11] Antonio Herman Benjamin, *Matters of judgement: The potential for sustainability at the local level is huge, fundamental and achievable*, (2017)19, https://wedocs.unep.org/bitstream/handle/20.500.11822/20259/Our%20Planet%20March%202017.pdf?sequence=1&isAllowed=y

[12] Governing Council of the United Nations Environment Programme, *Johannesburg Principles on the role of law and Sustainable Development adopted at the Global Judges Symposium held in Johannesburg, South Africa on 18–20 August 2002.*

[13] Roda Mushkat, *International Environmental Law and Asian Values: Legal Norms and Cultural Influences* § 50 (2004).

ever, is the nature of enabling laws and procedures on the environment laws in African states (taking Uganda, Kenya and Tanzania as examples) and what implication do these laws and procedures have for courts' role in the enforcement of environmental rights?

As already alluded to, in terms of the Johannesburg Principles on the role of law and Sustainable Development, courts are considered guardians of the rule of law in as far as environmental issues are concerned. There is therefore an expectation that the judiciaries in Africa should play a critical role in the enforcement of environmental rights. However, how effective can courts on the African continent be given the nature of the right to environment and the rules of procedure applicable to the enforcement of rights? The purpose of this chapter is to assess the implication of the nature of the right to the environment and procedures applicable to the adjudication on environmental-related matters for the judiciary's role in the fostering of environmental conservation. Environmental conservation for purposes of this discussion connotes any action or omission geared toward protecting the planet and conservation of its natural resources so that all living things can enjoy quality of life.[14] In resolving the issue raised, reference is made to Kenya, Uganda and Tanzania. As already alluded to, these three countries are randomly placed at the heart of the discussion because it is impracticable to exhaust the practice of all African states on these issues. Moreover, the frameworks of these three countries on aspects such as enabling laws and rules of procedure are very much comparable with those of other African countries. Consequently, the conclusions drawn may be relevant and applicable to other African states. That said, the chapter is divided as follows. Following the present introduction, Sect. 2 provides an overview of the environmental challenges in the three selected counties. Section 3 critically engages with the enabling laws of Uganda, Kenya and Tanzania with a view to answering the overarching question—do these laws have any implications for the judiciary's role in fostering environmental conservation? Section 4 discusses the rules of procedure, with particular emphasis on the notion of *locus standi*. Emphasis is placed on the notion of *locus standi* in light of

[14] Conserve Future, *Environmental Conservation,* https://www.conserve-energy-future.com/methods-and-importance-of-environmental-conservation.php

the literature alluding to the fact that "a significant barrier to effective judicial enforcement of international environmental norms are restrictive locus standi rules."[15] This discussion too is geared toward addressing the overarching issue—the implication of rules on *locus standi* in Kenya, Uganda and Tanzania for the judiciary's role in fostering environmental conservation. Section 5 draws on the theoretical framework on decision-making and how the approaches of the courts in Africa find force in this theoretical framework. Section 6 draws a conclusion and makes appropriate recommendations.

2 A Glimpse into the Environmental Challenges in the Three Selected African Countries

All African countries have their fair share of environmental challenges. In Tanzania, as is the case with many African states, majority of Tanzania's population is dependent on natural resources for survival.[16] Wood remains a major source of energy, the effect of this being increased deforestation. Air and water pollution is on the rise, particularly in major cities such as Dar es Salaam and Arusha, which constitute the hubs of industrialization in this country.[17] While Tanzania is endowed with multiple species of wildlife, the unsustainable and illegal wildlife exploitation of these species remains rampant.[18] Examples are indeed inexhaustible, but considered together, the manner in which the environment is generally being

[15] By Roda Mushkat, *International Environmental Law and Asian Values: Legal Norms and Cultural Influences*, § 51 (2004). See also Adem K Abebe, *Towards more liberal standing rules to enforce constitutional rights in Ethiopia*, 10 African Human Rights Law Journal. 407–431 (2010).

[16] Emmanuel Kallonga, Alan Rodgers Fred Nelson, Yannick Ndoinyo and Rugemeleza Nshala, *Reforming environmental governance in Tanzania: Natural resource management and the rural economy*, 1 (2002), Paper presented at the Inaugural Tanzanian Biennial Development Forum 24th–25th April 2003 at the Golden Tulip Hotel, Dar es Salaam Tanzania, http://www.ujamaa-crt.org/uploads/1/2/5/7/12575135/reforn_ujamaa.pdf

[17] Rahma Mohamed Msoffei, *Increase in deforestation: A key challenge to household charcoal supply – A case of Tanga urban, Tanzania*, 2 European Journal of Social Sciences Studies. 133–151 (2017).

[18] Jafari R Kideghesho1a, Julius W Nyahongo, Shombe N Hassan, Thadeo C Tarimold and Nsajigwa E Mbijele, *Factors and Ecological Impacts of Wildlife Habitat Destruction in the Serengeti Ecosystem in Northern Tanzania*, 11 *AJEAM-RAGEE*. 17–32 (2007).

handled and managed remains a far cry from the goals of ensuring the conservation of the environment and development that is sustainable.

In Kenya, natural resources are under increasing pressure of overexploitation and poor management. Water resources, for instance, continue to be threatened by industrial waste and agricultural chemical.[19] Kenya is known for the wide variety of species of wildlife; however, wildlife habitats are threatened by poachers, with reports suggesting that some species have since become extinct.[20] Deforestation remains rampant as reports indicate that unsustainable exploitation of forests over the past three decades has resulted into considerable reduction in forest cover.[21] The main drivers for deforestation include conversion of forest timber into charcoal for fuel, unsustainable methods of production and the cutting don of trees to secure land for agriculture.[22] The failure to adhere to environmental laws and regulations has also exacerbated the vulnerable state of the environment in Kenya. Notably, illegal dumping of hazardous waste and rubbish remains a serious challenge in both rural and urban Kenya.[23] Reports suggest that most, if not all, urban towns in Kenya have poor or inadequate garbage disposal and collection systems.[24] These realities point to the increasing need for enforcement of existing laws on environment by organs in Kenya, with the judiciary being one such organ.

In Uganda, the environmental challenges are not any different from those in Kenya and Tanzania. For instance, water catchment areas are increasingly being depleted.[25] With the population in Uganda on the rise, the pressure on natural resources has spiraled.[26] Natural resources are bur-

---

[19] Simeon Otieno Dulo, *Impact of Poor Solid Waste Management in Kenya on Groundwater*, § 39–40 (2009) Doctoral thesis, University of Nairobi.

[20] Sam Weru, *Wildlife protection and trafficking assessment in Kenya: Traffic report*, 1–49 (2016), https://portals.iucn.org/library/sites/library/files/documents/Traf-126.pdf

[21] Seton Stiebert, Deborah Murphy, Jason Dion and Scott McFatridge, *Kenya's Climate Change Action Plan: Mitigation, Chapter 4: Forestry*. 2–3 (2012), http://www.kccap.info/phocadownload/final/SC4/Chapter%204%20SC4_Forestry_Final.pdf

[22] Ibid.

[23] Dulo, *Impact of poor*.

[24] Ibid.

[25] Francis NW Nsubuga, Edith N Namutebi, Masoud Nsubuga-Ssenfuma, *Water resources of Uganda: An assessment and review*, 6 *Journal of Water Resource and Protection*. 1297–1315 (2014).

[26] Ibid., 1307.

dened as people search for food, wood for fuel and land for crop cultivation and business. This state of affairs has resulted into multiple environmental problems including soil degradation, deforestation, loss of biodiversity, wetland degradation and pollution of water, land and air. These challenges remain rife despite the host of legislation and constitutional provisions on effective management of environmental matters.[27]

3 The Nexus Between the Legal Framework on the Right to Environment and the Judicial Enforcement of Environmental Rights

Courts do not operate in a vacuum. In adjudicating over matters, including environmental ones, their point of reference is the law, which may either be written or be unwritten. Thus, courts may make reference to legislation and/or common law. Aside from legislation and common law, courts may also have recourse to constitutional provisions on environmental matters. Considered together, therefore, the enabling laws of a country on matters pertaining to the environment constitute entry points for the effective performance of the judiciary in the furtherance of environmental conservation. What form, therefore, do the laws in Kenya, Uganda and Tanzania take and do these laws help the judiciary advance environmental conservation through its decisions? The nature and scope of enabling laws are addressed country by country as below.

Tanzania

Tanzania has ratified/acceded to a host of international treaties, all of which place obligations on Tanzania to ensure that all interactions with the environment foster the goal of sustainable development and are alive

[27] Examples of such legislation include the Local Government Act Chapter 265, the Agricultural Seeds and Plants Act Chapter 28, the Uganda Wildlife Act Chapter 200, the National Environment Act Chapter153, the Land Act Chapter 227, the Water Act Chapter 152, the Fisheries Act Chapter 197 and the Penal Code Act Cap 120.

to the need to conserve the environment. Notable among these are the Convention on Biological Diversity, the Convention for the Protection, Management and Development of the Marine and Coastal Environment of the Eastern African Region and Related Protocols, the United Nations Convention to Combat Desertification, the Bamako Convention on Ban of the Import into Africa and the Control of Transboundary Movements of Hazardous Wastes Within Africa and the United Nations Framework Convention on Climate Change Ratified, the Vienna Convention on the Protection of Ozone Layer, the Montreal Protocol on Substances That Deplete the Ozone Layer and the Basel Convention on the Control of Transboundary Movements of Hazardous Wastes and Their Disposal.[28] By virtue of the binding nature of treaties under international law,[29] Tanzania is under obligation to enforce the obligations contained in these treaties in good faith. At the national level, Tanzania has several legislation which either directly or indirectly having a bearing on the protection and conservation of the environment. The most comprehensive law on environmental issues in Tanzania is the Environmental Management Act which was passed in 2004. This Act had the effect of replacing the National Environment Management Council Act of 1983. Amongst the issues addressed by the 2004 Act are environmental quality standards, risk and impact assessment, framework for institutions on management of the environment, control and prevention of pollution, enforcement and compliance measures and public participation.[30]

Aside from the above comprehensive legislation, Tanzania has a plethora of legislation which, though not exclusively devoted to environmental issues, contain provisions on environmental issues.[31] For instance, Tanzania's Mining Act, while focusing on issues of mining, makes Environmental Impact Assessment a legal requirement for all activities

[28] See status of ratification/accession of these treaties.

[29] In terms of article 26 of the Vienna Convention on the Law of Treaties 1969, the ratification of treaties creates binding obligations on states and as such, states are to implement provisions of treaties they are party to in good faith.

[30] See generally Environmental Management Act of 2004 of Tanzania.

[31] See, for example, The Mining Act, Chapter 123 R.E 2002 and Environmental Management Act 20 of 2004 of Tanzania.

pertaining to mining.[32] Generally, therefore, Tanzania has a comprehensive legislative framework on environmental matters. It would, therefore follow, logically so that recourse can effectively be made to these laws and the provisions contained therein to enforce the environmental rights and ultimately to ensure that the environment is conserved. Commentators, however, submit that the fragmented legislation on protection of the environment has made the protection of the environment illusive.[33] Since Tanzania's legal regime on issues of the environment is characterized by multiple legislation, various sectors are tasked to ensure the enforcement of the environmental provisions contained in these legislation. The enforcement of these provisions is therefore fragmented due to the fact that there is no coordination across sectors regarding the overlapping issues on protection of the environment.[34] The fragmented nature of the legislative framework has made both administrative and judicial enforcement illusive.

Apart from legislation, the Constitution of Tanzania contains some provisions with a bearing on environmental issues. As a starting point, the Constitution of Tanzania is divided into sections, two of these sections being: (1) fundamental objectives and directive principles of state policy (DPSP) and (2) basic rights and duties.[35] The section on "Basic rights and duties" essentially constitutes Tanzania's Bill of Rights. Suffice it to note, however, that this section is silent on environmental rights and to be exact, has no explicit provision on the right to a clean and healthy environment as is the case for some Constitutions in Africa. Rather, the provision that has some inkling on environmental matters is contained in the section that the Constitution of Tanzania refers to as "Fundamental objectives and directive principles of state policy." This provision reads as follows:

---

[32] Section 38(1) of the Mining Act, Chapter 123 of 2002. A similar requirement is entrenched in section 81 of the Environmental Management Act, No. 20 of 2004.

[33] Emmanuel Laban Kileo, *Enforcement and implementation of environmental laws and the protection of mining areas in Tanzania: a case study of Mererani*. § 56–57 (2013), Masters Degree Dissertation Mzumbe University.

[34] Ibid.

[35] See part II and part III of the Constitution of the Republic of Tanzania 1977.

> The object of this Constitution is to facilitate the building of the United Republic as a nation of equal and free individuals enjoying freedom, justice, fraternity and concord, through the pursuit of the policy of Socialism and Self Reliance which emphasizes the application of socialist principles while taking into account the conditions prevailing in the United Republic. Therefore, the state authority and all its agencies are obliged to direct their policies and programmes towards ensuring that activities of the Government are conducted in such a way as to ensure that the national wealth and heritage are harnessed, preserved and applied for the common good and also to prevent the exploitation of one person by another.[36]

The fact that the above provision makes mention of preservation and harnessing of national wealth could progressively be interpreted to encompass environmental protection. The question that remains unresolved, however, is—would this provision be relied on to advance environmental rights before courts of law? To answer this question, one would need to critically engage with the status of DPSP in national constitutions in light of the fact that the above provision rests not in Tanzania's Bill of Rights section, but rather, in the DPSP section.

Much has been written on the status of DPSP for enforcement of fundamental rights. Most constitutions contain a section on DPSP. Chinnappa has substantiated on the role that DPSP play, observing that "[they] specify the programs and the mechanisms of the state to attain the constitutional goals set out in the preamble."[37] In the words of Gebeye "they are instrumental in furthering aspirations of the people and implementing the very aims and objectives of DPSP."[38] There remains scholarly divergence on the legal status of provisions contained in the section on DPSP. According to one line of argument, "they are not legally enforceable. Instead, they are guidelines for creating a social order characterised by social, economic and political justice, liberty, equality

[36] Article 9(c) of the Constitution of the Republic of Tanzania 1977.

[37] Reddy Chinnappa, *The Court and the Constitution of India: Summit and Shallows*, § 73(2010).

[38] Berihun Adugna Gebeye, *The Potential of Directive Principles of State Policy for the Judicial Enforcement of Socio-Economic Rights: A Comparative Study of Ethiopia and India*, 4 ICL Journal. 4 (2015).

and fraternity as enunciated in the Constitutions' preamble."[39] Per this line of argument, therefore, they are devoid of an enforcement mechanism. Das, in fact, refers to them as "an appendix to the Constitution"[40] while Seervai considers them theoretical statements of "hopes, ideals and goals" rather than substantive provisions backed up by judicial mechanisms of enforcement.[41] Other commentators, however, take a contrary view. Kumar, for example, affirms the practical role that DPSP can play, underscoring that they have and continue to be a basis for groundbreaking jurisprudence.[42]

The debates of commentators aside, context is particularly pivotal in discussions pertaining to the status of DPSP. There are countries whose judiciaries have relied on DPSP to protect fundamental rights. India is a practical example in this regard, with the host of jurisprudence grounded in DPSP being testament to such viability.[43] This position has also been evident in the jurisprudence of courts in Africa, including Ghana's judiciary. In *New Patriotic Party v Attorney General*, the Court in Ghana rejected the objection regarding recourse to DPSP, ruling that "the Constitution as a whole is justiciable, if DPSP are not justiciable, the Constitution may say it so expressly and the very tenor of DPSP supports the view of justiciability."[44] But as alluded to already, context is critical in assessing the status of DPSP. Arguably, the reason why the Court in the *New Patriotic Party* case could so easily arrive at such a conclusion is because the Constitution of Ghana explicitly mandates the judiciary to

[39] Bikash Thapa, *Enforceability of Directive Principles with Reference to Judicial Decisions of Nepal*, 1(2013), https://papers.ssrn.com/sol3/papers.cfm?abstract_id=2221996

[40] Cited in Gebeye, *The potential*, 4.

[41] Cited in Bertus De Villiers, *Directive Principles of State Policy and Fundamental Rights: The Indian Experience*, 8 South African Journal of Human Rights. 34 (1992).

[42] Raj Kumar, *International human rights perspectives on the fundamental right to education-integration of human rights and human development in the Indian Constitution*, 12 Tulane Journal of International and Comparative Law. 265 (2004).

[43] See, for example, decisions in the cases of Sajjan Singh v State of Rajasthan 1965 SCR (1) 933; Kesavananda Bharati v State of Kerala and Another 1973 Supreme Court of India; Minerva Mills Ltd. & Others v Union of India & Others 1981 SCR (1) 206; Waman Rao and Ors v Union of India and Others (1981) 2 SCC 362; Chandra Bhavan Boarding and v The State of Mysore and Another 1970 SCR (2) 600. In these cases the courts drew profound insight from the DPSP in arriving at their decision to protect fundamental human rights.

[44] New Patriotic Party v Attorney-General (1996–97) Supreme Court of Ghana Law Reports.

draw guidance from DPSP when interpreting the Constitution. There are, however, constitutions which make it explicit that DPSP are not justiciable. Examples in Africa include Namibia, Nigeria, Lesotho, Sierra Leone and even Tanzania.[45] Suffice it to note that provisions directly speaking to environmental matters in most constitutions are located in the DPSP section. Again here, Tanzania as already explained is one such country with no explicit environmental rights provision under the Bill of Rights section, but with provisions on environment contained in the DPSP section. Article 7(2) of Tanzania's Constitution buttresses the non-enforceability of provisions in the DPSP section by reading as follows:

> The provisions of this Part of this Chapter are not enforceable by any court. No court shall be competent to determine the question whether or not any action or omission by any person or any court, or any law or judgment complies with the provisions of this Part of this Chapter.

The added advantage of having environmental rights explicitly entrenched in the Constitution is that it elevates their status in light of the fact that constitutions constitute supreme laws of countries. Such status arguably directly impacts on the seriousness with which these rights are taken and ultimately, their protection and enforcement. However, the fact that there are provisions on environmental rights under Tanzania's Constitution does not mean that the weight due these rights is enforced, let alone by the judiciary. The implication of article 7(2) above for the enforcement of environmental rights and ultimately, for environmental conservation cannot be emphasized enough. Courts in Tanzania are essentially barred from anchoring their adjudication of disputes in DSPS provisions on environmental matters. The courts in Tanzania are therefore having to rely on other general provisions to address environmental issues. In *Festo Balegele v Dar es Salaam City Council*,[46] for example, the High Court of Tanzania interpreted article 14 of the Constitution on the right to life as encompassing the right to live in a clean and healthy environment. The

[45] See generally the DPSP of the Constitutions of Namibia, Nigeria, Lesotho, Sierra Leone and Tanzania.

[46] Festo Balegele v Dar es Salaam City Council, Misc. Civil Case No. 90 of 1991, High Court of Tanzania.

creativity of the judiciary in this regard is indeed praise-worthy. However, such outcomes may not always be guaranteed as only creative judiciaries may be prepared to go to such lengths to progressively interpret the Bill of Rights to encompass environmental rights. There is no guarantee that matters pertaining to the environment will always be adjudicated upon by progressive judiciaries. Moreover, the very fact that environmental rights are categorized under the DPSP section, that is, the non-enforceable section of Tanzania's Constitution, could be interpreted to mean that these rights are yet to be taken seriously. In so far as the judiciary's role in enforcing environmental rights is concerned, there is no doubt that this role is stifled by the nature and form of environmental rights under Tanzania's laws.

Kenya

Kenya has a host of legislation on environmental matters.[47] As a former British colony, Kenya derives most of its legislation on environmental issues from the former British colonial regime. These laws were inherited by Kenya even after independence. The legislation on environmental matters were and still are mainly sectoral. The sectors encompassed include mining and geology, wildlife conservation, livestock and husbandry, waste disposal, forestry conservation and water conservation. While some of these legislation are not devoted to environmental protection, they contain provisions on environmental matters and these provisions are to be implemented by the sectors responsible for these sectors.

[47] See, for example, Environmental Management and Coordination Act (EMCA) 1999; Environmental Management and Coordination (Environmental Impact Assessment and Audit) Regulations 2003; Environmental Management and Coordination (Water Quality) Regulations 2006; Environmental Management and Coordination (Waste Management) Regulations 2006; Environmental Management and Coordination (Conservation of Biodiversity, Access to Genetic Resources and Benefit Sharing) Regulations 2006; Environmental Management and Coordination (Fossil Fuel Emission Control) Regulations 2006; The Wildlife (Conservation and Management) Act, Cap. 376; National Museums and Heritage Act 2006; Fisheries Act, Cap 378 1989; Forest Act 2005; Plant Protection Act Cap 324; Suppression of Noxious Weeds Act Cap 325; Water Act 2002; Agriculture Act Cap 318; Public Health Act Cap 242; Local Government Act Cap 265; Penal Code Cap 63; Traffic Act Cap 403; Radiation Protection Act Cap 243; The Maritime Zones Act Cap 371; Maritime Authority Act 2006; Continental Shelf Act Cap 312; Occupier's Liability Act Cap 34.

However, the fragmented nature of sectoral enforcement posed challenges of effective enforcement as there was hardly any coordination across sectors. Moreover, the fact that such enforcement was in some sectors dependent on the discretion of sector heads, many provisions on environmental matters were left unenforced. In the words of Bosek, "Kenya's post-colonial environmental law regime was somewhat obscure and largely scattered in a patchwork of sectoral legislation."[48] These inadequacies paved the way for the enactment of the Environmental Management and Co-ordination Act (EMCA) in 1999. The Act does not repeal the various sectoral legislation. Rather, it seeks to ensure coordination across sectors in the protection of the environment.

The enactment of the EMCA, therefore, brought about a shift in the enforcement of environmental rights. Amongst the novel developments in the EMCA was the explicit provision of the right to a clean environment.[49] The "rights" terminology in effect presented environmental issues as public law issues and not merely private law issues. Not surprisingly, when the Court, in *Rodgers Muema Nzioka & 2 Others v Tiomin Kenya Limited* (*Muema* case),[50] was confronted with a petition on pollution by the applicants on behalf of ordinary farming inhabitants, it did not miss its words on the public law nature of environmental matters, going ahead to rule that

> [e]nvironmental degradation is not necessarily individual concern or loss but public loss so in a matter of this kind the convenience not only of the parties to the suit, but also of the public at large is to be considered so that if the injunction is not issued it means that any form of feared degradation, danger to health and pollution will be caused to the detriment of the population.

Up to this point, however, the Constitution of Kenya had no explicit substantive provisions on environmental rights. The Constitution of Kenya of 2010 ushered in a new era on environmental protection,

---

[48] Joel Kimutai Bosek, *Implementing environmental rights in Kenya's new constitutional order: Prospects and potential challenges*, 14 *African Human Rights Law Journal*. 492 (2014).

[49] Section 3 of the Environmental Management and Coordination Act 1999 of Kenya.

[50] Rodgers Muema Nzioka & 2 Others v Tiomin Kenya Limited (*Muema* case) HCCC 97 of 2001.

making provision for a number of provisions on environmental rights, including a substantive provision on these issues. The Constitution of Kenya has multiple provisions on environmental matters. Article 42 establishes the main substantive provision on environmental rights in Kenya. This right falls within the "Rights" section of Kenya's Constitution, thus, placing the right to environment in the rank of fundamental rights in Kenya's democracy. This provision reads as follows:

> **Environment**
>
> **42.** Every person has the right to a clean and healthy environment, which includes the right—
>
> (*a*) to have the environment protected for the benefit of present and future generations through legislative and other measures, particularly those contemplated in Article 69; and
>
> (*b*) to have obligations relating to the environment fulfilled under Article 70.[51]

As already alluded to, prior to the enactment of the 2010 Constitution, the EMCA had already made explicit provision for the right to a clean environment. However, the Constitution's recognition of this right under the "rights" section not only elevated the status of this right but also signaled the seriousness with which environmental matters are taken. Moreover, the fact that rights have corollary obligations; the inclusion of this right in the "Rights" section automatically triggers obligations on the part of duty-bearers which may be state organs or privates persons. In Tanzania where no such substantive right is entrenched in the "rights" section of the Constitution, the courts in Tanzania have had to rely on other peripheral rights such as the right to life to enforce environmental rights and this of itself may be limited in some cases. The explicit entrenchment of the right to environment as a substantive right under the "rights" section as is the case under Kenya's Constitution presents courts with effective weapons for enforcement of environmental rights and ultimately allows courts to play a role not only in environmental conservation but also, sustainable development. Courts can equally rely

[51] Article 42 of the Constitution of the Republic of Kenya 2010.

on other provisions on environmental matters within the Constitution to buttress the substantive right and give meaningful content to it. As one can garner from article 42 above, reference is made to article 69 of the Constitution which is also a provision on environmental issues. In terms of this provision, the state is to

> (*a*) ensure sustainable exploitation, utilisation, management and conservation of the environment and natural resources, and ensure the equitable sharing of the accruing benefits;
> (*b*) work to achieve and maintain a tree cover of at least ten per cent of the land area of Kenya;
> (*c*) protect and enhance intellectual property in, and indigenous knowledge of, biodiversity and the genetic resources of the communities.
> (d) encourage public participation in the management, protection and conservation of the environment;
> (e) protect genetic resources and biological diversity;
> (f) establish systems of environmental impact assessment, environmental audit and monitoring of the environment;
> (g) eliminate processes and activities that are likely to endanger the environment; and
> (h) utilise the environment and natural resources for the benefit of the people of Kenya.
> (2) Every person has a duty to cooperate with State organs and other persons to protect and conserve the environment and ensure ecologically sustainable development and use of natural resources.[52]

Overall, it can be said that the scope and nature of the right to environment in Kenya's environmental law regime has thus far impacted positively on the role of the judiciary in the enforcement of environmental rights. One can garner that in the case of Kenya, the right to a clean environment was first ushered into Kenya's legal regime in 1999 through the EMCA. This was over ten years prior to the enactment of Kenya's 2010 Constitution. Yet, even before the entrenchment of this right under Kenya's Bill of Rights, the courts in Kenya were already enforcing it as such as evident in the *Muema* case. Already, its entrenchment in

[52] Article 69 of the Constitution of the Republic of Kenya 2010.

legislation had the effect of shifting the atmosphere of adjudication, from merely a private affair to a public law one. It goes without saying that this impacted positively on the judiciary's role in the enforcement of environmental rights in Kenya. With the coming into force of the 2010 Constitution, this standpoint was further buttressed. The increased engagement of courts with environmental matters subsequent to the enactment of the 2010 Constitution has even been confirmed by the praiseworthy jurisprudence of the Kenya's Environment and Land Court.[53] What lessons then does this state of affairs offer states such as Tanzania whose Constitution still does not recognize the right to a clean environment as a substantive right?

It is an indisputable fact that unlike legislation, constitutions are instruments which are not easily amended or re-enacted given the procedural rigor surrounding such a process. Though desirable, for constitutions that do not have substantive provisions on the right to a clean environment, it would be ambitious to expect immediate reforms. The same may, however, not be said of national legislation. For countries like Tanzania, more effective judicial enforcement of environmental rights could be triggered by reforms to legislation on environmental matters with explicit provisions on environmental rights. Moreover, a legislation akin to the EMCA in Kenya would not necessarily be a bad idea for Tanzania in light of the fact that Tanzania's environmental legal regime is also a product of British colonial law, with the enforcement of most of these laws remaining extremely fragmented and uncoordinated.

Uganda

As a former British colony, just like Kenya and Tanzania, the environmental law regime of Uganda finds its roots in the British legal system.

---

[53] See, for example, Joseph Leboo & 2 others v Director Kenya Forest Services & another Environment and Land No. 273 of 2013 eKLR; Moffat Kamau & 9 others v Aelous Kenya Limited & 9 others Constitutional Petition No.13 of 2015 [2016] eKLR; African Centre for Rights and Governance (ACRAG) &3 others v Municipal Council of Naivasha Petition No. 50 of 2012 [2017] eKLR. For commentary on some of the jurisprudence of the Environment and Land Court, see Caiphas B Soyapi, *Environmental Protection in Kenya's Environment and Land Court,* 31 Journal of Environmental Law. 151–161 (2019).

On attainment of independence, most of the laws left behind by the British were retained by Uganda and some continue to be applied. As these laws are mainly sectoral, their enforcement is fragmented and, in many respects, extremely weak. After independence, Uganda enacted some new legislation including the National Environment Act.[54] This Act contains a substantive provision on the right to a clean a health environment.[55] The Act also has constitutional backing in that the Constitution has an explicit substantive provision on the right to a clean and healthy environment.[56] The enforcement of this right in Uganda is made practical by the fact that the Constitution empowers the courts to adjudicate over matters pertaining to the rights guaranteed under the Constitution, amongst which are environmental rights.[57] The Courts have duly breathed life into these provisions, going as far as to give meaningful content to this right. In light of the fact that the right is couched as "clean and healthy environment," it could so easily be restricted to health from a medical perspective. The courts in Uganda have, however, given meaningful content to this right, observing in the case of *Uganda Electricity Transmission Co Ltd v De Samaline Incorporation Ltd*, that

> the right to a clean and healthy environment must not only be regarded as a purely medical matter. It should be regarded as a holistic socialcultural phenomenon because it is concerned with physical and mental well-being of human beings … a clean and healthy environment is measured in both ethical and medical context. It is about linkages in human well-being. These may include social injustice, poverty, diminishing self-esteem, and poor access to health services. That right is not restricted to a clinical model.[58]

Generally, rights are couched in very precise terms. It would follow therefore that even with the explicitness of substantive environmental rights, the mandate rests on courts to breathe life on these provisions with a view

[54] The National Environment Act Chapter153.

[55] Section 3 of the National Environment Act Chapter153.

[56] Article 39 of the Constitution of the Republic of Uganda 1995.

[57] Article 50(1) of the Constitution of the Republic of Uganda 1995.

[58] Uganda Electricity Transmission Co Ltd v De Samaline Incorporation Ltd, Miscellaneous Cause No. 181 of 2004, High Court of Uganda.

to enforcing them. One can garner here that the Court did exactly that by ensuring that the right to a clean and healthy environment is given a progressive interpretation, enough to ensure that the environment is not deprived of protection.

It is also interesting to note that despite the fact that both the Constitution of Uganda and the National Environment Act recognize the right to a clean and healthy environment, the courts have not unduly restricted themselves to this right in theory advancement of environmental rights. The courts have creatively relied on other substantive provisions under the "Rights" section including the right to life and the right to access to information to enforce environmental rights. Notably, in *Greenwatch v Attorney General and National Environment Management Authority*,[59] the Court underscored the critical role of access to information in possession of the state in the enforcement of environmental rights. Overall, this case affirmed the indispensability of the Bill of Rights as a whole in the enforcement of environmental rights. By the same token, the decision is testament to the fact that other rights under the Constitution constitute entry points for the enforcement of environmental rights. There is no doubt that having substantive provisions on environmental rights in the "Rights" section of constitutions is to be encouraged to ensure effective judicial enforcement of environmental rights. However, this, of itself, does not take away the fact that other rights can be relied on to complement substantive provisions on environmental rights. And for states such as Tanzania, recourse to other rights becomes inevitable in light of the absence substantive rights on environmental matters.

In a more recent 2015 landmark decision of the Supreme Court of Uganda in *Nyakaana v National Environment Management Authority (NEMA)* the critical role that courts can play in environmental conservation was brought to the fore. The facts of this case were that Nyakaana, the appellant, owned land taking the form of leasehold tenure. In his capacity as owner, he commenced to construct on the said land. The NEMA, a government organ mandated under legislation to ensure that

[59] Greenwatch v Attorney General and National Environment Management Authority Miscellaneous Application No. 371 of 2002, High Court of Uganda.

wetlands are preserved, issued a restoration order, directing the appellant to restore the land to its original condition prior to the encroachment by way of construction. The appellant's' failure to comply prompted the NEMA to demolish the appellant's structure on account of its location in wetland. The appellant instituted proceedings in the Constitutional Court of Uganda, arguing, amongst others, that the powers of the NEMA were unconstitutional as they unreasonably limited his right to a fair trial. The Constitutional Court dismissed the petition. The appellant subsequently appealed the Constitutional Court's decision in the Supreme Court of Uganda which court upheld the Constitutional Court's decision. The Supreme Court made a number of interesting observations touching on the role of judiciaries in the enforcement of environmental rights. Underscoring the critical role of land owners in conservation of the environment, the Supreme Court of ruled that "although one has a right to own land through one of the systems of land tenure listed in the Constitution, there may be situations which necessitate the government either to take over that land, or to regulate its use for purposes of promoting and protecting the environment for the common good of all the people of Uganda." The court added that "a person cannot degrade a wetland and cause pollution to other citizens simply because he owns the land. This would defeat the whole purpose of the Constitution which requires that citizens may own land, but not cause pollution or degradation of the environment which may affect other people and the country as a whole." The government's intervention in the manner in which the appellant was dealing with land under his ownership, in the court's view, did not take away his proprietary interest. Rather, the intervention was to prevent him from misusing the land.

Turning to the issue of the nature of the right to environment under Uganda's laws, the Supreme Court provided an interesting perspective to the legal status of DPSP in as far as judicial enforcement of environmental rights are concerned. The Court made reference to multiple DPSPs including Objective XIII of Uganda's Constitution which provides that "the state shall protect important natural resources, including Land, water, wetlands, minerals, oil, fauna and flora on behalf of the people of Uganda." As consistently demonstrated in this chapter, in some instances, in the absence of a substantive environmental rights, the right to environment remains non-justiciable. However, in this regards, Justice Katureebe observed that "these objectives have gone beyond merely guiding us in interpreting the Constitution, but may in themselves be justiciable. The directives on protection of the environment must in my view be read together with Article 39 of the Constitution on the right to a clean and healthy environment to which every Ugandan has a right to." The approach of the Court in this regard underscores the critical role that courts could play in progressively interpreting and applying legislation and the constitution to enforce environmental rights.

4 The Nexus Between the Rules of Procedure on Access to Courts and the Judiciary's Role in Enforcement of Environmental Rights

The nature of the rules of procedure applicable to proceedings before courts has implications for the judiciary's role in the enforcement of environmental rights. Rules of procedure can either ensure access of applicants to courts or deny them such access. This section discusses the rules of procedure, with particular emphasis on the notion of *locus standi*. This discussion assesses the nature of this rule in Uganda, Kenya and Tanzania with a view to assessing its implication for the judiciary's role in fostering environmental conservation.

It would be inchoate to discuss rules on standing in environmental matters without getting to grips with the law of tort. This is because most principles of environmental law find their roots in the law of tort. A tort is a civil wrong and generally, the law of tort is an aspect of private law.[60] Not surprisingly, in most common law jurisdictions, environmental matters were traditionally addressed as private matters. There are multiple examples of torts including trespass, nuisance and negligence. The law of trespass addresses issues of unauthorized or forcible entry on land. The

[60] Edwin Jaggard *Handbook of the law of torts* (1895).

core aim of the law of trespass is to protect private property.[61] A nuisance entails any action or omission by a person which causes interference with the quiet possession by other land users or neighbors.[62] Negligence, on its part, proceeds on the premise that a person is expected to exercise reasonable care to avoid harm to others.[63] Failure to exercise such care to the extent that it results into harm to others constitutes negligence. In all these torts, a person had to be aggrieved and only such aggrieved person had standing before court for purposes of a cause of action.[64] As the law of tort is generally a private law matter, it was viewed as such rather than a public law issue. Where public rights were infringed upon, the only person with standing before courts was the Attorney General. A person could therefore not purport to approach the courts claiming to be acting on behalf of others. These rules continued to be applied for the better part of African countries' post-independence era, the inevitable consequence of this being that courts were deprived of the platform to adjudicate over environmental matters for want of standing. This profoundly limited the role of the judiciary in enforcing environmental rights.

In Kenya, prior to the enactment of the EMCA in 1999 and the 2010 Constitution, these rules were very much alive and well. There were established precedents on matters of standing, thus, when confronted with environmental matters, these same rules came into play. These rules barred Kenyan courts from engaging with environmental matters on their merits in cases where the person who instituted proceedings was not aggrieved. In *Wangari Maathai v Kenya Times Media Trust Ltd*, the High Court of Kenya buttressed the view that only the Attorney General had standing before the Court and that the mandate of suing on behalf of the public was only vested in the Attorney General.[65] The Court ruled that Wangari (the plaintiff) had no cause of action against the respondent. The decision in *Gouriet v the National Union of Post Office Workers* so

[61] Ibid., 660–663.

[62] Jaggard, *Handbook*, 744–748.

[63] Jaggard, *Handbook*, 810–815.

[64] Jaggard, *Handbook*, 737–738.

[65] Wangari Maathai v Kenya Times Media Trust Ltd (1989) 1 KLR (E&L).

vividly traced these rigid rules of standing to English common law, in which it was ruled as follows:

> It was a fundamental principle of English law that public rights could only be asserted in a civil action by the Attorney-General as an officer of the Crown representing the public. Except where statute otherwise provided, a private person could only bring an action to restrain a threatened breach of the law if his claim was based on an allegation that the threatened breach would constitute an infringement of his private rights or would inflict special damage on him.[66]

In the main, the issue of standing in Kenya prior to 1999 was anchored in English common law. The implication of these rules for enforcement of environmental rights cannot be overstated. Unless the plaintiff established that they were aggrieved by the alleged action or omission, proceedings before courts would not be heard on their merits. Although the Attorney General bore the mandate to institute proceedings on behalf of the public, often this power was not exercised because in many instances the violations were committed by the state—the very organ that the Attorney General stood for.

With the enactment of the EMCA in 1999, the right to a clean and healthy environment was recognized and as already alluded to, cases such as the *Muema* one set the record straight regarding standing on environmental matters before the court. In this case, the plaintiff's quest to sue on behalf of the public was recognized by the Court. With the coming into force of the 2010 Constitution, the flexibility on the rules of standing was further buttressed and this further paved the way for public interest litigation on environmental issues. In terms of article 70(1) of Kenya's Constitution, "[i]f a person alleges that a right to a clean and healthy environment recognised and protected under Article 42 has been, is being or is likely to be, denied, violated, infringed or threatened, the person may apply to a court for redress in addition to any other legal remedies that are available in respect to the same matter." From the foregoing provision, it can be gleaned that standing before court is not

[66] Gouriet v the National Union of Post Office Workers (1977) 3 All ER 71–72.

restricted to persons who have been aggrieved but rather "a person who alleges."

Suffice it to note that for courts to adjudicate over disputes, a person has to institute proceedings before courts. Thus, provisions on standing can either stifle or advance environmental protection. Kenya's current regime in so far standing is concerned is progressive in that it makes room for public interest litigation and ultimately, for more environmental matters to be addressed by courts. It is also significant to note that article 70 (1) is not merely reactive but also proactive in that it anticipates future harm to the environment. This can be gleaned from the fact standing is granted to persons who allege that the right to a clean and healthy environment is likely to be infringed, threatened or violated (emphasis added). The anticipatory nature of this provision suggests that court orders in the form of injunctions, mandamus and so on have the potential of preventing environmental harm even before such harm happens. In so doing, the judiciary is given an opportunity to play a role in not only conserving the environment but also ensuring that developments are sustainable.

In Uganda, prior to the enactment of the Constitution of Uganda in 1995, the rules on standing required that the plaintiff demonstrates to court that he/she had suffered harm resulting from the alleged conduct or omission. This state of affairs limited the courts' role to entertain environmental matters. Uganda's current rules on standing can be credited to the enactment of its Constitution in 1995. With the coming into force of this Constitution, a platform has been created for the judiciary to play a role in the enforcement of environmental rights and to the environment's preservation. In terms of article 50(1), "[a]ny person who claims that a fundamental or other right or freedom guaranteed under this Constitution has been infringed or threatened, is entitled to apply to a competent court for redress which may include compensation."[67] The rights envisaged by the foregoing provision encompass the right to a clean and healthy environment. Article 50(2) buttresses the notion of public interest litigation even further by providing that "[a]ny person or organisation may bring an action against the violation of another person's or group's

[67] Article 50(1) of the Constitution of the Republic of Uganda 1995.

human rights."[68] While Uganda's provisions on enforcement of environmental rights deserves praise to the extent that it redeems Uganda from the unpalatable past of restricting standing before court to aggrieved persons, it does appear limited when compared to Kenya's provision. Notably, Kenya's article 70(1) envisages "likely" harm to the environment. Uganda's article 50(1), however, only grants standing to persons where environmental rights have been threatened or violated already. Such an approach seems reactive and in fact restrictive as there are likely/possible threats to the environment which can be prevented were standing to be granted before the threat, violation or infringement.

Tanzania's Constitution also makes room for public interest litigation. Article 30(3) of Tanzania's Constitution[69] is similar to Kenya's article 70(1)[70] to the extent that it envisages likely violations to the environment for purposes of determining standing. As the case is for Uganda and Kenya, one does not have to be aggrieved to have standing before court. This has made room for public interest litigation and the role of this development has been affirmed by Tanzania's courts. In *Mtikila v The Attorney General*, for example, the court underscored the role of public interest litigation particularly among the poor population in society who may not be able to marshal the necessary resources to institute proceedings, let alone to recognize that their rights are threatened and as such, warranting redress.[71] In so far as Tanzania is concerned, however, the benefits of public interest litigation appear to be limited in as far as environmental matters are concerned. The right to a clean and healthy environment, as consistently noted is not explicitly guaranteed under Tanzania's Bill of Rights. Rather, it is located in the judicially non-enforceable section of DPSP. Therefore, public interest litigation cannot be exercised directly with specific regard to this right. Litigants and the courts can however have recourse to the other rights guaranteed under the Constitution such as the right to life. Considered together therefore, it is not enough for a legal regime to have flexible rules on standing. These

[68] Article 50(2) of the Constitution of the Republic of Uganda 1995.

[69] The Constitution of the Republic of Tanzania 1977.

[70] Article 70(1) of the Constitution of the Republic of Kenya 2010.

[71] Mtikila v The Attorney General, Tanzanian Civil Suit No.5 of 1993.

rules work hand in hand with effective substantive provisions on environmental matters, absence of which the flexible rules on standing are rendered illusory.

5 Theories on Judicial Decision-Making: Making Sense of Courts' Approaches to Environmental Issues in Africa

Much has been written on the role of courts in the enforcement of rights.[72] Various theories also abound on how judges arrive at decisions with particular regard to the various theories that inform judicial decision-making.[73] However, very little research has linked these theories to the enforcement of environmental rights in Africa. Theories on judicial decision-making are multiple, however, two shall inform the current discussion and these are the formalistic theory of judicial decision-making and the realist theory of decision-making. The formalistic theory, also sometimes referred to as the positivist theory, proceeds on the premise that the principles of law are the ultimate guide for judges to arrive at decisions.[74] The principles of law spoken of in this regard are derived from statutes, common and precedent. In terms of this theory, once the law is determined, it is henceforth applied to the facts in issue and such application leads to a decision.[75] A decision arrived at using in reliance of this process is predictable and according to commentators, a judicial officer confronted by a similar set of facts should be able to arrive at a similar decision as application of the law to the facts is akin to a mathematical formula.[76] This theory is associated with positivist thinkers such as Hart

[72] See, for example, Ngang C. Carol, *Judicial enforcement of socio-economic rights in South Africa and the separation of powers objection: The obligation to take 'other measures*, 14 African Human Rights Law Journal. 655–680 (2014); Albie Sachs, *The Judicial Enforcement of Socio-Economic Rights: The Grootboom Case,* 56 Current Legal Problems. 579–601 (2003).

[73] Timothy Capurso, *How Judges Judge: Theories on Judicial Decision Making,* 9 University of Baltimore Law Forum. 5–16 (1998).

[74] Ibid., 9–12.

[75] Ibid.

[76] Ibid.

who draw a distinction between law and morality. As such the role of judges is not to supplant the law with morality but to apply the law as it is based on the facts the court is presented with. In as far as judicial enforcement of environmental rights is concerned, this theory could help explain why judges strictly apply provisions of the law. Worthy to note, some of these laws are a legacy of colonialism, with many of them being out of step with current developments in environmental law. With a formalistic approach to some of these obsolete laws, judicial officers may be deterred from playing a role in environmental protection. This is because progressive interpretation of legal provisions would be out of step with the formalistic theory of judicial decision-making. This possibly explains why judiciaries with legal frameworks such as that in Tanzania are left with no option but rely on other existing human rights provisions to enforce environmental rights as in the absence of an explicit environmental rights such as that the right to environment is not justiciable.

The realist theory, associated with jurists such as Joseph Bingham and Eugene Elrlich, is premised on the argument that prior to decision-making, judges are influenced by prejudices.[77] Principles of law are therefore merely invoked to rubberstamp conclusions drawn based on prejudice. Commentators' views, "judges determine the outcome of a lawsuit before deciding whether the conclusion is, in fact, based on an established legal principle."[78] Upon reaching such a conclusion, a judge merely explores principles to support preconceived decisions. Preconceived decisions are informed by multiple factors including political affiliation, legal experience, legal education, personal and family associations. What stands out in this theory therefore is the premise that judges are not always guided by law in arriving at decisions.[79] This theory could help make sense of the practice of judges, through judicial activism, to progressively interpret the law with a view to ensuring environmental conservation and sustainable development despite there being no explicit statutory provision or precedent on the environmental issues before such a court. This, for example, could explain why Kenya, despite not having

[77] Ibid., 5–8.

[78] Ibid.

[79] Ibid.

an explicit provision on the right to environment under the Constitution prior to 2010, proceeded to invoke the right to environment in as entrenched in the EMCA with a view to affording protection to the environment.[80] The same holds true for Uganda and Tanzania. As apparent in the decision of the court in the case of *Nyakaana v NEMA* and others, the court's willingness to elevate DSPS to a level of justiciable rights is testament to the non-formalistic approach of the court to decision-making.

Pertinent to note, the reality on ground in most African countries is that most laws are a legacy of colonialism. Although states have carried out amendments in some respects, many of the laws introduced during the colonial era remain alive and well, with not many of them taking cognizance of developments in environmental law. Not many constitutions in Africa, for example, have explicit provisions on the right to environment. This, of itself, makes this right non-justiciable. It is in these circumstances that the rubber meets the road, with judges being confronted with two options: either to invoke the formalistic approach to decision-making or the realist one. When the formalistic approach is invoked, in which case the law is applied as it is, environmental conservation has always remained a far-reaching goal. Judicial activism can only rest comfortably alongside a realist approach to decision-making. Such an approach, however, cannot always be guaranteed, thus underscoring a need for the right to environment to be explicit and for the rules of procedure to be flexible.

6 Conclusion

This chapter set out to assess the implication of the nature of the right to the environment and procedures applicable to the adjudication on environmental-related matters for the judiciary's role in the fostering of environmental conservation. This assessment has been conducted with specific reference to Uganda, Kenya and Tanzania. It is indisputable that courts play a critical role in the enforcement of environmental rights. Judges across African countries have indeed been granted this platform

[80] See *Muema* case above.

and if effectively utilized, it goes a long way in fostering environmental conservation. The analysis conducted in this chapter reveals that it is not enough for courts to be granted such a mandate. Significantly, the manner in which the right to environment is couched as well the nature of a country's rules on access to court play a critical role in ensuring that courts effectively execute their mandate. We garner from the framework in Tanzania that while the rules of standing have been relaxed to ensure that Courts increasingly engage with environmental matters with a view of redress, the fact that this right is not explicitly guaranteed under Tanzania's Constitution continues to pose challenges. Uganda long addressed this gap upon its enactment of its 1995 Constitution, however, the relative datedness of Uganda's environmental law provisions on standing would appear to suggest that it is time for Uganda to step up to the plate and to get in step with current developments on environmental law. This can be gleaned from the fact that the provisions on access to courts do not envisage "likely" threats to the environment. Kenya's Constitution is a more recent 2010 enactment and not coincidentally, it is extremely meticulous on issues of substantive environmental rights and access to courts. Traces in the environmental legal regimes of Uganda, Kenya and Tanzania can be located in the frameworks of countries such as South Africa, Ghana and Nigeria. This therefore suggests that the issues highlighted in this discussion are of relevance to other African countries. Overall, if states are really keen on ensuring that environmental rights are effectively enforced, it is critical for their environmental legal regimes to constantly be updated with a view to bringing them up to speed with more current trends on environmental conservation.

References

Adem K Abebe, "Towards more liberal standing rules to enforce constitutional rights in Ethiopia." *African Human Rights Law Journal* 10, (2010): 407–431.

Agbor, Avitus, "The Ineffectiveness and Inadequacies of International Instruments in Combatting and Ending the Transboundary Movement of Hazardous Wastes and Environmental Degradation in Africa" *African Journal of Legal Studies* 9, (2016): 235–267.

Antonio Herman Benjamin, *Matters of judgement: The potential for sustainability at the local level is huge, fundamental and achievable*, (2017). Accessed November 29, 2029. https://wedocs.unep.org/bitstream/handle/20.500.11822/20259/Our%20Planet%20March%202017.pdf?sequence=1&isAllowed=y.

Bertus De Villiers, "Directive Principles of State Policy and Fundamental Rights: The Indian Experience" *South African Journal of Human Rights* 8, (1992): 29–49.

Bikash Thapa, *Enforceability of Directive Principles with Reference to Judicial Decisions of Nepal*, (2013). Accessed November 29, 2019. https://papers.ssrn.com/sol3/papers.cfm?abstract_id=2221996.

Capurso Timothy, "How Judges Judge: Theories on Judicial Decision Making" *University of Baltimore Law Forum* 29, (1998): 5–16.

Bosek, Joel Kimutai "Implementing environmental rights in Kenya's new constitutional order: Prospects and potential challenges." *African Human Rights Law Journal* 14, (2004): 489–508.

Chinnappa, Reddy, *The Court and the Constitution of India: Summit and Shallows*. New Delhi, New York: Oxford University Press, 2010).

Christian Nellemann, Rune Henriksen, Arnold Kreilhuber, Davyth Stewart, Maria Kotsovou, Patricia Raxter, Elizabeth Mrema and Sam Barrat (eds) *The rise of environmental crimes*. Norway: United Nations Environmental Program, 2016.

Gebeye Berihun Adugna "The Potential of Directive Principles of State Policy for the Judicial Enforcement of Socio-Economic Rights: A Comparative Study of Ethiopia and India" *ICL Journal* (2015):1–63.

Jafari R Kideghesho, Julius W Nyahongo, Shombe N Hassan, Thadeo C Tarimold and Nsajigwa E Mbijele, "Factors and Ecological Impacts of Wildlife Habitat Destruction in the Serengeti Ecosystem in Northern Tanzania" *AJEAM-RAGEE* 11, (2007):17–32.

Jaggard Edwin, *Handbook of the law of torts*. Minnesota, West Publishing, 1895.

Kallonga, Emmanuel, Alan Rodgers, Fred Nelson, Yannick Ndoinyo and Rugemeleza Nshala, "Reforming environmental governance in Tanzania: Natural resource management and the rural economy" (2002), Paper presented at the Inaugural Tanzanian Biennial Development Forum at the Golden Tulip Hotel, Dar es Salaam Tanzania, 24th–25th April 2003. Accessed November 29, 2019. http://www.ujamaa-crt.org/uploads/1/2/5/7/12575135/reforn_ujamaa.pdf.

Kesavananda Bharati v State of Kerala and Another 1973 Supreme Court of India.

Kileo, Emmanuel, *Enforcement and implementation of environmental laws and the protection of mining areas in Tanzania: a case study of Mererani,* Masters Degree Dissertation Mzumbe University, 2013.

Kumar, Raj, "International human rights perspectives on the fundamental right to education-integration of human rights and human development in the Indian Constitution." *Tulane Journal of International and Comparative Law* 12, (2004): 237–265.

Mabogunje, Akin, "The Environmental Challenges in Sub Saharan Africa." African technology Forum (1998). Accessed 29 November 2019. http://web.mit.edu/africantech/www/articles/EnvChall.htm.

Mushkat, Roda, *International Environmental Law and Asian Values: Legal Norms and Cultural Influences*. Vancouver: UBC Press, 2004.

Ngang Carol, "Judicial enforcement of socio-economic rights in South Africa and the separation of powers objection: The obligation to take 'other measures." *African Human Rights Law Journal* 14, (2004): 655–680.

Nsubuga, NW, Namutebi, Edith, Nsubuga-Ssenfuma, Masoud, "Water resources of Uganda: An assessment and review" *Journal of Water Resource and Protection* 6, (2014): 1297–1315.

Rahma Mohamed Msoffei, "Increase in deforestation: A key challenge to household charcoal supply – A case of Tanga urban, Tanzania." *European Journal of Social Sciences Studies* 2, (2017):133–151.

Sachs Albie, "The Judicial Enforcement of Socio-Economic Rights: The Grootboom Case." *Current Legal Problems* 56, (2003): 579–601.

Seton Stiebert, Deborah Murphy, Jason Dion and Scott McFatridge, *Kenya's Climate Change Action Plan: Mitigation, Chapter 4: Forestry*, (2012). Accessed November 29, 2019. http://www.kccap.info/phocadownload/final/SC4/Chapter%204%20SC4_Forestry_Final.pdf.

Simeon Otieno Dulo *Impact of Poor Solid Waste Management in Kenya on Groundwater*, Doctoral thesis, University of Nairobi, 2009.

Soyapi Caiphas, "Environmental Protection in Kenya's Environment and Land Court." *Journal of Environmental Law* 31, (2019):151–161.

United Nations Environment Program, "UN Environment's Africa Office is supporting the continent's journey towards sustainable development" (2019). Accessed 29 November 2019. https://www.unenvironment.org/regions/africa.

Weru Sam, *Wildlife protection and trafficking assessment in Kenya: Traffic report*, (2016). Accessed November 29, 2019. https://portals.iucn.org/library/sites/library/files/documents/Traf-126.pdf.

8

Experiences from African Regional Human Rights Courts and Tribunals in the Protection of Environmental Rights

Michael Addaney and Chantelle Gloria Moyo

1 Introduction

The unparalleled dilapidation of the earth's essential ecosystems is among the most challenging issues facing the international community. Despite the adoption of numerous legal instruments to address global environmental problems, clashes between rich countries in the Global North and poor nations in the Global South have compromised the efficacy of international environmental law, resulting in stalemates in environmental treaty negotiations and non-compliance with existing multilateral

M. Addaney (✉)
Research Institute of Environmental Law, Wuhan University, Wuhan, China

University of Energy and Natural Resources, Sunyani, Ghana
e-mail: michael.addaney@uenr.edu.gh

C. G. Moyo
North-West University, Potchefstroom, South Africa

M. Addaney, A. Oluborode Jegede (eds.), *Human Rights and the Environment under African Union Law*, https://doi.org/10.1007/978-3-030-46523-0_8

environmental agreements.[1] The historic roots of the divide between the Global North and Global South in European expansionism manifest today in a range of issues including energy, justice, trade and investment, extractive industries, human rights, hazardous waste, and climate change.[2] The exclusive geographies of the environment in Africa and the continent's economic development trajectory such as considerably worse per capita incomes, high reliance on mineral resource extraction and the exposure of rural communities to environmental catastrophes such as mining-induced displacement and pollution from extractive activities exert pressure on the protection of the natural environment and associated human rights violations in Africa.[3] The activities of extractive industries are also corroding the value of ecological treasures such as green spaces, forests and water resources that minorities and other vulnerable groups depend on for survival and development.[4] This is a damaging path for the promotion and enjoyment of environmental rights, particularly the human rights of vulnerable communities and individuals to health, food, adequate housing, culture and water.

Human rights-based approaches are relatively new additions to the environment and development field.[5] The concept of development first emerged in the human rights field through considerations on the 'right to development'.[6] A human rights-based approach is substantially about having a 'social guarantee' that is linked to the way the collaborations between citizens, states and corporations are structured as well as how

[1] See, generally, Shawkat Alam, Sumudu Atapattu, Carmen G. Gonzalez and Jona Razzaque (eds) International Environmental Law and the Global South (Cambridge University Press 2015).

[2] Ibid.

[3] Michael Addaney, Elsabe Boshoff, and Michael Gyan Nyarko, 'Protection of environmental assets in urban Africa: Regional and Sub-Regional Human Rights and Practical Environmental Protection Mechanisms', (2018) 24(2) *Australian Journal of Human Rights* 182–200; John Bosco Isunju, Christopher Garimoi Orach, and Jaco Kemp 'Hazards and vulnerabilities among informal wetland communities in Kampala, Uganda', (2016) 28(1) Environment and Urbanization 275–93, at 275.

[4] Addaney, Boshoff, and Nyarko (n 3 above).

[5] Hans-Otto Sano, 'Development and human rights: The necessary, but partial integration of human rights and development', (2000) 22(3) Human Rights Quarterly 734–52, at 734.

[6] Magdalena Sepúlveda Carmona, 'The obligations of international assistance and cooperation under the International Covenant on Economic, Social and Cultural Rights: A possible entry point to a human rights-based approach to Millennium Development Goal 8', (2009) 13(1) *International Journal of Human* Rights 86–109 at 86.

these interactions affect the most marginalized sectors of society.[7] Uvin argues that, for decades, the development endeavor was sequestered from the human rights system and its implications for development.[8] However, this began to change in the 1990s when development theorists embarked on the quest to appreciate development as being about more than just economic growth, to include the realization of human rights and environmental protection.[9] There is, thus, a widespread acceptance that human rights and environmental protection ought to play a greater role in development. However, the extent of this role and its meaning for developing countries in the Global South, especially those in Africa, has remained both ambiguous and contested. For instance, studies show that environmentally harmful substances particularly from private multinational enterprises harm the environment and human health through polluting soil, air, groundwater, atmosphere, surface and marine waters and as a result violate numerous human rights of individuals and communities.[10] Pollutants usually find their way into the environment through accidents and acts of negligence that may occur during the production, transportation or use of harmful substances by these enterprises.

At the international level, such accidents habitually lead to the adoption of new or revision of existing environmental protection instruments with the aim of preventing future occurrence or, at least, effectively addressing accidents.[11] It has been noted that instruments adopted during or in the aftermath of accidents may serve to address operational pollution from the same source.[12] This chapter therefore explores the approach by the African Union (AU) and the experiences of African regional and sub-regional tribunals in addressing environmental

[7] Colin Brown and Léo Heller, 'Development cooperation in water and sanitation: Is it based on the human rights framework?' (2017) 22(7) Ciência and Saúde Coletiva 2247–56, at 2247.

[8] Peter Uvin, 'From the right to development to the rights-based approach: How 'human rights' entered development', (2007) 17(4/5) Development in Practice 597–606, at 597.

[9] Ibid.

[10] Chairman Okoloise 'Contextualising the corporate human rights responsibility in Africa: A social expectation or legal obligation?' (2017) 1 African Human Rights Yearbook 191–220 at 194; See, generally, Michael Addaney, Hlengiwe Dube and Samrawit Getaneh, 'Light through the Storm: Safeguarding the Human Right to Water in Challenging Landscapes in Africa', (2018) 5(1) *Journal of Comparative Law in Africa* 37–72.

[11] See, generally, Ellen Hey, *Advanced Introduction to International Environmental Law* (Edward Elgar Publishing 2016).

[12] Ibid.

pollution in Africa. It further assesses whether there exists some uniqueness in terms of how African human rights platforms address issues of environmental protection. In conducting this examination, the chapter adopts a doctrinal method and the human rights-based approach as analytical approach. The human rights-based approach establishes that since international human rights law places the protection and realization of fundamental human and group interests at its core, its legal threshold necessitates that all actions which can impact on human rights, including obnoxious corporate conduct, are bound by its rules.[13] This approach further adds greatly to the budding human rights and environmental law literature in Africa in making international environmental law more universally applicable and practically effective. The chapter is divided into four sections. Following this introduction, the second section discusses the relationship between environmental protection and the realization of human rights at the international and regional level. The third section examines the experiences of African regional and sub-regional human rights bodies in protecting environmental human rights in Africa, emphasizing African union (AU) law. The fourth section concludes by summarizing key findings and recommendations.

2 The Growth of Environmental Human Rights at the International and Regional Level in Africa

Since the 1972 Stockholm Conference on the Human Environment[14] through the Rio Conference on Environment and Development (the Earth Summit) to 2012 Rio +20 United Nations Conference on

[13] Andrea Cornwall and Celestine Nyamu-Musembi, 'Putting the "rights-based approach" to development into perspective' (2004) 25 *Third World Quarterly* 1417–1418, at 1415.

[14] UN, Report of the United Nations Conference on the Human Environment, Stockholm, Sweden, 5–16 June 1972, UN Doc. A/Conf.48/14/Rev. 1.

Sustainable Development,[15] the international community has taken action to address the global environmental crisis through adopting several treaties, declarations, UN General Assembly resolutions, customary rules and judicial decisions that respond to specific ecological threats. Areas that have received the attention of international law include the creation of new institutions to handle these environmental matters, adoption of new concepts to give them a firm theoretical foundation and a greater public awareness, and participation in all these matters by the people of the Global North and Global South.[16] Conversely, most countries in the Global South were under colonial rule when the Global North created the existing international legal architecture for globalization. For instance, international organizations such as the International Monetary Fund (IMF) and the 1947 General Agreement on Tariffs and Trade (GATT) were aimed at eroding the sovereignty of the newly decolonized states in order to ease the free flow of goods, services and capital across continents and national borders.[17] Therefore, this legal framework aided the Global North to fuel its economic expansion through the continued utilization of the Global South's natural resources, trapping the countries in the Global South in rancorous cycles of poverty and environmental degradation. This consequently extended the economic divide between the Global North and Global South.[18] Additionally, the economic policies pursued by the Global North resulted in global environmental harms such as ozone depletion, acid rain and climate change.[19]

Despite the varied political and economic trajectories of the Global South several decades after attaining political independence, most countries in the Global South were assimilated into the global economy as exporters of raw materials and importers of industrial goods.[20] This economic specialization has rendered countries in the Global South

[15] UN, Report of the United Nations Conference on Sustainable Development, Rio de Janeiro, Brazil, 20–22 June 2012, UN Doc. A/Conf.216/16.

[16] Ibid.

[17] Ibid.

[18] Clive Ponting, A Green History of the World: The Environment and the Collapse of Great Civilizations (New York: Penguin Books, 1991) 194–212.

[19] Ibid, at 383–392.

[20] Mary Eming Young, World Hunger (New York: Routledge, 1997) 41.

susceptible to deteriorating terms of trade for primary commodities compared to manufactured goods[21] and to the efforts of foreign investors to erode national sovereignty to protect the profitability of their investments in resource-extractive industries.[22] The unwillingness of states in the Global North to regulate the extraterritorial activities of their corporations in the Global South and the challenges in holding parent companies liable for the actions of their subsidiaries have resulted in corporate impunity for human rights and environmental abuses in the countries located in the Global South.[23] Multinational corporations headquartered in the Global North have been responsible for many of the environmental and human rights violations in the Global South,[24] as proven by the *Ogoni* case in Nigeria[25] and the litigation against Chevron in Ecuador.[26] Since these corporations operate in the gray zone between international law and domestic law, they have traditionally escaped scrutiny and accountability at the international level.[27]

The first direct recognition of the principle of environmental protection was the adoption of the Declaration of the United Nations Conference on the Human Environment,[28] which established a foundation for linking human rights and environmental protection in law. Principle 1 declares that 'man has the fundamental right to freedom, equality and adequate conditions of life in an environment of a quality that permits a life of dignity and well-being'.[29] Since the 1972 Stockholm

---

[21] Ibid, p. 42.

[22] Kate Miles, The Origins of International Investment Law: Empire, Environment and the Safeguarding of Capital (Cambridge University Press, 2013) 78–100.

[23] Carmona, note 6, 92–94.

[24] Beth Stephens, 'The Amorality of Profit: Transnational Corporations and Human Rights' (2002) 20 Berkeley Journal of International Law 45 at 49–53.

[25] See, Social and Economic Rights Action Center and Center for Economic and Social Rights v. Nigeria (the Ogoni Case), Case No. ACHPR/COMM/AO44/1,OAUDoc.CAB/LEG/67/3 (2001).

[26] Manuel A. Gómez, 'The Global Chase: Seeking Recognition and Enforcement of the Lago Agrio Judgment Outside of Ecuador' (2013) 1 Stanford Journal of Complex Litigation 429.

[27] Penelope Simons, 'International Law's Invisible Hand and the Future of Corporate Accountability for Violations of Human Rights' (2012) 3 Journal of Human Rights and the Environment 5.

[28] United Nations Conference on the Human Environment (Stockholm Declaration). Declaration of the United Nations Conference on the Human Environment. Stockholm, 5–16 June 1972. http://www.undocuments.net/unchedec.htm

[29] Stockholm Declaration, Principle 1.

Conference, there has been a plethora of international environmental policies at the global level, including the Rio Declaration on Environment and Development in 1992, the Millennium Development Goals (MDGs) and the Sustainable Development Goals (SDGs). The SDGs were adopted in 2015 and represent a more comprehensive and ambitious agenda than the MDGs.[30] As opposed to the 10 MDGs, the SDGs set out 17 major goals with 169 detailed strategic targets associated with the goals. The post-2015 development agenda sought to

> build on the Millennium Development Goals and complete what they did not achieve. They seek to realize the human rights of all and to achieve gender equality and the empowerment of all women and girls. They are integrated and indivisible and balance the three dimensions of sustainable development: the economic, social and environmental.[31]

Currently the international juridical order is organized in a manner that there are no special international environmental courts exercising jurisdiction over the hundreds of multilateral treaties that deal with environmental protection.[32] Often these treaties establish non-compliance or review mechanisms and procedures with quasi-judicial features through facilitative approaches, thereby providing transparency, financial and technological assistance and capacity-building rather than sanctions, such as the 'compliance committees' established by both the Kyoto Protocol and the Aarhus Convention on Access to Justice in Environmental Matters.[33] Despite the lack of a dedicated court, a large body of jurisprudence relating to environmental issues has originated from issue-specific judicial bodies, including regional human rights courts and human rights treaty bodies.[34] These bodies thus present opportunities for pursuing

---

[30] U.N. Transforming our world: The 2030 Agenda for Sustainable Development. UNGA, October 2015.

[31] Ibid, para 3.

[32] Christina Voigt and Evadne Grant, 'The legitimacy of human rights courts in environmental disputes', (2015) 6(2) Journal of Human Rights and the Environment 131–38, at 131.

[33] Addaney, Boshoff and Nyarko (n 3 above) at 4.

[34] See, also, Edith Brown Weiss, 'The Evolution of International Environmental Law' (2011) 54 Japanese Yearbook on International Law 1–27; Peter H Sand, *The History and Origin of International Environmental Law* (Edward Elgar Publishing 2015).

environmental claims at the international and regional levels that otherwise do not exist. Due to the increasing recognition of the strong interconnection between human rights and environmental protection especially in the Global South,[35] a growing number of current disputes adjudicated in regional human rights courts involve the environmental aspects of human rights claims.

In contrast to the rest of international environmental law, a human rights perspective directly addresses environmental impacts on the life, health, private life and property of individuals rather than on states or the environment in general. Boyle contends that it may serve to secure higher standards of environmental quality, based on the obligation of states to take measures to control pollution or degradation affecting health and private life.[36] Furthermore, a human rights-based approach to environmental protection helps to promote the rule of law by making governments directly accountable for their failure to regulate and control environmental nuisances such as those caused by private corporations as well as for facilitating access to justice and enforcing environmental laws and judicial decisions.[37] The expansion of economic and social rights to cover elements of the public interest in environmental protection has breathed new life to the notion that there is, or should be, a right to a safe and decent environment.[38] The growing environmental caseload of human rights courts and treaty bodies further indicates the importance of environmental protection in mainstream human rights law. However, it is self-evident that the environmental dimensions of rights found in the substantive human rights treaties are not direct, and therefore any reference to such implies a 'greening' of existing human rights law rather than the addition of new rights to existing treaties.[39]

---

[35] Ibid.

[36] Alan Boyle, 'Human rights and the environment where next?' in Ben Boer (ed) *Environmental law dimensions of human rights* (Oxford: Oxford University Press 2015) 202.

[37] Ibid.

[38] Francesco Francioni, 'International Human Rights in an Environmental Horizon', (2010) 21(1) *European Journal of International* Law 41–55, at 41.

[39] Boyle, 202.

3 African Regional and Sub-Regional Human Rights Tribunals and the Protection of Environmental Human Rights: African Union Law and Policy in Practice

Human rights protection depends on domestic Constitutions and international human rights mechanisms for their protection and enforcement as well as on each country's socioeconomic and legal systems and the varied levels of its development.[40] In Africa, the consequences of post-independence which include weak institutions, poverty and the inability to forge a common national identity, values and ethos, warranted an objective basis for human rights guarantees, other than the voluntary commitment of its organs of society. The law has been the most potent basis for the promotion and protection of human rights in Africa. Zeleza contends that human rights in Africa have historically been 'products of concrete social struggles and not simply textual or legal discourse',[41] and, thus, the law has been the only neutral source for securing their protection and their contemporary conceptualization on the continent. In Africa, where collective values supersede individual interests, the African human rights and criminal justice systems protect both individual and group concerns, recognize duties as correlative of rights, and increasingly anticipate a broader regime of accountability that captures appalling corporate conduct.[42] Although businesses are generally obliged to pursue policies that are desirable in terms of the objectives and values of society, this is hardly so in Africa where many transnational corporations (TNCs) have come under heavy criticism for their inadequate compensation for

---

[40] Lina Marcinkutė, 'The role of human rights NGO's: human rights defenders or state sovereignty destroyers?' (2011) 4 *Baltic Journal of Law & Politics* 52 74; Ryan Goodman and Derek Jinks 'How to influence states: socialization and international human rights law' (2004) 54 *Duke Law Journal* 623, 634.

[41] Paul Tiyambe Zeleza, 'Rethinking Rights in Africa: The Struggle for Meaning and the Meaning of the Struggle' (2007) 41 (3) Canadian Journal of African Studies 474–506.

[42] African Charter on Human and Peoples' Rights; article 31 of the African Charter on the Rights and Welfare of the Child; article 46C of the Protocol on Amendments to the Protocol on the Statute of the African Court of Justice and Human Rights.

land resource acquisition and use, poor consultations with local communities, forced displacement, environmental pollution, and the destruction of the sacred heritages of communities.[43]

The African Charter on Human and Peoples' Rights (the African Charter) is one of the few international human rights instruments that unequivocally recognizes and guarantees the human right to a satisfactory environment. Article 24 of the African Charter provides that 'all peoples shall have the right to a generally satisfactory environment favourable to their development'.[44] The Protocol to the African Charter on Human and People's Rights on the Rights of Women in Africa (the Maputo Protocol) also provides that 'women shall have the right to live in a healthy and sustainable environment'.[45] These provisions are reiterated in the 2003 Revised African Convention on the Conservation of Nature and Natural Resources (Revised African Convention). In article 3 concerning the Principles underlying the Convention, the Revised African Convention provides that 'in taking action to achieve the objectives of this Convention and implement its provisions, the Parties shall be guided by…the right of all peoples to a satisfactory environment favourable to their development'.[46] These rights and guarantees have also constituted the object of a complaint in terms of the individual complaints procedure[47] before the African Commission on Human and Peoples' Rights (African Commission).[48] Thus, as argued, the links between the

[43] Jernej Letnar Černič, 'European Perspectives on the Business and Human Rights Treaty Initiative' in Jernej Letnar Cernic and Nicolás Carrillo-Santarelli (eds) The Future of Business and Human Rights: Theoretical and Practical Considerations for a UN Treaty (Intersentia 2018) 229–250; Nicola M. Jagers and Marie-Jose van der Heijden 'Corporate human rights violations: the feasibility of civil recourse in the Netherlands' (2008) 33 *Brooklyn Journal of International Law* 833,833–834; Bennett Freeman, Maria B. Pica and Christopher N. Camponovo 'A new approach to corporate responsibility: the Voluntary Principles on Security and Human Rights' (2001) 24 *Hastings International and Comparative Law Review* 423,426.

[44] African Charter on Human and Peoples' Rights, 27 June 1981, OAU Doc CAB/LEG/67/3 rev 5, 21 ILM 58.

[45] Protocol to the African Charter on Human and Peoples' Rights on the Rights of Women in Africa (Maputo Protocol) AHG/Res.240 (XXXI) June 1995, Addis Ababa, Ethiopia, article 18.

[46] African Convention on the Conservation of Nature and Natural Resources, OAU Doc CAB/LEG/24.1 (2003), article 3(1).

[47] The individual complaints procedure resulted from article 55 of the African Charter in conjunction with chapter XVII of the Rules of Procedures of the African Commission, 10 June 1995: http://www.achpr.org/english/_info/rules_en.html (accessed 13 July 2017).

[48] Frans Viljoen, *International human rights law in Africa* (Oxford University Press 2012) 420.

environment and human rights have been recognized by regional courts and tribunals in Africa.

The African Court/Commission on Human and Peoples' Rights remain the most important regional monitoring bodies in relation to the rights guaranteed in the African Charter. Both bodies, particularly the African Commission, through its dual mandate of protection and promotion, present an opportunity for the advancement of human rights and development of standards for the region to support the achievement of the African Union's Agenda 2063 as well as the Sustainable Development Goals.[49] This is because human rights conventions and multilateral environmental agreements set standards for development goals to ensure that specific groups such as indigenous peoples, women, children and persons with disabilities are included in the development planning process and implementation. Given the challenges to governance and rule of law in the region, regional human rights tribunals supplement access to remedies where national human rights mechanisms are non-existent and/or inadequate.[50] In this regard, the African Court and Commission on Human and Peoples' Rights provide important platforms of ensuring accountability of the member states of the AU to their international and regional human rights obligations, including the rights-based development objectives contained within the Sustainable Development Goals and the AU Agenda 2063.[51]

Although environmental legislation is continuing to develop on the African continent and the field of human rights is gaining more focus, there continue to be many instances of environmental degradation and associated violations of basic human rights caused by unsustainable

[49] Ben Boer, 'Environmental law and human rights in the Asia Pacific' In Ben Boer (ed) *Environmental law dimensions of human rights* (Oxford: Oxford University Press 2015) 135.

[50] Michael Addaney, Elsabe Boshoff, and Michael Gyan Nyarko, 'Protection of environmental assets in urban Africa: Regional and Sub-Regional Human Rights and Practical Environmental Protection Mechanisms', (2018) 24(2) *Australian Journal of Human Rights* 182–200.

[51] See *Michelot Yogogombaye v Senegal*, Application 001/2008 Judgment (jurisdiction), 15 December 2009 (2009) 1 AfCLR 1; *Soufiane Ababou v Algeria*, Application 002/2011 Decision (jurisdiction), 16 June 2011 (2011) 1 AfCLR 24; *Efoua Mbozo'o Samule v The Pan African Parliament*, Application 010/2011 Decision (jurisdiction), 30 September 2011 (2011) 1 AfCLR 95; *Evodius Rutechura v Tanzania*, Application 004/2016 Order (provisional measures), 18 March 2016 (2016) 1 AfCLR 596; *Deogratius Nicholaus Jeshi v Tanzania*, Application 017/2016 Order (provisional measures), 3 June 2016 (2016) 1 AfCLR 649.

development practices.[52] The African Commission, being the foremost human rights institution of the AU with supervisory jurisdiction to oversee the implementation of the African Charter within member states, plays an important role in ensuring the realization of the right to environment in Africa. Notably, article 24 of the African Charter, which guarantees to peoples the 'right to a general satisfactory environment favourable to their development' is quite vague as it does not elaborate on the scope and content of the appropriate measures that states must adopt to realize this right. Therefore, the African Commission has, over the years, tried to elaborate on the exact scope and content of the right to satisfactory environment. The African Commission's first attempt at elaborating on the scope and meaning of the right to satisfactory environment dates back to 1989 when it adopted its Guidelines for National Periodic Reports.[53] The Commission herein elaborated that states were required to 'establish a system to monitor effective disposal of waste in order to prevent pollution' and 'to prohibit and penalise disposal of waste on the African soil by any company'.[54] Consequently, states are required to indicate in their initial state reports the 'legislation and other measures' adopted in preventing the dumping of toxic waste from industrialized countries in Africa and 'scientific and efficient methods utilised for effective disposal of locally produced wastes'.[55] In subsequent periodic reports, states are also required to report on the 'continuation of development to curb wastes and removal of pollution on land, in water and in the air'.[56]

Apart from the African Commission's elaboration on the reporting requirements on the right to satisfactory environment, under article 24, the Commission also clarified that in terms of the right to health under article 16 of the African Charter, states must indicate in their initial report,

[52] Paul Collier, 'Africa: New opportunities, old impediments' (2015) 35(2) Economic Affairs 169–77, at 413.

[53] African Commission on Human and Peoples' Rights, Guidelines for National Periodic Reports (1989).

http://www.achpr.org/files/instruments/guidelines_national_periodic_reports/achpr_guide_periodic_reporting_1989_eng.pdf

[54] Ibid, at para 11.

[55] Ibid, at para 12.

[56] Ibid, at para 13.

'measures taken to protect and improve all aspects of environmental and industrial hygiene, to prevent air, land and water pollution, to overcome the adverse effects of economic development and industrialization'.[57] Additionally, in terms of the right to enjoy the benefits of scientific progress, which the Commission considers to be an integral part of the right to take part in cultural life, states are required to report on the 'measures taken to ensure the application of scientific progress for the benefit of everyone, including measures to promote a healthy and pure environment'.[58] The most notable intervention of the African Commission with regards to environmental rights is its decision in the *Social and Economic Rights Action Centre and Another v Nigeria* (*SERAC* case),[59] which was decided in 2001. The facts emanate from the exploration of oil in the lands of the Ogoni people by a consortium between the Nigerian National Petroleum Company (NNPC) and Shell Petroleum Development Corporation (SPDC). The applicants alleged that the operations of the consortium had led to environmental degradation and health problems as a result of contamination of the lands of the Ogoni people. These included the disposal of oil waste into the environment and local waterways and numerous oil spillages near villages resulting from the failure of the consortium to maintain facilities. These had resulted in the pollution of land, air and water bodies of the Ogoni people, leading to infections and other health problems. The applicants also alleged that the government of Nigeria had failed to monitor the operations of the oil companies and had not required them to adhere to relevant safety standards. Additionally, the government had refused to give the Ogoni communities the relevant information on the dangers created by the activities of the oil companies and did not consult them on the impacts of these activities.

The African Commission held that the right to satisfactory environment 'requires the state to take reasonable [legislative] and other measures to prevent pollution and ecological degradation, to promote conservation, and to secure an ecologically sustainable development and use of natural resources'.[60] This right and the right to health require states

[57] Ibid, at para 35(c).

[58] Ibid, at para 15(b).

[59] *Social and Economic Rights Action Centre (SERAC) & Another v Nigeria (2001) AHRLR 60 (ACHPR 2001). Gbemre v. Shell*, Suit No FHC/B/CS/153/05.

[60] Ibid, para 52.

to refrain from any acts that 'directly threaten the health and environment of their citizens' or tolerating any act that has the same or similar effect.[61] This according to the Commission is the state's obligation to respect the right to satisfactory environment.[62] In addition to this, the state has the obligation to order or at least permit 'independent scientific monitoring of threatened environments, requiring and publicizing environmental and social impact studies prior to any major industrial development, undertaking appropriate monitoring and providing information to those communities exposed to hazardous materials and activities and providing meaningful opportunities for individuals to be heard and to participate in the development decisions affecting their communities'.[63] This can be likened to the positive obligation to protect and fulfill the right to satisfactory environment.

On these terms the African Commission concluded that even though the government of Nigeria has the right to extract oil for the socioeconomic benefits of Nigerians, this must be done in accordance with the relevant safeguards elaborated above to ensure that the rights of host communities are protected.[64] Therefore, the Commission found the Nigerian government in violation of the right to satisfactory environment of the Ogoni people and among others requested that environmental and social impact assessment should be conducted prior to the commencement of future oil explorations. The Nigerian government was also requested to provide information on health and environmental risks to affected communities and allow meaningful participation of the affected communities in decision-making.[65] Most importantly, the court referred to the International Court of Justice in conveying what the term 'environment' entails and pointed out that it must be considered as 'an indivisible whole, comprising the biotic and abiotic natural resources, notably air, water, land, fauna and flora and the interaction between these same factors'.[66] It was the conclusion of the court that there had been a

[61] Ibid.

[62] Ibid.

[63] Ibid.

[64] Ibid, para 54.

[65] Ibid, para 71.

[66] Ibid, para 100.

failure on the part of the Federal Republic of Nigeria to adopt any of the 'other' measures required by article 1 of African Charter to ensure the enjoyment of the right laid down in article 24 of the same instrument.[67] Van der Linde and Louw contend that the obligations elaborated by the African Commission contain both procedural and substantive aspects.[68] Procedural aspects include the 'right to access environmental information or information relating to a possible adverse impact on the natural environment' and the opportunity to seek redress for the violation of one's environmental rights. Substantive aspects relate to the obligation on the governments to prevent pollution and ecological degradation and to 'promote conservation and sustainable development'.[69] These obligations, they further contend, reflect the values of international environmental law such as the 'preventive principle and duty of care principle'.[70] Some of these obligations have been reiterated in subsequent soft law instruments adopted by the African Commission.[71]

The African Commission throughout these various attempts has always recognized the interdependence and invisibility of the right to satisfactory environment with other rights, particularly the right to health. As a continent that relies heavily on the exploitation of natural resources, it is important that the African Commission recognizes that the right to satisfactory environment envisages measures aimed at countering the detrimental effects of extractive activities. Again, the African Commission recognizes that the right to satisfactory environment entails both negative and positive aspects.[72] The negative aspects require states to refrain from polluting the environment or condoning its pollution by third parties while the positive aspects require the state to manage and/or conserve the

[67] Ibid, para 107.

[68] Morne Van Der Linde and Lirette Louw, 'Considering the interpretation and implementation of article24 of the African Charter on Human and Peoples' Rights in light of the SERAC communication', (2003) 3(1) African Human Rights Law Journal 167–78, at 167.

[69] Ibid.

[70] See, generally, Michael Kidd, *Environmental law: A South African perspective* (Juta 2011).

[71] See, The State Party Reporting Guidelines for Economic, Social and Cultural Rights in the African Charter on Human and Peoples' Rights (Tunis Reporting Guidelines) (2012a); and The Principles and Guidelines on the Implementation of Economic, Social and Cultural Rights in the African Charter on Human and Peoples' Rights (2012b).

[72] Ibid.

environment in a manner that is conducive for development. States are also required to provide relevant information about environmental risks to affected communities and involve them in decision-making processes relating to the management of the environment.[73] Where any of these obligations are violated, affected persons or communities must be afforded the consequent procedural right of access to appropriate redress. The African Court on Human and Peoples' Rights has not yet dealt with the right to environment in direct terms but has heard matters which threaten rights to life in circumstances where they were closely linked to the right to a healthy environment, as was evidenced in the *Ogoni* case.

In addition to the continental human rights frameworks and institutions that operate under the auspices of the African Union (AU), various regional economic communities (RECs) have human rights frameworks and courts that enforce these frameworks. The most developed of these RECs are the Economic Community of West African States (ECOWAS), East African Community (EAC) and the Southern African Development Community (SADC). In particular the ECOWAS Community Court of Justice (ECCJ) and the East African Court of Justice (EACJ) have taken active roles in the protection of human rights.[74] The most important environmental human rights cases decided by these Courts are the *SERAP v Nigeria*[75] and *African Network for Animal Welfare v Tanzania*,[76] respectively. The SERAP case arose from the same factual background as the SERAC case involving the destructive effects of oil extraction in the Niger

[73] Ibid.

[74] Karen J. Alter, Laurence Helfer and Jacqueline R. McAllister, 'A new international human rights court for West Africa: The ECOWAS Community Court of Justice', (2013) 107(4) *American Journal of International Law* 737–79, at 737; Solomon T. Ebobrah, 'A rights-protection goldmine or a waiting volcanic eruption? Competence of, and access to, the human rights jurisdiction of the ECOWAS Community Court of Justice', (2007) 7(2) *African Human Rights Law Journal* 307–329; Ally Possi, 'Striking a balance between community norms and human rights: The continuing struggle of the East African Court of Justice', (2015) 15(1) *African Human Rights Law Journal* 192–213, at 192.

[75] Socio-Economic Rights & Accountability Project (SERAP) v Nigeria, Judgment Number ECW/CCJ/JUD/18/12, ECOWAS (2012).

[76] *African Network for Animal Welfare v The Attorney General of Tanzania,* Reference Number 9 of 2010 (2014).

Delta.[77] Just like the SERAC case, the applicant alleged the violation of several rights including the right to health and healthy environment.[78] The ECCJ held that the duty imposed by article 24 of the African Charter is 'both an obligation of conduct and an obligation of result', which requires states to maintain the quality of the environment to enhance sustainable development.[79] Not only does this require the adoption of legislative, administrative and other measures but also 'it must ensure that vigilance and diligence are being applied and observed towards attaining concrete results'.[80] Further, these concrete measures must aim at preventing the occurrence of damage or ensuring accountability for such damage and effectively repairing environmental damage that occurs.[81] The ECCJ held that even though the Nigerian government had passed legislation and set up institutions relating to environmental governance, the failure to effectively enforce the laws amounted to a failure to take action to prevent damage to the environment and that the failure to hold offenders of environmental rights to account was in violation of article 24 of the African Charter. This case thus confirmed the African Commission's decision in the SERAC case on the obligations of states to ensure that they refrain from polluting the environment or condoning its pollution, provide affected communities with relevant information and ensure their participation in decision-making, and ensure that violators of environmental human rights are held to account.

In the *African Network for Animal Welfare v Tanzania*[82] case, the applicant challenged the proposed construction of a main road that ran through the Serengeti National Park. It was ANAW's submission that the proposed road would have deleterious environmental and ecological effects and most likely to cause irreparable and irreversible damage to the

[77] Evadne Grant, 'International Courts and Environmental Human Rights: Re-Imagining Adjudicative paradigms', in Anna Grear, and Louis J. Kotze (eds) Research Handbook on Human Rights and the Environment (Edward Elgar Publishing 2015) 379.

[78] *SERAP v Nigeria 2012*, para 14.

[79] Ibid, at paras 100–101.

[80] Ibid, at para 101.

[81] Ibid, at para 105.

[82] *African Network for Animal Welfare v The Attorney General of Tanzania*, Reference Number 9 of 2010 (2014).

delicate ecosystem of the Serengeti and adjoining national parks such as the Masai Mara in Kenya.[83] These included the 'disruption in animal migration; driving and scaring away wildlife from the game controlled areas; fragmentation of animal habitats and weakening or disappearance of an entire generation of a given animal population; disruption of the wildlife corridor; loss of scenic and visual quality; increased and disruptive vehicular traffic; enhanced and disruptive human activity; increased wildlife mortality due to road kill from speeding vehicles; deterioration of air quality; surface water and soil pollution and increased poaching activities'.[84] Therefore, the applicant sought a declaration that the proposed road was unlawful and in violation of the provisions of the East African Community Treaty.[85] The court noted that both parties agreed that construction of the road would have negative impacts on the Serengeti environment and ecosystem. Subsequently, the court held that while there was nothing unlawful about the Government of the Republic of Tanzania within its territory, where is found wanting in its obligations in the conservation and protection of the environment then challenges arise.[86] Moreover, the court noted that it was convinced that if the road project was implemented as originally intended, the effects would have a devastating effect on both the Serengeti and neighboring Parks like the Masai Mara in Kenya.[87] As such, the court ordered the respondents to stop the future 'degradation without taking away the respondent's mandate towards economic development of its people'.[88] In essence, the court observed that while nothing prevented States to act in a sovereign manner in the development of their territories, they had to ensure that such actions did not violate any of their international or regional obligations where the conservation and protection of the environment was con-

[83] Ibid, para 11.

[84] Ibid.

[85] Article 114 (1) (a) of the Treaty enjoins all Partner States to conserve, protect and manage the environment and natural resources and articles 5 (3) (c), 8 (1) (c) and 111 (2) of the Treaty which obligate Partner States to cooperate in the management and utilization of natural resources within the Community and to abstain from any measures that would jeopardize the attainment of the objectives of the Treaty.

[86] *African Network for Animal Welfare v Tanzania*, para 69.

[87] Ibid, para 89.

[88] Ibid.

cerned. These cases are a clear demonstration of the obligations that the AU imposes on State parties to not only protect their citizens from harm associated with environmental pollution, but to also protect the environment from pollution and degradation. The SERAP case is a clarion call to African States that where a State displays lucklustre measures in preventing environmental pollution, it is legally expected of the State to take measures to remedy such pollution and prevent it from worsening. Moreover, these cases show the seriousness with which the AU considers cases which violate people's socio-economic rights through environmentally harmful practises.

4 Conclusion

This chapter recognizes the intimate connection between protecting human rights and securing the environment upon which humans depend not only for their physical existence but for their economic, social and cultural prosperity. It observed the progress made in the development of closer links between human rights and the environment in the AU. The African Charter was the first human rights instrument to expressly recognize the right to satisfactory environment as a human right. From the above discussion, it is clear that there is a role for human rights law in the protection of the environment in Africa, since the protection of ecological treasures are often correlated to the violation of human rights, not least the right to a satisfactory environment. Even though the protection of the environment is not an issue that the African regional human rights tribunals have ever been called upon to address directly, the instruments and case law discussed above from the African Commission, the ECOWAS Court and the EAC Court provide general principles on the right to satisfactory environment in Africa, which are applicable in all situations including in the context of extractive activities by multinational companies. Since all African states have ratified the African Charter, they are generally obliged to adhere to these standards. Nonetheless, the African Commission has provided further elaboration on this right through resolutions, guidelines and jurisprudence through individual complaints. These efforts have been supplemented by sub-regional courts such as the

ECOWAS Court. The jurisprudence from these tribunals has directly addressed rights violation relating to environmental degradation in the extractive sector.

References

Addaney, M., Boshoff, E., and Nyarko, M.G., 'Protection of Environmental Assets in Urban Africa: Regional and Sub-Regional Human Rights and Practical Environmental Protection Mechanisms', (2018) 24(2) *Australian Journal of Human Rights* 182–200.

Addaney, M., Dube, H., and Getaneh, S., 'Light through the Storm: Safeguarding the Human Right to Water in Challenging Landscapes in Africa', (2018) 5(1) *Journal of Comparative Law in Africa* 37–72.

Alam, S., Atapattu, S., Gonzalez, C.G., and Razzaque, J. (eds) International Environmental Law and the Global South (Cambridge University Press 2015).

Alter, K.J., Helfer, L. and McAllister, J.R. 'A New International Human Rights Court for West Africa: The ECOWAS Community Court of Justice', (2013) 107(4) *American Journal of International Law* 737–79.

Boer, B. 'Environmental Law and Human Rights in the Asia Pacific' in Boer, B. (ed) *Environmental Law Dimensions of Human Rights* (Oxford University Press 2015).

Boyle, A.E., 'Human Rights and the Environment Where Next?' in Boer, B. (ed) *Environmental Law Dimensions of Human Rights* (Oxford University Press 2015).

Brown, C., and Heller, L., 'Development Cooperation in Water and Sanitation: Is It based on the Human Rights Framework?' (2017) 22(7) Ciência and Saúde Coletiva 2247–56.

Carmona, M.S., 'The Obligations of International Assistance and Cooperation under the International Covenant on Economic, Social and Cultural Rights: A Possible Entry Point to a Human Rights-based Approach to Millennium Development Goal 8', (2009) 13(1) *International Journal of Human* Rights 86–109.

Černič, J.L. 'European Perspectives on the Business and Human Rights Treaty Initiative' in Cernic, J.L, and Carrillo-Santarelli, N. (eds) *The Future of Business and Human Rights: Theoretical and Practical Considerations for a UN Treaty* (Intersentia 2018).

Collier, P. 'Africa: New Opportunities, Old Impediments' (2015) 35(2) Economic Affairs 169–77.

Cornwall, A., and Nyamu-Musembi, C., 'Putting the "Rights-based Approach" to Development into Perspective' (2004) 25 *Third World Quarterly* 1417–1418.

Ebobrah, S.T. 'A Rights-protection Goldmine or a Waiting Volcanic Eruption? Competence of, and Access to, the Human Rights Jurisdiction of the ECOWAS Community Court of Justice', (2007) 7(2) *African Human Rights Law Journal* 307–329.

Francioni, F., 'International Human Rights in an Environmental Horizon', (2010) 21(1) *European Journal of International* Law 41–55.

Freeman, B, Pica, M.B. and Camponovo, C.N., 'A New Approach to Corporate Responsibility: The Voluntary Principles on Security and Human Rights' (2001) 24 *Hastings International and Comparative Law Review* 423, 426.

Gómez, M.A., 'The Global Chase: Seeking Recognition and Enforcement of the Lago Agrio Judgment Outside of Ecuador' (2013) 1 Stanford Journal of Complex Litigation 429.

Goodman, R., and Jinks, D., 'How to Influence States: Socialization and International Human Rights Law' (2004) 54 *Duke Law Journal* 623, 634.

Grant, E. 'International Courts and Environmental Human Rights: Re-Imagining Adjudicative paradigms', in Grear, A. and Kotze L.J. (eds) Research Handbook on Human Rights and the Environment (Edward Elgar Publishing 2015).

Hans-Otto, S., 'Development and Human Rights: The Necessary, but Partial Integration of Human Rights and Development', (2000) 22(3) Human Rights Quarterly 734–52.

Hey, E., *Advanced Introduction to International Environmental Law* (Edward Elgar Publishing 2016).

Isunju, J.B., Orach, C.O., and Jaco Kemp 'Hazards and Vulnerabilities among Informal Wetland Communities in Kampala, Uganda', (2016) 28(1) Environment and Urbanization 275–93.

Jagers, N.M. and van der Heijden, M-J, 'Corporate Human Rights Violations: The Feasibility of Civil Recourse in the Netherlands' (2008) 33 *Brooklyn Journal of International Law* 833 833–834.

Kate, Miles, The Origins of International Investment Law: Empire, Environment and the Safeguarding of Capital (Cambridge University Press, 2013).

Kidd, M. *Environmental law: A South African perspective* (Juta 2011).

Marcinkutė, L., 'The Role of Human Rights NGO's: Human Rights Defenders or State Sovereignty Destroyers?' (2011) 4 *Baltic Journal of Law and Politics* 52–74.

Okoloise, C., 'Contextualising the Corporate Human Rights Responsibility in Africa: A Social Expectation or Legal Obligation?' (2017) 1 African Human Rights Yearbook 191–220.

Ponting, C., A Green History of the World: The Environment and the Collapse of Great Civilizations (Penguin Books, 1991).

Possi, A. 'Striking a balance between Community Norms and Human Rights: The continuing struggle of the East African Court of Justice', (2015) 15(1) *African Human Rights Law Journal* 192–213.

Sand, P.H., *The History and Origin of International Environmental Law* (Edward Elgar Publishing 2015).

Simons, P., 'International Law's Invisible Hand and the Future of Corporate Accountability for Violations of Human Rights' (2012) 3 Journal of Human Rights and the Environment 5.

Stephens, B., 'The Amorality of Profit: Transnational Corporations and Human Rights' (2002) 20 Berkeley Journal of International Law 45.

UN Transforming our world: The 2030 Agenda for Sustainable Development. UNGA, October 2015.

Uvin, P., 'From the Right to Development to the Rights-based Approach: How 'Human Rights' entered Development', (2007) 17(4/5) Development in Practice 597–606.

Van Der Linde, M and Louw, L. 'Considering the Interpretation and Implementation of Article 24 of the African Charter on Human and Peoples' Rights in light of the SERAC Communication', (2003) 3(1) African Human Rights Law Journal 167–78.

Viljoen, F., *International Human Rights Law in Africa* (Oxford University Press 2012).

Voigt, C. and Grant, E., 'The Legitimacy of Human Rights Courts in Environmental Disputes', (2015) 6(2) Journal of Human Rights and the Environment 131–38.

Weiss, E.B., 'The Evolution of International Environmental Law' (2011) 54 Japanese Yearbook on International Law 1–27.

Young, M.Y., World Hunger (Routledge, 1997).

Zeleza, P.T., 'Rethinking Rights in Africa: The Struggle for Meaning and the Meaning of the Struggle' (2007) 41 (3) Canadian Journal of African Studies 474–506.

9

International Environmental Governance: A Case for Sub-regional Judiciaries in Africa

Eghosa Osa Ekhator

1 Introduction

There has been a rise in the use of international, regional and sub-regional judiciaries in promoting and upholding of human rights. Recently, many of these international courts, especially regional and sub-regional judiciaries or courts (including quasi-judicial organs), have been at the forefront of promoting environmental governance around the world especially in Africa. Examples include the African Commission on Human and Peoples' Rights, African Court of Human and Peoples' Rights and the different sub-regional judiciaries such as the ECOWAS Court of Justice (ECCJ) and East African Court of Justice (EACJ). The ECCJ has been utilized by non-governmental organizations (NGOs) to seek redress for victims of environmental injustices in Nigeria.[1] For example,

[1] SERAP v Federal Government of Nigeria Judgment No. ECW/CCJ/JUD/18/12. Also see Eghosa Ekhator, 'Improving Access to Environmental Justice under the African Charter on Human and People's Rights: The Roles of NGOs in Nigeria' (2014) 22 (1) AJICL 63.

E. O. Ekhator (✉)
University of Derby, Derby, UK

M. Addaney, A. Oluborode Jegede (eds.), *Human Rights and the Environment under African Union Law*, https://doi.org/10.1007/978-3-030-46523-0_9

Socio-Economic Rights and Accountability Project (SERAP), a Nigerian-based NGO, has been at the forefront of filing cases on socio-economic rights (including right to environment and education among others) at the ECCJ.[2] Furthermore, the East African Court of Justice stopped the Tanzanian government from constructing a road across the Serengeti National Park because of its potential adverse environmental impacts.[3] Thus, Professor Gathii has contended that the recent decisions of the sub-regional courts (especially ECCJ and EACJ) have led to an 'expansion towards international judicial environmentalism'[4] in Africa. Here, the sub-regional courts (ECCJ and EACJ) in *SERAP v. Federal Republic of Nigeria*[5] and *the African Network for Animal Welfare (ANAW) v. The Attorney General of the United Republic of Tanzania*[6] have added to the emergent international environmental governance by extending the remit of sub-regional judiciaries to include environmental protection or right to environment-related issues.

In West Africa, regional integration initiatives or measures predated the post-independence era or period.[7] These regional initiatives in the colonial period were influenced by the need for colonial powers to engage in trade or exchange between the colonies and Western countries in Europe.[8] Some of the notable colonial regional measures in the West African sub-region included the West African Currency Board (WACB) and the West African Airways Corporation (WAAC) in English-speaking colonies (Nigeria, Ghana, Sierra Leone and the Gambia).[9] However, many of these regional measures (especially in the English-speaking countries) collapsed in the post-independence era. Hence in 1970s, West

---

[2] Generally, see Ekhator (n 1).

[3] African Network for Animal Welfare (ANAW) v. The Attorney General of the United Republic of Tanzania, Ref. No. 9 of 2010, Judgment, East African Court of Justice at Arusha First Instance Div. 64 (Jun. 20, 2014). Cited in James Gathii, 'Saving the Serengeti: Africa's new international judicial environmentalism' (2016) 16 (2) CJIL 386.

[4] Gathii ibid.

[5] SERAP v Federal Government of Nigeria (n 1).

[6] *ANAW* (n 3).

[7] Mike Obadan, 'Introduction' in Ladi Hamalai and Mike Obadan (eds) 40 Years of ECOWAS (1975–2015) (NILS 2015) 26.

[8] ibid.

[9] Obadan (n 7).

African leaders came together to develop new initiatives in the sub-region. The setting up of the Economic Community of West African States (ECOWAS) in May 1975 is a culmination of these efforts by the leaders.[10]

ECOWAS is a regional group of 15 states founded in 1975 and its mission is the attainment of regional and economic integration of the member states.[11] One of the key institutions of the ECOWAS is the Community Court of Justice (ECCJ). A major aim of the ECCJ is the promotion and protection of human rights and peoples' rights in accordance with the tenets of the African Charter on Human and Peoples' Rights.[12] Similarly, formal regional integration in East Africa can be traced to 1967 with the founding of the EAC by Kenya, Tanzania and Uganda. By 1977, the EAC was dissolved following disagreements among its member states. Efforts to revive the EAC began in 1991 and culminated in the development of a new EAC Treaty in 1999.[13] The East African Court of Justice is 'the judicial arm of the East African Community and is vested with the primary mandate of interpreting and applying the Treaty for the Establishment of the East African Community.'[14]

Due to the non-justiciability of the right to environment doctrine and lack of adequate access to environmental justice in many African countries, NGOs, activists, communities and individuals[15] now utilize the sub-regional judiciaries in accessing justice in human rights issues or litigation. For example, the ECCJ has been utilized by NGOs to seek redress

[10] Obadan (n 7).

[11] ECOWAS Website https://www.ecowas.int/about-ecowas/basic-information/ accessed 12 November 2019.

[12] Revised ECOWAS Treaty, Article 4(g).

[13] Solomon Ebobrah, 'Sub-Regional Judicial Enforcement of Economic, Social and Cultural Rights' in Danwood Chirwa and Lilian Chenwi (eds), *The Protection of Economic, Social and Cultural Rights in Africa: International, Regional and National Perspectives* (CUP 2016) 281.

[14] Victor Lando, 'The domestic impact of the decisions of the East African Court of Justice' (2018) 18 (2) *African Human Rights Law Journal* 463, 463.

[15] The Supplementary Protocol (2005) amended 1991 Protocol of the Court to give individuals a direct right of access to the ECCJ in human rights issues or litigation. However, see Solomon Ebobrah, 'The Uneven Impact of International Human Rights Law in Africa's Subregional Courts' in Martin Scheinin, (ed). *Human Rights Norms in 'Other' International Courts* (Cambridge University Press 2019) 307–308. Also see Osaghae and others v Republic of Nigeria ECW/CCJ/JUD/O3/17.

for victims of environmental injustices in Nigeria.[16] The main question this chapter seeks to address is whether the rise of environmental governance or litigation in sub-regional judiciaries will lead to better environmental protection for the victims and communities.

Due to the different institutions, norms and frameworks developed under the African Union (AU), some scholars have argued that the concept of 'African Union Law' is emerging or evolving.[17] AU law has been defined 'as the bodies of treaties, resolutions and decisions that have direct and indirect application to the member States of the Union.'[18] Human rights issues are at the center of the development of AU bodies and the emergent African Union Legal order (AU Law).[19] The African Charter on Human and Peoples Rights[20] is at the center of the human rights architecture or regime in Africa. The African Charter establishes a system or framework for the promotion and protection of human rights in Africa within the framework of the Organisation of African Unity (now AU).[21] The African Charter promotes a plethora of human rights such as civil and political, socio-economic and cultural, individual and collective rights.[22] It is the first regional mechanism in the world to incorporate the different generations of human rights in a single document.[23] The African Charter makes no distinctions between the different types of rights in its provisions (e.g., socio-economic rights, civil and political rights, and group rights). Hence, the various categories of rights enshrined in the African Charter are all enforceable and justiciable under the African

[16] See Ekhator (n 1).

[17] Olufemi Amao, African *Union Law: The Emergence of a Sui Generis Legal Order* (Routledge 2018).

[18] Amao, ibid. 22.

[19] Amao (n 17).

[20] African Charter on Human and Peoples' Rights, adopted 27 June 1981, entered into force on 21 October 1986, OAU Doc. CAB/LEG/67/3 Rev. 5. However, there are other relevant AU mechanisms on human rights. See, Amao (n 18) 100–101.

[21] African Charter of Human and Peoples' Rights OAU CAB/LEG/67/3 rev. 5, 21 I.L.M. 58 entered into force October 21, 1986.

[22] Manisuli Ssenyonjo (ed), The African Regional Human Rights System: 30 Years after the African Charter on Human and Peoples' Rights (Martinus Nijhoff Publishers 2012).

[23] ibid.

human rights system including the sub-regional judiciaries.[24] The African Charter is expressly mentioned in the treaties of the EAC and ECOWAS.[25]

This chapter is divided into five parts including this introduction. The second part discusses the roles of sub-regional judiciaries in the emerging environmental governance architecture in Africa. This chapter focuses on the ECCJ and EAC because they are among the most active sub-regional judiciaries in Africa.[26] The third part focuses on the ECCJ. Here, some of the relevant ECCJ cases on the environment will be discussed. This part of the chapter considers the barriers militating against the successful implementation of ECCJ judgments in the sub-region. Part four of the chapter focuses on the EACJ and its relevance to the environmental protection discourse in the East African sub-region. The fifth part of the chapter which is the conclusion focuses on the limitations of the sub-regional judiciaries. Also, in this section, some suggestions or recommendations are discussed.

2 Regional Economic Communities (RECs) in Africa

Currently, there are over 14 economic groupings in Africa that qualify as regional economic communities. However, the African Union (AU) only recognizes eight of these organizations as the building blocks of the AU and the African Economic Community (AEC).[27]

These regional economic communities (RECs) were originally set up to serve as vehicles of regional integration on the continent. Also, many of these RECs as a means or vehicle of fostering regional and trade

[24] However, Ebobrah (n 13) 289 aver that socio-economic rights are recognized indirectly in the treaties.

[25] Ebobrah (n 13) 286. For example, Article 6(d) of the EAC Treaty and Article 4(g)(h) of the ECOWAS Treaty refers specifically to the African Charter.

[26] See Gathii (n 3). However, the SADC Tribunal is currently suspended and many of its decisions were never enforced or implemented.

[27] African Union website https://au.int/en/organs/recs contains names of the eight AU-recognized RECs. Lucyline Murungi and Jacqui Gallinetti, 'The role of sub-regional courts in the African human rights system' 13 *SUR-Int'l J. on Hum Rts.* 119, (observing that RECs are also referred to as sub-regional economic communities).

relations established judicial organs or courts within their existing frameworks or treaties.[28] These judicial organs were initially set up to settle inter-state differences and to interpret treaties and other legal mechanisms within their respective frameworks.[29] A major impact of the adoption of the continent-wide treaty for the creation of an AEC is the emergence of a new era in regionalism in Africa.[30] And this coincided with what is termed the 'new regionalism,' which led to the expansion of the human rights mandate of many sub-regional judiciaries in Africa.[31] The EACJ and ECCJ have been at the forefront in the expansion of human rights mandate under their various treaties and sub-regional orders. Hence 'both the EACJ and the ECOWAS Court are now recognised as critical players in the African human rights system (AHRS).'[32]

Roles of Sub-regional Judiciaries in Environmental Governance Architecture in Africa

In the global order on international environmental law, there are no explicit international environmental courts exercising jurisdiction over the thousands of multilateral treaties dealing with environmental protection.[33] These treaties tend to create non-compliance or review measures encompassing quasi-judicial features via facilitative approaches thereby enhancing capacity building, transparency, financial and technological assistance instead of sanctions.[34] Some examples include the Compliance

[28] Solomon Ebobrah, 'Courts of Regional Economic Communities in Africa and Human Rights Law' in Stefan Kadelbach, Stefan Rensmann and Thilo Rieter (eds), *Judging International Human Rights* (Springer 2019) 223.

[29] Ebobrah, ibid. 224, (arguing that new regionalism 'takes regional integration beyond mere trade liberalisation').

[30] Ebobrah ibid.

[31] Ebobrah (n 28); Daniel Abebe, 'Does International Human Rights Law in African Courts make a difference' (2016) 56(3) *Virginia Journal of International Law* 527.

[32] Ebobrah (n 28) 225.

[33] Christina Voigt and Evadne Grant, 'The Legitimacy of Human Rights Courts in Environmental Disputes' (2015) 6 (2) *Journal of Human Rights and the Environment* 131.

[34] Michael Addaney, Elsabé Boshoff and Michael Gyan Nyarko, 'Protection of environmental assets in urban Africa: Regional and Sub-Regional Human Rights and Practical Environmental Protection Mechanisms' (2018) 24 (2) *Australian Journal of Human Rights* 182, 185.

Committee under the Aarhus Convention on Access to Justice and the Kyoto Protocol.[35] Despite the absence of an explicit environment court on the international plane, a large body of jurisprudence emanating or touching upon environmental issues has arisen from issue-specific judicial bodies not limited to human rights courts and treaties.[36] Hence, Addaney et al. posit that these international bodies or mechanisms 'present opportunities for pursuing environmental claims at the international and regional levels that otherwise do not exist.'[37]

Furthermore, there has been a rise in the use of international, regional and sub-regional judiciaries in the promotion and upholding of human rights. Some of these international courts especially regional and sub-regional judiciaries or courts (including quasi-judicial organs) have been at the forefront of promoting environmental governance around the world especially in Africa. Examples include the African Commission on Human and Peoples' Rights (African Commission),[38] African Court of Human and Peoples' Rights (African Court) and the different sub-regional judiciaries such as ECCJ and EAC. It is arguable that reasons for the burgeoning environmental governance under sub-regional judiciaries in Africa are the lack of access to environmental justice in some parts of the continent and delays in the judicial process in many countries.[39]

3 The ECOWAS Court of Justice

CSOs, individuals and communities have relied on the ECCJ to seek redress for victims of environmental injustice in Nigeria and other parts of the sub-region. One of the key institutions of the ECOWAS is the ECCJ. A major aim of the ECCJ is the promotion and protection of human and peoples' rights in reliance on the tenets of the African Charter on Human and Peoples' Rights.[40] By virtue of Article 76(2) of the Revised

[35] Voight and Grant (n 33).

[36] Voight and Grant (n 33) 131.

[37] Addaney et al. (n 34) 186.

[38] However, it should be noted that the African Commission is a part-time quasi-judicial body.

[39] Ekhator (n 1). Also see Gathii (n 3).

[40] Article 4(g) of the Revised ECOWAS Treaty.

Treaty of the Economic Community of West African States[41] (Revised ECOWAS Treaty), the decisions of the ECCJ are final and not subject to appeal. These decisions or judgments are binding on Member States, the Institutions of the Community and on individuals and corporate bodies who are subject to the jurisdiction of the ECCJ.[42] Parties that can institute actions at the ECCJ include Member States and the Authority of Heads of State or Government.[43] By virtue of Article 4(g) of the Revised ECOWAS Treaty and 9(4) of the 2005 Supplementary Protocol of the ECOWAS Court of Justice[44] the ECCJ can entertain claims by individuals and corporate bodies for relief for violation of their human rights. The ECCJ adjudicates cases filed by individuals and NGOs alleging violations of human rights. ECOWAS judges have an express mandate to hear such cases—under the Supplementary Protocol which grants 'jurisdiction to determine cases of violation of human rights that occur in any Member State' in response to complaints by private litigants.[45]

Article 4(g) of the Revised ECOWAS Treaty of 1993 enjoins Member States of the regional group to adhere to the 'recognition, promotion and protection of human rights in accordance with the provisions of the African Charter on Human and Peoples' Rights.' This obligation is also reflected in the preamble to the Revised Treaty as well as in Article 56(2) where Member States commit themselves to cooperate for the realization of the mandates or aims of the African Charter.[46] Notwithstanding that Nigeria is a signatory to the 1993 ECOWAS Revised Treaty, the Treaty is yet to be domesticated or incorporated into its national laws.[47] Nigeria,

---

[41] Signed on 24 July 1993 and entered into force on 23 August 1995, (1996) 25 I.L.M. 663.

[42] Article 15(4) of the Revised ECOWAS Treaty.

[43] Article 76(2) of the Revised ECOWAS Treaty. See also ECOWAS Community Court of Justice, Protocol A/P.1/7/91, (Adopted July 6, 1991, Came into Force Nov. 5, 1996, Amended by Supplementary Protocol A/SP.1/01/05 in 2005), at Art. 9(1).

[44] Supplementary Protocol A/SP.1/01/05.

[45] Ebobrah (n 28) 242.

[46] Solomon Ebobrah, 'Human Rights Realisation in the Africa Sub-Regional Institutions' in Manisuli Ssenyonjo (ed), *The African Regional Human Rights System: 30 Years after the African Charter on Human and Peoples' Rights* (Martinus Nijhoff Publishers, 2012) 287.

[47] This is replicated in countries that are dualist in nature in the sub-region. See Muyiwa Adigun 'Enforcing ECOWAS judgments in Nigeria through the common law rule on the enforcement of foreign judgments' (2019) 15 (1) *Journal of Private International Law* 130, 133.

Ghana, Sierra Leone, Liberia and Gambia are common law countries in the sub-region that operate dualist system wherein treaties are not applied domestically unless incorporated via the machinery of legislation.[48]

There have been a few cases on environmental protection in the ECCJ and this chapter focuses on two of these cases.[49] The cases that this chapter discusses are *SERAP v Federal Government of Nigeria*[50] and *Osaghae and others v. Republic of Nigeria*.[51] In *SERAP v. Federal Republic of Nigeria*, SERAP, (a Nigerian-based NGO) instituted an action against the Nigerian government. SERAP contended that the Niger Delta region despite its natural endowments has borne the negative consequences arising from the activities of companies operating in that region. Some of these negative externalities include oil spills and environmental degradation among others. Also, SERAP alleged violations by the government on the right to health, economic and social development, adequate standard of living and the inability of the Nigerian government to adequately enforce the environmental law and regulations in the oil and gas industry among others. The plaintiffs maintained that Federal Government of Nigeria has been culpable for environmental degradation in the Niger Delta. The core of the reliefs sought by the applicant (SERAP) was a declaration that the Nigerian government had violated the tenets of the African Charter (and other relevant international standards) and that the Niger Delta communities should have a right to a clean or general satisfactory environment.[52] For example, in regard to right to health, SERAP relied on Articles 16 and 24 of the African Charter and Article 12(1) of the International Covenant on Economic, Social and Cultural Rights (ICESCR) to argue that the Nigerian government has failed to promote environment or conditions to live a healthy life due to its failure to stop the widespread pollution arising from the oil industry which has impacted

[48] See Amos Enabulele, 'Reflections on the ECOWAS Community Court Protocol and the Constitutions of Member States' (2010) 12 (1) INT'L COMM L REV 111, 121.

[49] Also see, Marie Molmon & 114 Ors. v Guinea ECW/CCJ/JUD/16/16, May 17, 2016 http://prod.courtecowas.org/wp-content/uploads/2019/01/ECW_CCJ_JUD_16_16.pdf accessed 12 November 2019.

[50] *SERAP* case (n 1).

[51] Osaghae and others v Republic of Nigeria (n 15).

[52] *SERAP* (n 1) para. 63.

negatively on the quality of life of the affected communities in the Niger Delta.[53]

The ECCJ held that the Nigeria violated Articles 1 and 24 of the African Charter and ordered that the Nigerian government to take effective measures within the shortest possible period to restore or remediate the environment of the Niger Delta.[54] The ECCJ further held that the Nigerian government must take steps to prevent the occurrence of damage to the environment in the Niger Delta and take measures to hold the architects of environmental damage responsible for their actions.[55] The ECCJ posited that the Nigerian government is expected to comply and enforce this decision by virtue of Article 15 of the Revised Treaty and Article 24 of the ECCJ Supplementary Protocol.[56]The ECCJ further held that the Nigerian government must take steps to prevent the occurrence of damage to the environment in the Niger Delta and take measures to hold the architects of environmental damage responsible for their actions. Unfortunately, till date, the government of Nigeria is yet to enforce or implement this decision.[57]

In *Osaghae v others v Republic of Nigeria*,[58] the case was instituted by four individuals from the Niger Delta in Nigeria. The first plaintiff who is from Edo State in Nigeria avers that he has suffered marginalization from the Nigerian government and its agents.[59] The second plaintiff who is from Delta State in Nigeria avers that he has been a victim of environmental injustice arising from the activities of oil companies and the third plaintiff who is from Edo State claims that the communal fishing water in the Niger Delta region has been destroyed due to the activities of oil firms in the region.[60] Finally, the fourth plaintiff who is from Edo State avers that he is a victim of the improper takeover of its communal or community natural resources and environmental deprivation. The

---

[53] Ibid. para. 67.
[54] SERAP (n 1).
[55] *SERAP* (n 1).
[56] *SERAP* (n 1).
[57] Ekhator (n 1). The court also held that SERAP has the requisite *locus standi* to institute the case.
[58] Osaghae and others v Republic of Nigeria (n 15).
[59] Osaghae and others v Republic of Nigeria (n 15) para. 7.
[60] ibid.

plaintiffs aver that they are 'suing for themselves and on behalf of the Niger Delta people of Nigeria.'[61] The plaintiffs based their complaint or application on the following:

(i) A Declaration that the allocation of oil concessions or blocs to private companies and individuals in Nigeria is unlawful and violates the provisions of Articles 21, 22 and 24 of the African Charter; Article 1 (1–3) of the International Covenant on Civil and Political Rights (ICCPR); and Article 1 (1–3) and 11 of the International Covenant on Economic, Social and Cultural Rights (ICESCR)
(ii) A Declaration that the unrestrained oil exploration, gas flaring and attendant pollution and deaths in the Niger Delta arising from the activities in the oil sector constitutes threats to right to life, health and right to self-determination of the Niger Delta people as enshrined in Articles 1, 2, 4, 16 and 24 of the African Charter; Article 1 and 6 of the ICCPR; and Article 1 and 12 of the ICESCR.
(iii) To declare a moratorium on oil bloc transactions in respect of Article 1 and 21.2 of the African Charter.
(iv) To mandate the Government of Nigeria to redistribute the ownership of all onshore and offshore oil blocs in the Niger Delta back to the indigenous oil communities in lines with Articles 21 and 22 of the African Charter and Article 11 of the ICESCR.
(v) An order to direct the Nigerian Government to pay the sum of $30 billion to repair the environmental damage caused by over 9 million barrels of spilt crude oil in the Niger Delta. This complaint is based on Articles 1, 21 and 24 of the African Charter and Article 12 of the ICESCR.
(vi) An order compelling the Nigerian Government to conduct a self-determination referendum for the people of the Niger Delta. This is based on Articles 1.1, 1.2 and 1.3 of the ICCPR & ICESCR.

On the other hand, the Nigerian Government averred that the ECCJ had no jurisdiction in the case because the plaintiffs lacked the requisite *locus standi* to institute the case, lack of reasonable cause of action and the suit

[61] ibid.

was an abuse of the court process.[62] In terms of the substance of the case, the ECCJ stated that the plaintiffs have led evidence to corroborate their claims against the defendants.[63] However, the ECCJ held that

> it is important to distinguish the capacity upon which the parties act, i.e. as non-natural and natural persons. While in SERAP supra, the Plaintiff by virtue of its registration under the Laws of Nigeria is recognized to represent the People of Niger Delta without the need to produce any proof of authorization. The Plaintiffs in this case are natural persons claiming to appear on behalf of the People of Niger Delta without authorization. The proof of authorization in the case of natural persons acting on behalf of a group cannot be dispensed with. The Niger Delta is so vast that an action brought for and on behalf of the said people without authorization sounds questionable. The Plaintiffs have failed to attach a mandate if any, given to them to clear the air in this regard.[64]

This decision has been criticized by some scholars.[65] For example, Giacomini argues that this decision in *Osaghae* represents a missed opportunity by the ECCJ for the ventilation or enforcement of the rights of the Niger Delta people due to the various procedural errors highlighted in the judgment.[66] Arguably, this decision will have negative impacts on the burgeoning environmental governance architecture under the ECCJ. However, this case evidences the important role of ECCJ in environmental protection in the sub-region.

Enforcing ECCJ Judgments

A major conundrum inherent in the ECCJ is the enforceability of its decisions in the Member States. There are conflicting views on the enforceability of ECCJ's rulings; the views include that decisions are advisory or persuasive, and enforceability depends on the legal structure

[62] Osaghae and others v Republic of Nigeria (n 15) para. 9.3.

[63] Ibid. para. 9.2.

[64] Ibid. para. 9.3.

[65] See Ebobrah (n 15).

[66] Giada Giacomini, 'Niger Delta People v Nigeria: A Missed Occasion before the ECOWAS Court of Justice' Focus Africa (2018) < https://www.federalismi.it/nv14/articolo-documento.cfm?artid=35993 accessed 12 November 2019.

or the mode of domesticating international treaties in the different Member states,[67] directly enforceable[68] or not enforceable in the Member States.[69] Despite the barriers militating against the implementation or domestication of the ECCJ judgments in Member States, the recent decisions by the ECCJ have created opportunities for victims of environmental injustice or abuses to by-pass the justice machinery and attempt to get justice for victims of environmental abuses or injustice.[70]

However due to the constitutional impediments to the implementation of decisions of the ECCJ, diverse suggestions have been proffered by scholars.[71] Enabulele suggests that there should be synergy or integration among the ECOWAS countries in implementing the ECCJ decisions.[72] This is because some of the countries in the ECOWAS sub-region are dualist in nature and the problems 'associated with the differences between the ECOWAS treaty and the domestic law of Community States, and between the community court and the national courts.'[73] Furthermore, article 24 of the ECOWAS Supplementary Protocol states that decisions of the ECCJ should be executed via the judicial machinery of the states. All ECOWAS states are mandated to set up a national authority for the implementation of its judgments. Unfortunately, very few Member States have established such national authorities. Thus,

---

[67] Enabulele (n 48). Thus, recently in Ghana (which is also dualist state), the high court (in Ghana) refused to enforce a judgment of the ECCJ. *In the Matter of an Application to Enforce the Judgment of the Community Court of Justice of the ECOWAS against the Republic of Ghana and In the Matter of Chude Mba v. The Republic of Ghana*, Suit No. HRCM/376/15 (High Court, Ghana, 2016) (unreported), cited in Richard Oppong, 'The High Court of Ghana declines to Enforce an ECOWAS Court Judgment' (2017) 25 (1) *African Journal of International and Comparative Law* 127, 128.

[68] See Enyinna Nwauche, 'Enforcing ECOWAS Law in West African National Courts' (2011) 55(2) *Journal of African Law* 181. This is arguably the situation in monist countries in West Africa where international law applies directly.

[69] This is arguably the position in dualist countries wherein international treaties need to be domesticated into national laws by the relevant authorities for them to have effect. Generally, see Helen Chuma-Okoro, 'The Nigerian Constitution, the ECOWAS Treaty and the Judiciary: Interplay of roles in the Constitutionalisation of Free Trade' (2015) 4(1) *Global Journal of Comparative Law* 43.

[70] Ekhator (n 1).

[71] ibid.

[72] Enabulele (n 48).

[73] Enabulele (n 48) 113. The English-speaking countries in the sub-region such as Nigeria, Ghana, Sierra Leone, Liberia and Gambia are generally referred to as dualist countries in the ECOWAS sub-region. Generally, see Nwauche (68) for the constitutional provisions in their various constitutions.

enforcing ECCJ judgments in the ECOWAS states is a difficult task.[74] Notwithstanding that Nigeria has set up the machinery of enforcement, the impact of the ECCJ judgments have been insignificant in the country.[75] In 2011, Nigeria appointed the Attorney General of the Federation as the national authority responsible for implementing the ECCJ's decisions.[76] Also, only five Member States including Nigeria, Guinea, Mali, Burkina Faso and Ghana have appointed competent nationality authority for the implementation or enforcement of the ECCJ judgments in accordance with Article 24(4) of the Protocol as amended.[77] Despite the various statistics on the enforcement of ECCJ decisions, the consensus is that some of the decisions have not been enforced in the various countries.[78] In a recent academic paper by the Chief Registrar of the ECCJ, he posits that since the adoption of the Supplementary Protocol on the ECCJ, only 35 out of the 64 enforceable decisions by the ECCJ have been complied with.[79]

In Ghana, akin to the position in Nigeria, the ECOWAS Protocol and the treaty establishing the ECOWAS are not in force and the Parliament is yet to exercise its powers under Article 72(2) of the Constitution of the Republic of Ghana.[80] Recently in Ghana (which is also dualist state), the high court refused to enforce a judgment of the ECCJ.[81] Also, in *Republic v High Court (Commercial Division) Accra, Ex parte Attorney General,*

[74] Jadesola Lokulo-Sodipo and Abiodun Osuntogun, 'The Quest for a Supranational Entity in West Africa: Can the Economic Community of West African States Attain the Status?' (2013) 16(3) *PER/PELJ* 255, 259.

[75] Ekhator (n 1).

[76] Chuma-Okoro (n 69). However, Tony Anene-Maidoh, 'Enforcement of Judgments of ECOWAS Court of Justice' (2018) 1 (1) *Journal of Law Review National Institute for Legislative and Democratic Studies* 58, 59 suggests that 'it is noteworthy that some countries that are yet to appoint national authorities have complied with decisions and judgements of the Court.'

[77] Guardian Newspaper 'ECOWAS court records highest number of Judgment, ruling in 2018/2019' (12 July 2019) https://guardian.ng/news/ecowas-court-records-highest-number-of-judgment-ruling-in-2018-2019/ accessed 29 November 2019.

[78] Anene-Maidoh (n 76).

[79] Anene-Maidoh (n 76) 59.

[80] Oppong (n 67) 128.

[81] See *Mba* case (n 67).

NML Capital and the Republic of Argentina,[82] which involved an action to enforce the decision of the International Tribunal for the Law of the Sea (ITLOS). The Ghanaian Supreme Court held that Ghana is a dualist country and even if a treaty is ratified by the Parliament until it is incorporated or domesticated into municipal law by the appropriate legislation, it has no legal effect in the country.[83] However, in Africa, it is only the Member States of the East African Community that have incorporated the Treaty Establishing the East African Community into their national laws.[84]

Many of the cases brought to the ECCJ originate from Nigeria and till date, the precise number of ECCJ judgments implemented or enforced in Nigeria is unknown.[85] Also, despite the slow rate of enforcement of ECCJ judgments in the sub-region, it has been contended that no country has declined to comply at all, probably due to the fact that the 'ECOWAS legal regime makes provision for sanctions where a state fails to comply with a decision of the ECCJ.'[86] Notwithstanding the apparent lack of enforcement of many ECCJ judgments in Nigeria, it is probable that in the future, ECCJ judgments might be implemented in Nigeria. A major reason for this is that contracting countries to treaties cannot rely on the basis of its domestic laws as reasons or justification for not performing its expected obligations under such treaties.[87] In the ECOWAS sub-region, where Nigeria is the major economic and military power, thus, if ECOWAS is to be successful, Nigeria must be seen to respect and implement the various ECOWAS treaties it has ratified. Furthermore, by virtue of section 19(d) of the Constitution of Nigeria (as amended)

[82] Civil Motion No J5/10/2013 (Supreme Court, Ghana, 2013). Cited in Richard Oppong and Lisa Niro, 'Enforcing Judgments of International Courts in National Courts' (2014) 5(2) *Journal of International Dispute Settlement* 344.

[83] Oppong and Niro (n 82).

[84] Oppong (n 67) 130–131.

[85] This point is debatable because the Chief Registrar of the ECCJ has posited that Nigeria has implemented some ECCJ decisions. Generally, see Anene-Maidoh (n 76).

[86] Ebobrah (n 13) 299. Generally, see Article 77 of the Revised ECOWAS Treaty and Article 24 of the 2005 Supplementary Protocol.

[87] See Article 27 of the Vienna Convention on the Law of Treaties.

'respect for international law' is one of the foreign policy objectives of the Nigerian government enunciated in the constitution.[88] Thus, Nigeria should strive to observe and enforce international law in the country.

4 The East African Court of Justice

The EACJ is the judicial arm or organ of the East African Community (EAC) created under Article 9 of the Treaty of the Establishment of the East African Community.[89] The EACJ consists of a First Instance Division and an Appellate Division.[90] The jurisdiction of the EACJ as exemplified by Articles 23 and 27 of the 1991 EAC Treaty (as amended) to hear and determine disputes on the interpretation and application of the Treaty.[91] Notwithstanding the absence of an explicit human rights mandate, the EACJ has claimed jurisdiction over matters arising from violation of human rights.[92] Also, it has been suggested that 'the judges of the EACJ have been proactive in encouraging human rights cases to come before the Court.'[93]

A recent case on environmental protection in EACJ is *African Network for Animal Welfare (ANAW) v. Attorney General of Tanzania*.[94] ANAW, the applicant in this case is a registered NGO in Kenya and it averred that the actions of the Tanzanian government was in violation of Articles 114 (1) and Articles 5(3) (c), 8(1) (c) and 111 (2) of the Treaty for the

[88] This provision falls under Chapter II of the Nigerian constitution, making it non-justiciable and, in effect, unenforceable. Despite its non-enforceability, arguably this provision evidences the fact that Nigerian government should respect its international obligations.

[89] EAC website http://eacj.org/ accessed 12 November 2019.

[90] Ebobrah (n 13).

[91] Also see Ebobrah (n 28) 224.

[92] Article 27(2) of the EAC Treaty suspends the human rights jurisdiction of the EACJ until a Protocol is adopted to explicitly extend its jurisdiction to human right issues. In James Katabazi & 21 Others v the Secretary-General of the EAC & Another and numerous cases, the EACJ 'through a mix of judicial activism and creative interpretation has claimed for itself limited human rights jurisdiction.' See Lando (n 14) 467.

[93] James Gathii, 'Variation in the use of Subregional Integration Courts between Business and Human Rights Actors: The Case of the East African Court of Justice' (2016) 79(1) *Law & Contemp. Probs* 37.

[94] *ANAW* case (n 3).

Establishment of the EAC.[95] In this case, the applicants argued that Tanzania's government decision to build a road across Serengeti National Park constituted a violation of treaty provisions that require EAC Partner States to protect and preserve the quality of the environment. The court further held that it was a violation and accordingly, the EACJ issued an injunction restraining Tanzania from implementing that decision.[96] Thus, the EACJ stopped the government of Tanzania from building a 'road across Serengeti National Park because of its potential adverse environmental impacts.'[97] In this case, the First Instance Division of the EACJ in 2014 issued a permanent order stopping the Tanzanian government from building a road through the Serengeti National Park and this decision was mainly upheld by the Appellate Division of the EACJ in July 2015.[98]

Analogous to enforcement of ECCJ judgments, the EAC Treaty requires Partner States to take prompt measures to enforce the judgments of the EACJ.[99] Some scholars have contended that there has been a fair rate of enforcement of the EACJ decisions.[100] Ebobrah contends that this is due to the fact that judgments from the EACJ have not generally resulted in declarations or orders that have huge policy and financial impacts.[101] On other the hand, Possi argues that not much 'is known about the extent of compliance with the EACJ's decisions. One could correctly characterize the EACJ's decisions as being academic.'[102] Notwithstanding the various criticisms against the EACJ,[103] it has

[95] *ANAW* case (n 3) para. 13.

[96] Ebobrah (n 13) 294–295.

[97] Gathii (n 3) 386.

[98] ibid.

[99] Ebobrah (n 13).

[100] Generally, see Lando (n 14).

[101] Ebobrah (n 13).

[102] Ally Possi, 'An appraisal of the functioning and effectiveness of the East African Court of Justice' (2018) 21(1) *Potchefstroom Electronic Law Journal/Potchefstroomse Elektroniese Regsblad* 1, 20.

[103] However, the EACJ has remained proactive in its judgments despite the backlash against it by Partner States. See Tomasz Milej, 'Human Rights protection by international courts – What role for the East African Court of Justice?' (2018) 26(1) *African Journal of International and Comparative Law* 108.

impacted positively on the national judiciaries of the states in the resolution of disputes in national courts.[104]

On the part of the ECCJ in its exercise of its human rights mandate has consistently exercised jurisdiction in human rights cases and provided a sub-regional avenue for 'aggrieved Community citizens to seek redress for human rights violations.'[105] However, the implementation or enforcement of the ECCJ decisions has been a recurring issue notwithstanding that there is 'a provision in amended Protocol for national court of Member States to refer questions of interpretation of the Treaty, Protocol or other ECOWAS legal texts to the ECOWAS Court of Justice, no such referral has been received from any Member State.'[106] Thus, some of the key problems besetting the ECCJ include problems of enforcement, lack of referrals by member states, non-appointment of competent national authorities, and failure to domesticate the Revised Treaty and Protocol to the ECCJ.[107]

5 Conclusion

Despite the various Community frameworks on enforcement of judgments of sub-regional courts in Africa, sub-regional courts 'do not have the powers to enforce their judgments.'[108] Due to the difficulties in implementing the decisions of sub-regional judiciaries in Africa (especially the ECCJ), some academics[109] have suggested diverse range of strategies, such as the use of common law regime on the enforcement of foreign judgments.[110] Furthermore, some scholars have suggested that that one method of implementing ECCJ judgments in Nigeria is by

[104] Generally, see Lando (n 14).

[105] Muhammed Ladan, 'Appraisal of the Human Rights Mandate of the ECOWAS Court of Justice '(April 18, 2018). International Conference on the Protection of Human Rights as a Factor for Peace Building in West Africa, 2018. 2.

[106] Ladan ibid. 17.

[107] Ladan (n 105).

[108] Ebobrah (n 13) 298.

[109] Adigun (47) and Oppong (67) among other scholars suggest that the common law regime can successfully be adapted to enforce decisions of international courts or tribunals in African countries.

[110] Oppong (n 67); Adigun (n 47).

registering the ECCJ rulings as foreign judgments by virtue of the Foreign Judgements (Reciprocal Enforcement) Act[111] (this is one slant of the statutory regime for enforcing foreign judgments in Nigeria. However, the Foreign Judgements (Reciprocal Enforcement) Act is said to be inchoate and not in force in Nigeria).[112] Arguably, this is fraught with a lot of difficulties.[113] Some of these difficulties include the non-domestication of the Protocol to the ECCJ and ECOWAS Revised Treaty in Nigeria and whether ECCJ judgments can be considered to be 'foreign judgments' under the Foreign Judgments (Reciprocal Enforcement) Act in the country. This is also exemplified in Ghana in the *Mba* case, where the high court refused to enforce a judgment of the ECCJ.

South Africa has both a common law and statutory regime for enforcing foreign judgments.[114] The latter is regulated by the Enforcement of Foreign Civil Judgments Act 1988. Section 2(1) of the Act provides that this Act 'shall apply in respect of judgments given in any country outside the Republic which the Minister has for the purposes of this Act designated by notice in the Gazette.'[115] In *Government of the Republic of Zimbabwe v Louis Karel Fick*,[116] the Constitutional Court of South Africa (CCSA) became the first apex court to hold that the common law regime for enforcing judgments from the courts of a foreign state can also be used to enforce a decision of an international court or tribunal. This case involved the possible enforcement of a decision of the South African Development Community Tribunal (SADC).[117] The CCSA enforced the SADC Tribunal's judgment by developing the common law, treating the judgment as a foreign judgment and applying South African private international law rules. Hence, common law or dualist countries (via their judiciaries) in the ECOWAS sub-region should also develop their

[111] 1990 now Cap F35 LFN 2004. Generally, see Adigun (n 47), Anene-Maidoh (n 76).

[112] Adewale Olawoyin, 'Enforcement of Foreign Judgments in Nigeria: Statutory Dualism and Disharmony of Laws' (2014) 10(1) Journal *of Private International Law* 129.

[113] For a general overview of the foreign judgments' enforcement paradigm in Nigeria, see Olawoyin ibid.

[114] Generally, see Oppong and Niro (n 82). South Africa has a mixed legal system.

[115] ibid.

[116] [2013] ZACC 22.

[117] Oppong and Niro (n 82).

respective common law regime on enforcing foreign judgments to include the implementation or enforcement of ECCJ judgments in their respective jurisdictions. Thus, Adigun has persuasively argued that the 'common law on the enforcement of foreign judgments can be successfully adapted to give domestic effect to the judgments of ECOWAS Court as an international tribunal in Nigeria.'[118]

However, it has been suggested that the CCSA decision in *Fick* limited the evolution of common law in this regard solely to 'to the enforcement of judgments and orders of international courts or tribunals *based on international agreements that are binding on South Africa.*'[119] Notwithstanding the above assertion, common law evolves with time and arguably, judiciaries of dualist countries in the ECOWAS sub-region can extend the common law rule on enforcement of foreign judgments to ECOWAS judgments in their respective countries.[120] Despite the difficulties in relying on the common law regime of enforcement of foreign judgments, this chapter suggests that it can serve as one of the means of enforcing ECCJ judgments in the common law (dualist) countries in West Africa. Courts in South Africa which also has common law heritage have successfully relied on the common law regime of enforcement of foreign judgments to implement or enforce decisions arising from sub-regional judiciaries.[121]

Official data on the enforcement of ECCJ judgments is not readily available online. Therefore, compliance rates by countries are difficult to determine due to the lack of available and reliable data.[122] Abebe contends that despite that the EACJ has adjudicated on a plethora of human

---

[118] Adigun (n 47) 130.

[119] Government of the Republic of Zimbabwe v Louis Karel Fick para. 53. Also cited in Oppong and Niro (n 82)15.

[120] Christian Okeke, 'The use of International Law in the Domestic Courts of Ghana and Nigeria' (2015) 32 (3) *Ariz. J. Int'l & Comp. L.* 371, 383 states that common law in Ghana has evolved with time. Hence some scholars such as Nwauche advocates for the development of common law regimes unique to the specific countries in Africa (such as Ghana and Nigeria). Generally, see Enyinna Nwauche, 'The Constitutional Challenge of the Integration and Interaction of Customary and the Received English Common Law in Nigeria and Ghana' (2010) 25 (1) *Tul. Eur. & Civ. LF* 37.

[121] Generally, see Oppong and Niro (82).

[122] Abebe (n 31) 554.

rights cases, 'information is not readily available on whether the nations found in violation of human rights treaties have taken steps to comply with the EACJ's judgments.'[123] The ECCJ also suffers from official information deficit. Due to the fact, that no Member State in the ECOWAS has communicated to the ECCJ the status and compliance with of judgments and decisions so far, the ECCJ 'has been able to get unofficial information from lawyers and parties involved in some cases.'[124] To mitigate this information deficit, this chapter suggests that official online data-bases should be created by the sub-regional organizations (especially ECCJ) showing various cases and their implementation status in Member States. Also, this information should be readily and freely available online.

There is a chronic failure in the enforcement or implementation of the decisions or judgments of sub-regional judiciaries (especially ECCJ) in Africa.[125] One way of improving the implementation of ECCJ judgments is that the ECOWAS should apply political pressure on the dualist countries in the sub-region to domesticate the Revised Treaty and the Protocol on the ECCJ into their national laws. This will enhance the implementation of the ECCJ judgments in the sub-region. Arguably, the solution lies in the political will of the governments in the sub-regions—the governments need to be committed to the implementation of the decisions or judgments arising from the sub-regional judiciaries.

References

Abebe, D 'Does International Human Rights Law in African Courts make a difference?' (2016) 56(3) *Virginia Journal of International Law* 527.

Addaney M, Boshoff E and Nyarko M, 'Protection of Environmental Assets in Urban Africa: Regional and Sub-Regional Human Rights and Practical Environmental Protection Mechanisms' (2018) 24 (2) *Australian Journal of Human Rights* 182.

[123] ibid.

[124] Anene-Maidoh (n 76) 59.

[125] Generally, see Paul Kagame, 'The Imperative to Strengthen Our Union' *Report on the Proposed Recommendations for the Institutional Reform of the African Union* (2017).

Adigun M 'Enforcing ECOWAS judgments in Nigeria through the common law rule on the enforcement of foreign judgments' (2019) 15 (1) *Journal of Private International Law* 130, 133.

African Network for Animal Welfare (ANAW) v. The Attorney General of the United Republic of Tanzania, Ref. No. 9 of 2010, Judgment, East African Court of Justice at Arusha First Instance Div. 64 (Jun. 20, 2014).

Anene-Maidoh T, 'Enforcement of Judgments of ECOWAS Court of Justice' (2018) 1 (1) *Journal of Law Review National Institute for Legislative and Democratic Studies* 58.

Chuma-Okoro H, 'The Nigerian Constitution, the ECOWAS Treaty and the Judiciary: Interplay of roles in the Constitutionalisation of Free Trade' (2015) 4(1) *Global Journal of Comparative Law* 43.

Ebobrah S, 'Sub-Regional Judicial Enforcement of Economic, Social and Cultural Rights' in Chirwa D and Chenwi L (eds), *The Protection of Economic, Social and Cultural Rights in Africa: International, Regional and National Perspectives* (CUP 2016) 274–302.

Ebobrah S, 'Courts of Regional Economic Communities in Africa and Human Rights Law' in Kadelbach S, Rensmann S and Rieter T (eds), *Judging International Human Rights* (Springer 2019) 223–254.

Ebobrah S, 'The Uneven Impact of International Human Rights Law in Africa's Subregional Courts' in Martin Scheinin, (ed). *Human Rights Norms in 'Other' International Courts* (Cambridge University Press 2019).

Gathii J, 'Saving the Serengeti: Africa's new international judicial environmentalism' (2016) 16 (2) CJIL 386.

Gathii J, 'Variation in the use of Subregional Integration Courts between Business and Human Rights Actors: The Case of the East African Court of Justice' (2016) 79(1) *Law & Contemp. Probs* 37.

Giacomini G, 'Niger Delta People v Nigeria: A Missed Occasion before the ECOWAS Court of Justice' Focus Africa (2018) https://www.federalismi.it/nv14/articolo-documento.cfm?artid=35993 accessed 12 November 2019.

Government of the Republic of Zimbabwe v Louis Karel Fick and Others [2013] ZACC 22.

In the Matter of an Application to Enforce the Judgment of the Community Court of Justice of the ECOWAS against the Republic of Ghana and In the Matter of Chude Mba v. The Republic of Ghana, Suit No. HRCM/376/15 (High Court, Ghana, 2016) (unreported).

Lando V, 'The domestic impact of the decisions of the East African Court of Justice' (2018) 18 (2) *African Human Rights Law Journal* 463.

Milej T, 'Human Rights protection by international courts—What role for the East African Court of Justice?' (2018) 26(1) *African Journal of International and Comparative Law* 108.
Nwauche E, 'The Constitutional Challenge of the Integration and Interaction of Customary and the Received English Common Law in Nigeria and Ghana' (2010) 25 (1) *Tul. Eur. & Civ. LF* 37.
Obadan M, 'Introduction' in Hamalai L and Obadan M (eds) 40 Years of ECOWAS (1975–2015) (NILS 2015) 26–44.
Olawoyin A, 'Enforcement of Foreign Judgments in Nigeria: Statutory Dualism and Disharmony of Laws' (2014) 10(1) Journal *of Private International Law* 129.
Oppong R and Niro L, 'Enforcing Judgments of International Courts in National Courts' (2014) 5(2) *Journal of International Dispute Settlement* 344.
Oppong R, 'The High Court of Ghana declines to Enforce an ECOWAS Court Judgment' (2017) 25 (1) *African Journal of International and Comparative Law* 127.
Osaghae and others v Republic of Nigeria ECW/CCJ/JUD/O3/17.
Possi A, 'An appraisal of the functioning and effectiveness of the East African Court of Justice' (2018) 21(1) *Potchefstroom Electronic Law Journal/ Potchefstroomse Elektroniese Regsblad* 1.
Republic v High Court (Commercial Division) Accra, Ex parte Attorney General, NML Capital and the Republic of Argentina Civil Motion No J5/10/2013 (Supreme Court, Ghana, 2013).
SERAP v Federal Government of Nigeria Judgment No. ECW/CCJ/JUD/18/12.
Voigt C and Grant E, 'The Legitimacy of Human Rights Courts in Environmental Disputes' (2015) 6 (2) *Journal of Human Rights and the Environment* 131.

10

Climate Change as a 'constitutional essential': Trend and Significance for Africa

Ademola Oluborode Jegede

1 Introduction

That climate change is a major global challenge of the twenty-first century is no longer disputed.[1] For long, the findings from the scientific reports of the Intergovernmental Panel on Climate Change (IPCC) establish that climate change is real and due to human activities.[2] Especially in its 2013 report, the IPCC affirms that "the warming of the earth is unequivocal" and the vulnerability of population to its adverse

[1] Chris Wold, David Hunter and Melissa Powers, *Climate Change and the Law* (LexisNexis 2009) 1–3; *see* UNGA, 'Protection of global climate for present and future generations of mankind' UNG.A. Res. 43/53, 70th plenary meeting 6 December 1988 (UNGA Resolution 43/53).

[2] Thomas Stocker and others (eds), *The Physical Science Basis. Contribution of Working Group I to the 5th Assessment Report of the Intergovernmental Panel on Climate Change* (Cambridge University Press 2013).

A. Oluborode Jegede (✉)
Department of Public Law, University of Venda, Thohoyandou, South Africa
e-mail: ademola.jegede@univen.ac.za

M. Addaney, A. Oluborode Jegede (eds.), *Human Rights and the Environment under African Union Law*, https://doi.org/10.1007/978-3-030-46523-0_10

effects is increasing.[3] This position has been reinforced in subsequent findings by the Royal Society and Royal Academy of Engineering that it is nearly impossible for conventional climate response measures to achieve net zero carbon dioxide (CO_2) emissions and stabilize global temperatures below 2 °C on their own.[4] The human activities which put pressure on the global environment include the fossil-based economic development path of developed nations in the North,[5] and the pursuit of a similar development path associated with large-scale agriculture, mining, construction and logging, which as shown by scientific evidence, are a considerable driver of climate change and its adverse impacts across the world.[6]

Climate change will bring more negative consequences than positive implications to most parts of the world.[7] However, effects of climate change may differ in different parts of the world. Some parts of Africa, such as northern and southern Africa, as projected, will become drier; others, such as East Africa, may become wetter, with different results for food production and health conditions.[8] Effects also vary elsewhere in Europe where climate change is expected to hinder economic activity in Southern Europe more than in other sub-regions in different parts of Europe,[9] and due to their delicate physical characteristics, low-level

---

[3] Ibid., 8, 15.

[4] Royal Society & Royal Academy of Engineering, *Greenhouse Gas Removal* (Royal Society & Royal Academy of Engineering 2018) 7.

[5] World Bank, *World Development Report 2010: Development and Climate Change* (World Bank 2010) 1–35.

[6] Gilbert Rist, *The History of Development: From Western Origins to Global Faith* (Zed Books 2009) 21–24; on the negative impacts of these activities on the climate, see Ross Gorte and Pervase Sheikh, 'Deforestation and Climate Change' (Congressional Research Service Report for Congress March 2010) http://nationalaglawcenter.org/wp-content/uploads/assets/crs/R41144.pdf accessed 21 July 2019.

[7] Earth Observatory, 'How Will Global Warming Change Earth?' https://earthobservatory.nasa.gov/Features/GlobalWarming/page6.php accessed 21 July 2019.

[8] Paul Collier, Gordon Conway and Tony Venables, 'Climate Change and Africa' (2008) 24 *Oxford Review of Economic Policy* 337; Isabelle Niang and others, 'Africa' in Vicente Barros and others (eds), *Climate Change 2014: Impacts, Adaptation, and Vulnerability. Part B: Regional Aspects. Contribution of Working Group II to the Fifth Assessment Report of the Intergovernmental Panel on Climate Change* (Cambridge University Press 2014) 1199.

[9] Sari Kovats and others, 'Europe' in Vicente Barros and others (eds), *Climate Change 2014: Impacts, Adaptation, and Vulnerability. Part B: Regional Aspects. Contribution of Working Group II to the Fifth*

islands are highly vulnerable to climate change.[10] In general terms, established vulnerable sectors to the impact of climate change, actual and projected, are profiled as water resources, food security, natural resource management and biodiversity, human health, settlements and infrastructure, and desertification.[11]

Despite the foregoing, the global search for an appropriate legal response to addressing climate change is problematic. While a cohort of authors have demonstrated the potential in human rights as a response of law to climate change,[12] and elaborately made the linkages of human rights with climate change,[13] there are authors who question the approach by highlighting its anthropocentric (human interest) focus,[14] and weak amenability to litigation due to 'causation', 'extra-territoriality' and 'sovereignty' as problematic.[15] A whole range of literature has dealt generally

Assessment Report of the Intergovernmental Panel on Climate Change (Cambridge University Press 2014) 1267–1326.

[10] Leonard Nurse and others, 'Small Islands' in Vicente Barros and others (eds), *Climate Change 2014: Impacts, Adaptation, and Vulnerability. Part B: Regional Aspects. Contribution of Working Group II to the Fifth Assessment Report of the Intergovernmental Panel on Climate Change* (Cambridge University Press 2014) 1613–1654.

[11] Martin Parry and others (eds), *Climate Change 2007: Impacts, Adaptation and Vulnerability. Contribution of Working Group II to the Fourth Assessment Report of the Intergovernmental Panel on Climate Change* (Cambridge University Press 2007) 133–357.

[12] Sara Aminzadeh, 'A Moral Imperative: The Human Rights Implications of Climate Change' (2007) 39 *Hastings International & Comparative Law Review* 234.

[13] UN HRC, 'Human Rights and Climate Change' Res. 7/23, UN Doc. A/HRC/7/78 (2 July 2009); UN HRC, 'Human Rights and Climate Change,' Res. 10/4, 41st meeting, A/HRC/RES10/4 (25 March 2009); UN HRC, 'Human Rights and Climate Change,' A/HRC/RES/18/22 (24 March 2011); UN HRC, 'Human Rights and Climate Change,' A/HRC/26/L.33 (23 June 2014); UN HRC, 'Climate Change and Human Rights,' A/HRC/32/L.34 (30 June 2016); Ademola Oluborode Jegede, 'Climate Change in the Work of the African Commission on Human and Peoples' Rights' (2017) 31 (2) *Speculum Juris* 136; William Burns, *The Paris Agreement and Climate Geoengineering Governance: The Need for a Human Rights-Based Component* 111 (Centre for International Governance Innovation Papers 2016) 16–17.

[14] Daniel Bodansky, 'Customary (And not so customary) International Environmental Law,' (1995) 3 *Indiana Journal of Legal Studies* 116; Alan Boyle, 'Human Rights or Environmental Rights? A Reassessment' (2007) 18 *Fordham Environmental Law Review* 471.

[15] Eric Posner, 'Climate Change and International Human Rights Litigation: A Critical Appraisal' (2007) 155 *University of Pennsylvania Law Review* 1925; Shi-Ling Hsu, 'A Realistic Evaluation of Climate Change Litigation through the Lens of a Hypothetical Lawsuit' (2008) 79 *University of Colorado Law Review* 101; Armelle Gouritin, 'Potential Liability of European States under the ECHR for Failure to take Appropriate Measures with a view to Adaptation to Climate Change' in Michael Faure and Marjan Peeters (eds), *Climate Change liability* (Edward Elgar Publishing 2011) 134–152.

with constitutional environmentalism with passing reference to climate change.[16] However, academic interest on the issue of climate change with specific link to national constitutions in Africa is scanty. Existing literature does not consider the issue of climate change from the context of 'constitutional essentials' and is deficit of the trend in this regard. Whether or not climate change is an appropriate subject of the constitution is only an emerging discussion. Is there a basis and trend with regard to this development, and if so, what potentials does the constitutionalization of climate change hold for averting further change in climate and address its impact in Africa? This chapter responds to the foregoing questions by demonstrating climate change as an 'essential' for constitution, and with reference to direct and indirect evidence on the link of climate change to constitutions of some states in Africa, the paper further establishes the extent of this trend and its significance for Africa.

2 Constitutional 'Essentials': Climate Change in Context

There has been scholarly attention on the essentials that a constitution should embody to aid the fulfillment of its role as a guide to governance. According to Rawls, "constitutional essentials" should reflect the fundamental principles that specify the general structure of government, political process, and the equal basic rights and liberties of citizenship that majorities are to respect.[17] Michelman expresses commitment to the implementation of rights and laws and their effective administration as

---

[16] James May and Erin Daly, *Global Environmental Constitutionalism* (Cambridge University Press 2015) 269; Louis Kotzé, 'The Conceptual Contours of Environmental Constitutionalism' (2015) 21 *Widener Law Review* 187; David Boyd, *The Environment Rights Revolution: A Global Study of Constitutions, Human Rights and the Environment* (UBC Press 2012) 10; but for a preliminary attempt to examine a comprehensive link of climate change with environmentalism, *see* Ademola Oluborode Jegede, 'Climate Change and Environmental Constitutionalism: A Reflection on Domestic Challenges and Possibilities', in James May and Erin Daly (eds), *Implementing Environmental Constitutionalism* (Cambridge University Press 2018) 84–99.

[17] John Rawls, *Political Liberalism* (Columbia University Press 2005) 227.

'constitutional essentials' for a state.[18] Reinforcing this viewpoint, Elster argues that "essentials" of constitutions should, at the very least, "define and protect the rights of the citizens" and "establish the machinery of government".[19] Similarly, Rawls notes that the constitution should contain arrangements which are appropriate for reconciling conflicting opinions about rights and justice.[20] Viewing the protection of basic rights as essential to protect 'collective goods',[21] Raz explains that the court is particularly competent to deal with a conflict between rights affecting individuals and the functioning of organs of governance.[22] Another writing has demonstrated peace-promoting mechanism as a collective good that a constitution should embody as an essential.[23] However, the institutional case for the court as a suitable medium to safeguard rights in order to protect common goods rests on two bases. First, litigation before the court is a channel of political action in solving intractable conflicts concerning individual rights.[24] Secondly, courtrooms generally provide groups with 'insufficient economic power', 'little social prestige' or 'poor education' a forum which allows their voices to be heard.[25]

The foregoing analysis is significant in the context of climate change as an essential of a constitution because climate change generally relates to the environment which is a public good deserving of protection. The fact that the environment is a public good has been the subject of focus for authors including Erlich,[26] Hardin[27] and Catton,[28] whose writings also

[18] Frank Michelman, 'Constitutional Essentials' (paper delivered at conference on the 20th anniversary of Israel's human rights revolution, at the Knesset, Jerusalem, Israel, 19 December 2011) 6.

[19] Jon Elster, 'Constitutionalism in Eastern Europe' (1991) 58 (2) *The University of Chicago Law Review* 447, 465.

[20] Rawls (n 17) 197.

[21] Joseph Raz, *The Morality of Freedom* (Oxford University Press 1988) 258.

[22] Ibid.

[23] Ademola Oluborode Jegede, 'Bridging the Peace Gap in Nigeria: The Panel of the Wise as a Constitutional Essential' (2016) 60 (2) *Journal of African Law* 264.

[24] Joseph Raz, 'Rights and Politics' (1995) 71 (1) Indiana *Law Journal* 27, 42.

[25] Ibid., 42–43.

[26] Paul Ehrlich and Anne Ehrlich, 'The Population Bomb Revisited' (2009) 1 *Electronic Journal of Sustainable Development* 63.

[27] Garrett Hardin, 'The Tragedy of the Commons' (1968) 162 *Science* 1243.

[28] William Catton, *Overshoot: The Ecological Basis of Revolutionary Change* (University of Illinois Press 1982).

emphasize the interplay of population growth and need for political responsibility over environmental issues. The whole world has only 'one environment', hence, 'one climate', a reality that makes climate change a public good of global concern. Further reinforcing the global nature of the concern around climate change is the almost universal attention given to the phenomenon in the decisions and guidelines from yearly meetings of the implementing bodies of the UNFCCC,[29] Kyoto Protocol[30] and Paris Agreement[31] such as the Conference of Parties (COP)/Meeting of Parties (MOP),[32] as well as reports from organs such as the Subsidiary Body for Implementation (SBI)[33] and the Subsidiary Body for Scientific and Technological Advice (SBSTA).[34] The decisions of states through these institutions show that climate change is real and reflect a global trend toward acknowledging and addressing climate change as a challenge.

In the same way, if rights are valid essentials of a constitution, the activities at the international level under the aegis of United Nations (UN), and in the particular context of Africa, the African Union, which link climate change to human rights appear to justify climate change as a deserving component in national constitutions. For instance, the work of the Human Rights Council (HRC) under the UN Resolutions 10/4, 18/22 and 26/33 link climate change to human rights.[35] More specifically,

[29] United Nations Framework Convention on Climate Change (UNFCCC) ILM 851 (1992).

[30] United Nations Kyoto Protocol to the United Nations Framework Convention on Climate Change, 1998 (entered into force 16 February 2005).

[31] Paris Agreement under the United Nations Framework Convention on Climate Change 2015, adopted by Conference of the Parties, 21st Session Paris, 30 November-11 December 2015 FCCC/CP/2015/L.9/Rev.1. (Paris Agreement).

[32] The COP is established pursuant to art 7 of the UNFCCC while the MOP is established by virtue of art 13 of the Kyoto Protocol; these institutions are the highest decision-making bodies, see Daniel Bodansky, 'International Law and the Design of a Climate Change Regime', in Urs Luterbacher and Detlef Sprinz (eds), *International Relations and Global Climate Change* (MIT Press 2001) 213.

[33] UNFCCC (n 29) art 10; the SBI reviews policy aspects of national reports and help the COP in evaluating summative effects of implementation measure, see Bodansky (n 32) 201.

[34] UNFCCC (n 29) art 9; the SBSTA is created to provide expeditious information and advice on scientific and technological matters relating to the UNFCCC, see Fred Gale, 'A Cooling Climate for Negotiations: Intergovernmentalism and its Limits,' in Timothy Cadman (ed), *Climate Change and Global Policy Regime: Towards Institutional Legitimacy* (Palgrave Macmillan 2013).

[35] United Nations Human Rights Council (UNHRC), 'Resolution 10/4 – Human Rights and Climate Change', adopted on 25 March 2009 at the 41st meeting of the Human Rights Council Preamble; United Nations Human Rights Council (UNHRC), 'Resolution 18/22 – Human Rights

the OHCHR Report which responded to the request of the HRC calling for discussion on the link between climate change and human rights describes the effect of climate change on a range of rights, including right to life,[36] the right to adequate food,[37] the right to adequate water,[38] the right to health,[39] the right to adequate housing,[40] and the right to self-determination.[41] This position is reinforced in Resolution 10/4 which affirms that human rights obligations and commitments have the potential to inform and strengthen international and national policy-making in the area of climate change.[42] It is also reiterated in Resolution 18/22, emphasized in Resolution 26/33,[43] and buttressed by Resolution 32/34, which notes, in addition, that climate change is an 'existential threat' that has impacted negatively on the fulfillment of the Universal Declaration of Human Rights.[44] In fact, the Paris Agreement adopted by States in December 2015 leaves no doubt that human rights obligations are important in climate change activities. It provides:

> Parties should, when taking action to address climate change, respect, promote and consider their respective obligations on human rights, the right to health, the rights of indigenous peoples, local communities, migrants, children, persons with disabilities and people in vulnerable situations and

---

and Climate Change,' adopted 30 September 2011 at the 37th meeting (A/HRC/RES/18/22) Preamble; United Nations Human Rights Council (UNHRC), 'Resolution 26 L/33 – Human Rights and Climate Change,' adopted on 23 June 2014 at the 26th session A/HRC/26/L.33 para 1; see also Ademola Oluborode Jegede, 'Rights Away from Home: Climate-induced Displacement of Indigenous Peoples and the Extraterritorial Application of the Kampala Convention (2016) 16 *African Human Rights Law Journal* 58.

[36] United Nations Human Rights Council (UNHRC), 'Report of the Office of the United Nations High Commissioner for Human Rights on the Relationship between Climate Change and Human Rights,' A/HRC/10/61 15 January 2009 (OHCHR Report) paras 21–24.

[37] Ibid., paras 25–27.

[38] Ibid., paras 28–30.

[39] Ibid., paras 31–34.

[40] Ibid., paras 35–38.

[41] Ibid., paras 39–41.

[42] UNHRC Resolution 10/4 (n 35) preamble.

[43] UNHRC Resolution 18/22 (n 35) preamble; UNHRC Resolution 26/33 (n 35) para 1.

[44] UNHRC Resolution 32/34 (n 35) preamble.

the right to development, as well as gender equality, empowerment of women and intergenerational equity.[45]

At the regional level in Africa, the African human rights system which is based on the African Charter on Human and Peoples' Rights (African Charter),[46] and the activities of its treaty monitoring body, the African Commission on Human and Peoples' Rights (African Commission),[47] has generated Resolutions 153,[48] 148,[49] 271[50] which point out the link of human rights to climate change.[51] Raz's linkage of constitutional essential to the importance of the court as an avenue of justice for the voiceless is no less an important factor in the argument that links climate change to 'constitutional essentials'. The role of court is evident in the position of May and Daly who construct the meaning of environmental constitutionalism as

> a relatively recent phenomenon at the confluence of constitutional law, international law, human rights, and environmental law. It embodies the recognition that the environment is a proper subject for protection in constitutional texts and for vindication by constitutional courts worldwide.[52]

---

[45] Paris Agreement (n 31).

[46] OAU, African Charter on Human and Peoples' Rights, CAB/LEG/67/3 rev. 5, 21 ILM 58, 27 June 1981 (entered into force 21 October 1986).

[47] Makau Mutua, 'The African Human Rights System: A Critical Evaluation' http://hdr.undp.org/sites/default/files/mutua.pdf accessed 21 January 2020.

[48] AU, African Commission of Human and Peoples' Rights, ACHPR/Res153 (XLVI) 09: Resolution on Climate Change and Human Rights and the Need to Study its Impact in Africa (25 November 2009).

[49] AU, 'African Commission of Human and Peoples' Rights, 148: Resolution on the Establishment of a Working Group on Extractive Industries, Environment and Human Rights Violations in Africa', adopted at 46th ordinary session held in Banjul, The Gambia (11–25 November 2009).

[50] AU, 'African Commission on Human and Peoples' Rights '271: Resolution on Climate Change in Africa', adopted at the 55th ordinary session of the African Commission on Human and Peoples' Rights held in Luanda, Angola (28 April-12 May 2014).

[51] Jegede (n 13) 136.

[52] James May and Erin Daly, *Global Environmental Constitutionalism* (Cambridge University Press 2015) 3.

It is also linked to courts' role in interpreting and applying environmental laws and constitutional obligations.[53] Climate change affects vulnerable populations whose cause will be strengthened if the phenomenon is made a constitutionally recognized issue. As Mali indicates in its submission to the invitation by the HRC on the link of climate change to human rights: "it is almost impossible for populations in poor countries to identify and pursue channels of justice, to have their cases heard, or to prove responsibility".[54] Hence, the question to which one must now turn is to consider the trend in relation to the incorporation of climate change as an essential in the constitutions of states in Africa.

3 Trending the Link of Climate Change to Constitutions in Africa

While a range of constitutions of states in Africa have provisions on environmental protection,[55] the ideas of incorporating climate change as an 'essential' in the constitution is only recently emerging. Arguably, in this regard, a direct and indirect link of climate change to the constitution is noticeable.

Direct Linkage of Climate Change to the Constitution

By direct linkage of climate change to the constitution, it is meant that evidence exist among states in Africa of constitutions which embody the fight against climate change as a prominent feature. This is provable by section 44 of the Tunisia Constitution of 2014 which provides that the State shall *"provide the means necessary to guarantee a healthy and balanced*

---

[53] Louis Kotzé and Alexander Paterson (eds), *The Role of the Judiciary in Environmental Governance: Comparative Perspectives* (Kluwer Law International 2009).

[54] OHCHR, 'Submission of Mali to OHCHR Study "Human Rights and Climate Change"' (Mali Submission) http://www.ohchr.org/Documents/Issues/ClimateChange/Submissions/Mali.pdf accessed 18 July 2019.

[55] May and Daly (n 52).

environment and contribute to the climate's integrity".[56] Section 45 of the Constitution guarantees "the right to a healthy and balanced environment and the right to participate in the protection of the climate". In addition, it urges the government to "provide the necessary means to eliminate environmental pollution".[57] The significance of these provisions to Tunisia is not farfetched as it has been argued that tourism and agribusiness, two major sectors on which the Tunisian economy depends, will be adversely and directly impacted by climate change. Tourism will be adversely affected because Tunisia could lose more than one-third of its coastal areas to climate change; at the same time, the sustainability or growth in agricultural production is also being affected by climate instability.[58] In particular, the nation's nearly 3000-year history of olive farming is under considerable threat with indications that production is at risk of halving by 2030, owing to extremes of climate change, ranging from floods to droughts.[59]

To address the foregoing trend, the inclusion of articles 44 and 45 in the Tunisia Constitution has been considered in commentaries as a positive step.[60] Although the constitutional provision is recent, and so far untested in the courtroom, there is evidence that the provision is influencing policy choices of the government in addressing climate change. For instance, Tunisia is one of the pioneering countries to draft Nationally Appropriate Mitigation Actions, and one of the first non-Annex I countries to submit its first biennial report on December 31, 2014, in which it expressed an ambition to draw both on internal and external resources to mitigate its emissions of greenhouse gases.[61] Also, there is a reference to the constitution in its Intended Nationally Determined Contribution

[56] UNFCCC, 'Republic of Tunisia Intended Nationally Determined Contribution' (UNFCC 2015) 4 (Tunisia INDC).

[57] Tunisia Constitution of 2014 (as translated by UNDP and reviewed by International IDEA https://issafrica.org/ctafrica/uploads/TunisiaConstitution2014Eng.pdf accessed 20 January 2020.

[58] 'Leaders: Can Tunisia Afford to Care about Climate Change?' https://tn.usembassy.gov/leaders-cantunisia-afford-to-care-about-climate-change/ accessed 20 January 2020.

[59] Africa News, 'Climate Change Threatens Tunisia Olive Farming'. http://www.news24.com/Africa/News/cliatechange-threatens-tunisia-olive-farming-20151130-2 accessed 20 January 2020.

[60] Tunisia Live, 'Climate Change: The Threat No One's Talking About,' December 5, 2015 http://www.tunisia-live.net/2015/12/05/environment/ accessed 17 May 2019.

[61] Tunisia INDC (n 56).

of 2015, a document which it submitted in accordance with the requirement under the Paris Agreement.[62] As evident in the document, the inclusion of climate change in the Constitution is already driving the direction of the state in relation to its energy system. To combat the challenge of insecurity of energy supply and ensure economic sustainability, in line with the Constitution, the Tunisian authority has decided to strengthen its focus on energy efficiency and renewable energies. The energy transition agenda aims at reducing primary energy demand of 30% by 2030 and improving renewable energies in electricity production at the same rate.[63]

A similar trend on the direct incorporation of climate change in the constitution is also noticeable in the amended Constitution of Zambia in 2016 which imposes a duty on the State to "establish and implement mechanisms that address climate change".[64] This is not surprising as Zambia is not only one of the several states in Africa affected by adverse effects of climate change, it is also involved in climate change mitigation program. Zambia became one of the pilot countries for the REDD+ (Reducing emissions from deforestation and forest degradation, and fostering conservation, sustainable management of forests, and enhancement of forest carbon stocks) initiative since 2010.[65]

Climate change has negative implications for health profile in Zambia. For increased cases of malaria and indeed other major diseases are associated with floods and increased temperature regimes. Incidence of malaria is increasing due to the incursion of malaria.[66] A reason for this is that malaria is a climate-sensitive disease, that is, an illness that is sensitive to weather or climatic factors. Malaria is regarded as the leading killer

---

[62] Paris Agreement (n 30) art 3.

[63] Tunisia INDC (n 56) 4.

[64] Andrew Gilder and Olivia Rumble, 'South African Climate Change Policy Trajectory in light of the Paris Agreement' https://www.saiia.org.za/wp-content/uploads/2016/03/2016-02-01-COP21outcome-Andrew-Gilder-and-Olivia-Rumble.pdf accessed 17 May 2019.

[65] See generally on the adverse effect of climate change in Zambia and the implementation of REDD+, Ademola Oluborode Jegede, The *Climate Change Regulatory Framework and Indigenous Peoples' Lands in Africa: Human Rights Implications* (Pretoria University Law Press 2016).

[66] Ibid.

disease in Zambia.[67] The natural resources inclusive of wildlife and forestry sectors are vulnerable to adverse impacts of climate change in Zambia. As documented by the national adaptation plan of action (NAPA) of Zambia,[68] the 1992 drought left in its wake the death of several hippopotamuses in South Luangwa National Park and the migration of most animals from the Park.[69] Also, in 2005, it was reported that drier weather occasioned changes in condition of elephants.[70]

Zambia has abundance of water since it holds much of the water in Southern African Development Community (SADC).[71] Due to harsh climatic condition, however, it has experienced consistent droughts bringing about water scarcity in several parts of the country. Excessive rainfall has disrupted communities living in the valley and towns with high water table such as Lusaka.[72] As a result of droughts, ground water resources are giving way to diminishing water tables as well as boreholes and rivers.[73] The potential in water sources such as Kariba dam, and Kafue gorge for energy generation is under threat, due to increasing droughts traceable to climate change.[74]

With an approximately 49,468,000 ha amounting to 67% of land surface covered by forests, Zambia is one of the most forested countries in Africa.[75] The most common forest type is Miombo woodland, covering 42% of the land area.[76] Generally forests play a significant role in the livelihoods of the vast majority of people living in the rural area. It serves

---

[67] Republic of Zambia National Adaptation Programme of Action on climate change (September 2007) (Zambia NAPA).

[68] Ibid.

[69] Ibid.

[70] Ibid.

[71] Republic of Zambia *Initial National Communication under the United Nations Framework Convention on Climate Change* (August 2002) (Zambia Initial National Communication) 58; Zambia NAPA (n 67) 12.

[72] Zambia National Communication (n 71) 58.

[73] Ibid.

[74] Zambia NAPA (n 67) 10.

[75] Royd Vinya and others, 'Preliminary Study on the Drivers of Deforestation and Potential for REDD+ in Zambia' (A consultancy report prepared for the Forestry Department and FAO under the national UN-REDD+ Programme Ministry of Lands & Natural Resources. Lusaka, Zambia 2012) 2–3 (Zambia Preliminary Study).

[76] Ibid.

as a means of subsistence, generating income and employment.[77] However, deforestation remains a challenge. Estimated at a growing rate of approximately 1.5% per annum, Zambia is ranked as one of the countries with the highest rates of deforestation in the world.[78] While there is emerging evidence of it in the North-Western province, major corridors of deforestation cover four key provinces, namely Southern, Lusaka, Central and Copperbelt.[79] In 1996, for instance, the Food and Agricultural Organisation (FAO), referring to the findings of Alajarvi,[80] reported that the annual average rate of deforestation in Zambia is around 250,000 ha per annum. Higher rate has been confirmed by subsequent findings of researchers, including Chidumayo, who reported a deforestation rate of 300,000 ha per annum, signifying that deforestation is scaling up.[81] Analyzing the scenario in a more recent studies, Vinya et al. predicted that, if not halted, the trend of deforestation in Zambia would dramatically increase, with Copperbelt being the worst affected province.[82]

In addition, there are a number of socio-economic reasons for which forests are depleted in Zambia. In both rural and urban areas, the forests serve as a source of fuelwood directly contributing to deforestation.[83] Also associated with deforestation are activities of socio-economic benefits,[84] such as, investment in charcoal industry which accounts for no less than 80% of Zambia energy source[85] and agricultural expansion.[86] In the Central, Copperbelt, Northern and Western provinces, research findings have shown that agricultural expansion is the second most frequent

[77] Zambia Preliminary Study (n 75) 8.

[78] Matieu Henry and others, 'Implementation of REDD+ in sub-Saharan Africa: State of Knowledge, Challenges and Opportunities' (2011) 16 *Environment & Development Economics* 381.

[79] Zambia Preliminary Study (n 75) 8.

[80] Pekka Alajarvi, *Forest Management Planning and Inventory* (ZFAP, MENR 1996).

[81] Emmanuel Chidumayo, *Development of Reference Emission Levels for Zambia* (Report prepared for FAO 2012).

[82] Zambia Preliminary Study (n 75) 10.

[83] Zambia Preliminary Study (n 75) 21.

[84] Zambia Preliminary Study (n 75) 23.

[85] Thomson Kalinda and others, *Use of Integrated Land Use Assessment (ILUA) data for forestry and agricultural policy review and analysis in Zambia* (Integrated Land Use Assessment Project 2008) 22.

[86] Zambia Preliminary Study (n 75) 23.

driver of deforestation.[87] Similarly, the mining sector has greatly contributed to a declining forest cover.[88] Huge tract of lands are cleared as the need arises to make space for mining and its infrastructures.[89] It is estimated that at the Kalumbila Mining Concession, infrastructure preparations will result in the loss of more than 7000 ha of land before the concession becomes fully operational.[90] Similarly, demographic factor in the form of growth in populations contributes to increasing degradation of the forests.[91] Adverse effects of climate change are visible in Zambia's indigenous forests which have played a key role in providing timber and non-timber products for communities around forest reserves and the nation at large.[92] Due to increase in temperature, climate change impacts have been reported as reducing the capacity of regeneration of forests such as the Miombo forests, signifying fewer natural resources for communities that rely on them for livelihoods.[93] Hence, the incorporation of climate change in the Constitution of Zambia is not difficult to appreciate.

In fact, the foregoing development in Tunisia and Zambia shows that states in Africa undergoing the process of constitutional amendment which share similar experiences in terms of adverse effects of climate change and modest efforts on mitigation are likely to incorporate climate change in their constitutions. However, even states without an evidence of incorporating climate change as a component of their constitution have indirectly done so by referring to the constitution to ground the content of their submissions under the aegis of the UNFCCC. An example of such document is discernible under the Paris Agreement which requires states to make nationally determined contributions to greenhouse gas mitigation (NDCs) and to transparently report on their activities toward fulfilling those commitments.[94]

---

87 Zambia Preliminary Study (n 75) 24.

88 Zambia Preliminary Study (n 75) 25.

89 Ibid.

90 Zambia Preliminary Study (n 75) 26.

91 FAO, 'Global Outlook Study for Africa. Sub-regional Report: Southern Africa' (African Development Bank/European Commission/FAO. Rome 2005).

92 Zambia NAPA (n 67) 10.

93 Zambia NAPA (n 67) 51.

94 'UNFCCC' https://unfccc.int/news/new-era-of-global-climate-action-to-begin-under-paris-climate-change-agreement-0 accessed 20 December 2019.

Indirect Linkage of Climate Change to the Constitution

There are states in Africa which have not incorporated climate change in the constitution but refer to it in their official submissions on climate change at the international level. This development, which is most feasible in the NDCs of certain states of Africa, constitutes an indirect use of the constitution to advance the response to climate change. Using the constitution to ground their commitment to climate change, in its INDC submission, Chad refers to sections 47 and 52 of its Constitution which guarantee the protection of environment to reinforce its commitment to climate change.[95] The reference to climate change is not only important given the vulnerability of the country to climate change and the sensitivity of its economy to the phenomenon, it equally demonstrates Chad's desire to make an effective contribution to the global effort to combat global warming.[96] Chad also appeals to the Constitution in its INDC by referring to the realization of human rights as the basis on which the country is participating in mitigation projects in its territory notably by means of the Clean Development Mechanism (CDM) and the REDD+ program. Its constitutional provisions on human rights, according to Chad's INDC, will shape issues associated with the implementation of the programs such as sharing benefits with the local population, access to sustainable energy, generation of local jobs, reduction of inequalities and gender-based inequality. Emphatically the Chad INDC affirms that implementing climate change interventions will place particular emphasis on human rights and equality between the sexes.[97]

The Constitution of the Republic of Uganda[98] does not make reference to climate change but its relevance is evident in Uganda's submission on the NDC. According to the submission, the mitigation and adaptation intentions of the state are not only based on the country's National Climate Change Policy (NCCP),[99] and do not only reflect Uganda Vision

[95] UNFCCC, 'Intended Nationally Determined Contribution (INDC) for the Republic of Chad' (UNFCCC 2015) 3.

[96] Ibid.

[97] Ibid.

[98] The Constitution of the Republic of Uganda (1995, as amended in 2005 and 2015).

[99] National Climate Change Policy (NCCP) (2015).

2040,[100] it is premised on the Constitution. It is, therefore, not surprising that based on its constitutional aspiration, Uganda prioritizes the reduction of the vulnerability of its population, environment and economy by implementing adaptation actions. It also aspires to transform its society from "a peasant to a modern and prosperous nation" and "implement strategies, plans and actions for low greenhouse gas emission development" in the context of its development goals.[101] The Uganda NDC includes crosscutting respect for human rights and gender-responsive climate change actions. The protection of vulnerable groups, including women, is a crosscutting priority.[102] Hence, it urges that climate change adaptation and mitigation actions should observe human rights and include gender consideration in the interest of women.[103]

In its INDC, Kenya indicated that addressing the adverse effects of climate change will help the state to actualize the promises of its 2010 Constitution and the attainment of Vision 2030.[104] Furthermore, the Kenyan Climate Change Act passed in 2016 links with the constitutional provisions on the environment in section 42 and obligations of the state in creating institutional structures and the 'mainstreaming' of climate change actions within government. It creates three institutional mechanisms. The first is the National Climate Change Council which is chaired by the President, with the Cabinet Secretary responsible for Climate Affairs as its Secretary. The Secretary is supported by the Climate Change Directorate. Part V of the Act also deals with public participation and access to information, the latter which is guaranteed under section 35 of the Constitution. The expectation that public participation and consultation are key in climate change agenda is further reflected in the Kenya INDC document which emphasizes the enhancement of education, training, public awareness, public participation, public access to information across public and private sectors. These rights-related provisions are

[100] National Planning Authority, Uganda Vision 2040 (National Planning Authority 2013).

[101] Ministry of Water and Environment, 'Uganda's Intended Nationally Determined Contribution' (October 2015) 3.

[102] Ibid., 2.

[103] Ibid., 15.

[104] Ministry of Environment and Natural Resources, Kenya's Intended Nationally Determined Contribution (INDC) (UNFCCC 2015) 1.

a useful example on how states with no specific provision on climate change in their constitutions can through a national legislation link climate ambitions with other constitutional aspirations.[105]

In the submission of Malawi, human rights and gender issues are linked to climate change. According to the document, the Government of Malawi is committed to the promotion of gender equality in all government sectors, programs and projects associated with climate change. Consideration of these crosscutting issues is pertinent in the design and implementation of mitigation and adaptation actions.[106] It further documents that public entities are required to undertake public awareness and consultations, and safeguard gender mainstreaming, in line with the Constitution.[107]

Morocco refers to its 2011 Constitution in its INDC. As the submission indicates, the Constitution offers the government the necessary impetus to achieve sustainable development, a development that is further confirmed in other documents. These include the Framework Law on the National Charter for Environment and Sustainable Development (NCESD),[108] which asserts the rights and duties inherent to the environment and sustainable development accorded to natural and legal persons and urges for implementation by the state, local authorities, public institutions and businesses.[109] As the submission on the INDC further shows, the implementation of Morocco's NDC is part of an integrated approach that includes respect for human rights and gender equality, as enshrined in Morocco's 2011 Constitution.[110]

Eritrea identifies its Constitution[111] as one of the main instruments which help facilitate the integration of climate change mitigation and adaptation concerns into national development policy framework. In fact

105 UNFCCC, 'Kenya's Intended Nationally Determined Contribution' (UNFCCC 2015).

106 Ibid., 12.

107 UNFCCC, 'Republic of Malawi Intended Nationally Determined Contribution' (UNFCCC 2015) 7.

108 NCESD, 'Framework Law on the National Charter for Environment and Sustainable Development' (NCESD, 2014).

109 UNFCCC, 'Morocco Nationally Determined Contribution under the UNFCCC' (UNFCCC 2015) 1 and 2.

110 Morocco Constitution, 2011.

111 Constitution of Eritrea, 1997.

it indicates that the preparation of NDC is based on national documents including the Constitution.[112] In constructing the INDC, the Eritrea's constitution indicates the direction of the climate adaptation program 2020–2030 which seeks to strengthen its capacities to cope with climate change. It is the basis of including the civil society to expedite the implementation process of adaptation as well as develop the sense of belongingness while tackling the challenges associated with climate change. As it appears, the INDC reflects the constitutional value of equity where it indicates that the incorporation of the most vulnerable sectors of the society is inevitable in the fight against climate change. According to the INDC, this requires the need to reduce social and geographical inequality as well as narrow the gaps between women and men's rights in Eritrea.[113]

The content of the South Africa's INDC is informed by the environmental right set out in section 24 of the Constitution, and its National Development Plan (NDP),[114] which provides a '2030 vision' to guide the attainment of its National Sustainable Development Strategy. It is also in that context that the progress being made in implementing climate-compatible sectoral plans, such as the integrated energy and electricity plans, industrial policy action plans and the new growth path of South Africa is framed.[115]Similarly, the Zimbabwe INDC refers to the provisions of the 2013 Constitution which guarantees every person environmental rights that include the right to healthy environment, protection of the environment for the benefit of present and future generations, through reasonable legislative and other measures that prevents pollution and ecological degradation, but, promotes conservation; and secure ecologically sustainable development and use of natural resources in the context of economic and social development.[116] More importantly, it indicates that the

[112] UNFCCC, 'The State of Eritrea Nationally Determined Contributions Report to UNFCCC' (UNFCC 2015) 12.

[113] Ibid., 24.

[114] National Development Plan 2012.

[115] UNFCCC, 'South Africa's Intended Nationally Determined Contribution' (UNFCCC 2015) 2.

[116] UNFCCC, 'Zimbabwe's Intended Nationally Determined Contribution' (UNFCCCN 2015) 1.

Constitution shall guide generally its mitigation efforts and low carbon development pathways.[117]

The reference to the Constitution in terms of the content of the INDC of the Gambia is within the context of the right of everyone to quality basic education as contained in the 1997 Constitution of The Gambia. In what shows that education is a key response to climate change, the INDC uses the constitutional provision on the right to education to underscore the integration of climate change into the basic and higher education curricular and calls upon financial and technical support to achieve that purpose.[118] The link is arguably well made by the Gambia in the sense that it conforms to existing literature on the subject which affirms the relevance of education as a climate change intervention strategy.[119]

Thus the foregoing trend demonstrates that while few states in Africa have directly incorporated climate change in their constitution, a range of other states have shown indirect reference to the constitution, through their INDC to drive climate change Agenda. It remains to be examined the potential in this development for states in Africa.

4 The Significance of Constructing Climate Change as a Constitutional Essential

In every nation that has adopted one, the significance of a constitution can generally not be overstated. It is the source of other legislation in the land, and as the highest law, it contains important rules for governance and politics within a polity, sets out the rights of citizens and provides for the obligations of the state. The spirit and tenor of the constitution serves further value of not only governing the operation of the political organs of the state, it influences the functioning of the judiciary in the exercise of their powers and discretions. The foregoing significance has a range of bearing on the construction of climate change as a constitutional essential.

[117] Ibid., 3.

[118] UNFCCC, 'Intended Nationally Determined Contribution of the Gambia' (UNFCCC 2015) 19.

[119] Ibid.

First, incorporating climate change into the constitution is symbolic and reflective of how importance climate change is to the global community. The adverse effects of climate change as has been observed are not disputed. It poses an existential threat to states and communities, particularly in Africa.[120] Hence, just as constitutions have always been accommodative of existential issues of global concern such as peace, human rights and development, incorporating climate change into the constitution will allow all stakeholders within a state to engage with the phenomenon in a manner that reflects its importance. Also considering that the constitution is difficult to change, the link of climate change to the Constitution guarantees that the subject is beyond the vicissitudes of petty politics.[121] Hence, constructing climate change as a constitutional essential will not only remove it from the politics of government but will afford vulnerable populations a premise to argue their case on the adverse effects of climate change both to political organs of a state and in the court room in Africa. Arguably, to exclude climate change from such constitutional arrangement may in fact be disempowering to these populations who need the backing of courts for protection against the adverse effects of climate change.

Secondly, where established in the constitution, it can become a core issue for governance and political choice. For instance, it can afford the government an opportunity to ground its policy and direction on core issues that interface with climate change such as its economic developmental path, creation of new jobs and energy creation and consumption. Also, the parliament which plays an important oversight role in any government can on the basis of such incorporation engage with the executive on the extent to which addressing climate change has been a subject of consideration in their policy and decision-making process. The civil society is better able to engage political actors about the subject, using such a

---

[120] Niang and others (n 6); Collier, Conway and Venables (n 6); Toulmin (n 6).

[121] Ademola Oluborode Jegede, 'Climate Change and Environmental Constitutionalism: A Reflection on Domestic Challenges and Possibilities,' in James May and Erin Daly (eds), *Implementing Environmental Constitutionalism* (Cambridge University Press 2018) 84–99.

platform as a tool to test the genuineness of government commitment to the cause and in fact its performance in addressing the adverse effects of climate change.

Lastly, incorporating climate change in the constitution is very crucial to the functioning of the judiciary in terms of its interpretive and adjudicatory role. It will allow courts to be more proactive in pursuing enhanced access to justice by vulnerable or concerned persons and institutions in the civic space who may wish to challenge the government or non-state actors about the constitutionality of their actions. That this in fact possible is already emerging in the case law of certain states in Africa where courts are beginning to draw the link between climate change and the provisions of the Constitution. In *Gbemre* v. *Shell Petroleum Development Company Nigeria Limited and others*,[122] a Nigerian Court ordered the Shell, ExxonMobil, Chevron- Texaco, and the Nigerian National Petroleum Corporation companies to stop gas flaring in the Niger Delta on the ground that it has negative implication on the constitutional right to life communities in which they operate.[123] In the South Africa's case of *Earthlife Africa Johannesburg v. Minister of Environmental Affairs and others*,[124] it was made evident by the court decision that the rights to just administrative action provided for under section 33 and right of anyone to approach the court for the enforcement of rights guaranteed under section 38 of the Constitution are relevant to court in considering the implementation of project that may have a likely adverse effect on the climate.[125] The foregoing signifies that the incorporation of climate change in the constitution can embolden the approach of courts in addressing disputes related to climate change and thereby enrich the jurisprudence of states in Africa.

---

[122] Gbemre v. Shell Petroleum Development Company Nigeria Limited and Others [2005] AHRLR 151 (NgHC 2005).

[123] Ibid., paras 5–7.

[124] Earthlife Africa Johannesburg v. Minister of Environmental Affairs and others [2017] JOL 37526 (GP).

[125] Ibid., paras 28 and 94.

5 Conclusion

Scholarship on the appropriate legal response to climate change has been problematic, but whether the incorporation of climate change as an essential in national constitutions can constitute a legal response is the focus of this contribution. As shown in the chapter, there is an emerging evidence in the form of direct and indirect recognition of the need to address the climate change crisis as an essential subject of constitutions in Africa. As further shown, the emerging development has potentials for addressing the adverse effects of climate change in Africa. It can shape the approach of all stakeholders including the parliament, executive, non-state actors, civil society and the judiciary on the subject of climate change. Accordingly, the search for an appropriate legal response to address climate change should include the incorporation of climate change as an essential in national constitutions of states in Africa.

References

Africa News, 'Climate Change Threatens Tunisia Olive Farming'. www.news24.com/Africa/News/cliatechange-threatens-tunisia-olive-farming-20151130-2 accessed 20 January 2020.

Alajarvi P, *Forest Management Planning and Inventory* (ZFAP, MENR 1996).

Aminzadeh S, 'A Moral Imperative: The Human Rights Implications of Climate Change' (2007) 39 *Hastings International & Comparative Law Review* 234.

AU, 'African Commission of Human and Peoples' Rights, 148: Resolution on the Establishment of a Working Group on Extractive Industries, Environment and Human Rights Violations in Africa', adopted at 46th ordinary session Africa.

AU, 'African Commission on Human and Peoples' Rights '271: Resolution on Climate Change in Africa', adopted at the 55th ordinary session of the African Commission on Human and Peoples' Rights held in Luanda, Angola (28 April-12 May 2014).

AU, African Commission of Human and Peoples' Rights, ACHPR/Res153 (XLVI) 09: Resolution on Climate Change and Human Rights and the Need to Study its Impact in Africa (25 November 2009).

Bodansky D, 'Customary (And not so customary) International Environmental Law,' (1995) 3 *Indiana Journal of Legal Studies* 116.

Bodansky D, 'International Law and the Design of a Climate Change Regime', in Luterbacher U and Sprinz D (eds), *International Relations and Global Climate Change* (MIT Press 2001).

Boyd D, *The Environment Rights Revolution: A Global Study of Constitutions, Human Rights and the Environment* (UBC Press 2012).

Boyle A, 'Human Rights or Environmental Rights? A Reassessment' (2007) 18 *Fordham Environmental Law Review* 471.

Burns W, *The Paris Agreement and Climate Geoengineering Governance: The Need for a Human Rights-Based Component* 111 (Centre for International Governance Innovation Papers 2016).

Catton W, *Overshoot: The Ecological Basis of Revolutionary Change* (University of Illinois Press 1982).

Chidumayo E, *Development of Reference Emission Levels for Zambia* (Report prepared for FAO 2012).

Chris Wold, David Hunter and Melissa Powers, *Climate Change and the Law* (LexisNexis 2009).

Collier P, Conway G and Venables T, 'Climate Change and Africa' (2008) 24 *Oxford Review of Economic Policy* 337.

Earth Observatory, 'How Will Global Warming Change Earth?' https://earthobservatory.nasa.gov/Features/GlobalWarming/page6.php accessed 21 July 2019.

Ehrlich P and Ehrlich A, 'The Population Bomb Revisited' (2009) 1 *Electronic Journal of Sustainable Development* 63.

Elster J, 'Constitutionalism in Eastern Europe' (1991) 58 (2) *The University of Chicago Law Review* 447.

FAO, 'Global Outlook Study for Africa. Sub-regional Report: Southern Africa' (African Development Bank/European Commission/FAO. Rome 2005).

Gale F, 'A Cooling Climate for Negotiations: Intergovernmentalism and its Limits,' in Cadman T (ed), *Climate Change and Global Policy Regime: Towards Institutional Legitimacy* (Palgrave Macmillan 2013).

Gilder A and Rumble O, 'South African Climate Change Policy Trajectory in light of the Paris Agreement' https://www.saiia.org.za/wp-content/uploads/2016/03/2016-02-01-COP21outcome-Andrew-Gilder-and-Olivia-Rumble.pdf accessed 17 May 2019.

Gorte R and Sheikh P, 'Deforestation and Climate Change' (Congressional Research Service Report for Congress March 2010) http://nationalaglawcenter.org/wp-content/uploads/assets/crs/R41144.pdf accessed 21 July 2019.

Gouritin A, 'Potential Liability of European States under the ECHR for Failure to take Appropriate Measures with a view to Adaptation to Climate Change' in Faure M and Peeters M (eds), *Climate Change liability* (Edward Elgar Publishing 2011) 134–152.

Hardin G, 'The Tragedy of the Commons' (1968) 162 *Science* 1243.

Henry M and others, 'Implementation of REDD+ in sub-Saharan Africa: State of Knowledge, Challenges and Opportunities' (2011) 16 *Environment & Development Economics* 381.

Hsu S, 'A Realistic Evaluation of Climate Change Litigation through the Lens of a Hypothetical Lawsuit' (2008) 79 *University of Colorado Law Review* 101.

Jegede AO, *The Climate Change Regulatory Framework and Indigenous Peoples' Lands in Africa: Human Rights Implications* (Pretoria University Law Press 2016).

Jegede AO, 'Bridging the Peace Gap in Nigeria: The Panel of the Wise as a Constitutional Essential' (2016) 60 (2) *Journal of African Law* 264.

Jegede AO, 'Climate Change and Environmental Constitutionalism: A Reflection on Domestic Challenges and Possibilities', in May J and Daly E (eds), *Implementing Environmental Constitutionalism* (Cambridge University Press 2018) 84–99.

Jegede AO, 'Climate Change in the Work of the African Commission on Human and Peoples' Rights' (2017) 31 (2) *Speculum Juris* 136.

Jegede AO, 'Rights Away from Home: Climate-induced Displacement of Indigenous Peoples and the Extraterritorial Application of the Kampala Convention' (2016) 16 *African Human Rights Law Journal* 58.

Kalinda T and others, *Use of Integrated Land Use Assessment (ILUA) data for forestry and agricultural policy review and analysis in Zambia* (Integrated Land Use Assessment Project 2008).

Kotzé L and Paterson A (eds), *The Role of the Judiciary in Environmental Governance: Comparative Perspectives* (Kluwer Law International 2009).

Kotzé L, 'The Conceptual Contours of Environmental Constitutionalism' (2015) 21 *Widener Law Review* 187.

Kovats S and others, 'Europe' in Vicente Barros and others (eds), *Climate Change 2014: Impacts, Adaptation, and Vulnerability. Part B: Regional Aspects. Contribution of Working Group II to the Fifth Assessment Report of the Intergovernmental Panel on Climate Change* (Cambridge University Press 2014) 1267–1326.

'Leaders: Can Tunisia Afford to Care about Climate Change?' https://tn.usembassy.gov/leaders-cantunisia-afford-to-care-about-climate-change/ accessed 20 January 2020.

May J and Daly E, *Global Environmental Constitutionalism* (Cambridge University Press 2015).

Michelman F, 'Constitutional Essentials' (paper delivered at conference on the 20th anniversary of Israel's human rights revolution, at the Knesset, Jerusalem, Israel, 19 December 2011).

Ministry of Environment and Natural Resources, Kenya's Intended Nationally Determined Contribution (INDC) (UNFCCC 2015).

Ministry of Water and Environment, 'Uganda's Intended Nationally Determined Contribution' (October 2015).

Mutua W, 'The African Human Rights System: A Critical Evaluation' http://hdr.undp.org/sites/default/files/mutua.pdf accessed 21 January 2020.

National Climate Change Policy (NCCP) 2015.

National Development Plan 2012.

National Planning Authority, Uganda Vision 2040 (National Planning Authority 2013).

NCESD, 'Framework Law on the National Charter for Environment and Sustainable Development' (NCESD 2014).

Niang I and others, 'Africa' in Vicente Barros and others (eds), *Climate Change 2014: Impacts, Adaptation, and Vulnerability. Part B: Regional Aspects. Contribution of Working Group II to the Fifth Assessment Report of the Intergovernmental Panel on Climate Change* (Cambridge University Press 2014).

Nurse L and others, 'Small Islands' in Vicente Barros and others (eds), *Climate Change 2014: Impacts, Adaptation, and Vulnerability. Part B: Regional Aspects. Contribution of Working Group II to the Fifth Assessment Report of the Intergovernmental Panel on Climate Change* (Cambridge University Press 2014) 1613–1654.

OAU, African Charter on Human and Peoples' Rights, CAB/LEG/67/3 rev. 5, 21 ILM 58, 27 June 1981 (entered into force 21 October 1986).

OHCHR, 'Submission of Mali to OHCHR Study "Human Rights and Climate Change"'. http://www.ohchr.org/Documents/Issues/ClimateChange/Submissions/Mali.pdf accessed 18 July 2019.

Parry M and others (eds), *Climate Change 2007: Impacts, Adaptation and Vulnerability. Contribution of Working Group II to the Fourth Assessment Report of the Intergovernmental Panel on Climate Change* (Cambridge University Press 2007) 133–357.

Posner E, 'Climate Change and International Human Rights Litigation: A Critical Appraisal' (2007) 155 *University of Pennsylvania Law Review* 1925.
Rawls J, *Political Liberalism* (Columbia University Press 2005).
Raz J, 'Rights and Politics' (1995) 71 (1) Indiana *Law Journal* 27.
Raz J, *The Morality of Freedom* (Oxford University Press 1988).
Republic of Zambia *Initial National Communication under the United Nations Framework Convention on Climate Change* (August 2002) (Zambia Initial National Communication).
Republic of Zambia National Adaptation Programme of Action on climate change (September 2007) (Zambia NAPA).
Rist G, *The History of Development: From Western Origins to Global Faith* (Zed Books 2009).
Royal Society & Royal Academy of Engineering, *Greenhouse Gas Removal* (Royal Society & Royal Academy of Engineering 2018).
Stocker T and others (eds), *The Physical Science Basis. Contribution of Working Group I to the 5th Assessment Report of the Intergovernmental Panel on Climate Change* (Cambridge University Press 2013).
Tunisia Live, 'Climate Change: The Threat No One's Talking About,' December 5, 2015 www.tunisia-live.net/2015/12/05/environment/ accessed 17 May 2019.
'UNFCCC' https://unfccc.int/news/new-era-of-global-climate-action-to-begin-under-paris-climate-change-agreement-0 accessed 20 December 2019.
UNGA, 'Protection of global climate for present and future generations of mankind' UNG.A. Res. 43/53, 70th plenary meeting 6 December 1988 (UNGA Resolution 43/53).
United Nations Human Rights Council (UNHRC), 'Report of the Office of the United Nations High Commissioner for Human Rights on the Relationship between Climate Change and Human Rights,' A/HRC/10/61 15 January 2009.
United Nations Human Rights Council (UNHRC), 'Resolution 10/4 – Human Rights and Climate Change,' adopted on 25 March 2009 at the 41st meeting of the Human Rights Council.
United Nations Human Rights Council (UNHRC), 'Resolution 18/22 – Human Rights and Climate Change,' adopted 30 September 2011 at the 37th meeting (A/HRC/RES/18/22).
United Nations Human Rights Council (UNHRC), 'Resolution 26 L/33 – Human Rights and Climate Change,' adopted on 23 June 2014 at the 26th session A/HRC/26/L.33.

Vinya R and others, 'Preliminary Study on the Drivers of Deforestation and Potential for REDD+ in Zambia' (A consultancy report prepared for the Forestry Department and FAO under the national UN-REDD+ Programme Ministry of Lands & Natural Resources. Lusaka, Zambia 2012) 2–3 (Zambia Preliminary Study).

World Bank, *World Development Report 2010: Development and Climate Change* (World Bank 2010).

Part III

Environmental Protection and Human Rights: Regional and National Case Studies

11

Implementing Human-Rights-Related Environmental Obligations in Ethiopia

Desalegn Amsalu

1 Introduction

The formal origin of environmental rights is traced back to the United Nations Conference on the Human Environment that was held in Stockholm from 5–16 June 1972. The first Principle of the Declaration that ensued from this conference linked environmental protection to human rights norms.[1] The Declaration influenced the legal and institutional development of states for the next two decades.[2] For instance, the

[1] Principle one of the Declaration: 'Man has the fundamental right to freedom, equality and adequate conditions of life, in an environment of a quality that permits a life of dignity and well-being, and he bears a solemn responsibility to protect and improve the environment for present and future generations. In this respect, policies promoting or perpetuating apartheid, racial segregation, discrimination, colonial and other forms of oppression and foreign domination stand condemned and must be eliminated.'

[2] Günther Handl, 'Declaration of the United Nations Conference on the Human Environment (Stockholm Declaration), 1972, and the Rio Declaration on Environment and Development,

D. Amsalu (✉)
Institute of Ethiopian Studies, Addis Ababa University, Addis Ababa, Ethiopia
e-mail: desalegn.amsalu@aau.edu.et

M. Addaney, A. Oluborode Jegede (eds.), *Human Rights and the Environment under African Union Law*, https://doi.org/10.1007/978-3-030-46523-0_11

United Nations Environment Program (UNEP) was established in the same year the Stockholm Declaration was adopted.[3]

Two decades later, on 3–14 June 1992, the United Nations held another landmark conference in Rio de Janeiro, Brazil, that focused on promoting responsible development approaches to halt the problems on the human environment. The Rio Conference like that of its precursor resulted in the Rio Declaration consisting of twenty-seven principles. Principle one of the Declaration provides that human beings should be 'at the center of concerns for sustainable development. They are entitled to a healthy and productive life in harmony with nature.'

Since the Rio Conference, many countries have incorporated environmental rights into their regional and national legal systems. The first report of its kind on environmental rule of law by UNEP presents, until 2017, at least 176 countries had environmental framework laws; 150 enshrined environmental protection or the right to a healthy environment in their constitutions; and 164 set up cabinet-level bodies in charge of environmental protection. Moreover, over 350 environmental courts and tribunals were established in over 50 countries and over 60 countries had at least some legal provisions on the citizens' environmental information.[4]

The global trajectories on the environment and the human rights have influenced the African continent too.[5] African countries have increasingly recognized the right to a healthy environment derived from the obligation of states to adopt the principles reflected in the Rio Declaration.[6] The 1981 African Charter on Human and Peoples' Rights under Article 24 included environmental rights as part of human rights, and by way of this, Africa as a whole became the first continent to formally introduce

1992' (United Nations Audiovisual Library of International Law 2012) http://legal.un.org/avl/pdf/ha/dunche/dunche_e.pdf, accessed 5 July 2019.

[3] Donal Worster, 'The Vulnerable Earth: Towards a Planetary History' in Donal Worster (ed) *The Ends of the Earth* (Cambridge University Press 1989) 3.

[4] United Nations Environment Programme, *Environmental Rule of Law: First Global Report* (United Nations Environmental Program 2019). https://wedocs.unep.org/bitstream/handle/20.500.11822/27279/Environmental_rule_of_law.pdf?sequence=1&isAllowed=y, accessed 4 July 2019.

[5] Jan Glazewski, *Environmental Law in South Africa* (Butterworths 2000) 4.

[6] Handl (note 2 as above).

environmental rights.[7] Many individual countries in Africa have also committed themselves to bring in environmental rights into their constitutions or other subsidiary laws to address new challenges the continent is facing concerning human rights to the environment.[8] The 1996 Constitution of South Africa and constitutions of other countries such as Botswana,[9] Cameroon,[10] Democratic Republic of Congo,[11] and Ethiopia incorporated environmental rights to its citizens. Until 2017, more than thirty African countries have incorporated environmental protection into their laws and a few of them have incorporated environmental rights specifically into their constitutions.[12]

This chapter investigates the extent of recognition and coverage of environmental rights in the Ethiopian legal framework on the environment. Besides, it explores legal and institutional constraints in the implementation of those rights guaranteeing environmental rights to the citizens. The chapter employed an interdisciplinary research approach. The part of the paper that focuses on the legal analyses of environmental rights follows a legal analyses method while it uses both legal analyses and empirical research methods for investigation of the practical aspects. For empirical data collection, interviews were conducted with responsible government officials as well as residents of Addis Ababa, Ethiopia's capital, on practical aspects of environmental rights. Direct observation of some selected industries also complemented the interview.

---

[7] Michael Faure & Willmien Plessis, 'Introduction' in Michael Faure & Willmien Plessis (eds) *The Balancing of Interests in Environmental Law in Africa* (Pretoria University Law Press 2012) xxix.

[8] Mekete Bekele, 'The Scope of Citizens' Environmental Rights Protection under Ethiopian Law' in Michael Faure & Willmien Plessis (eds) *The Balancing of Interests in Environmental Law in Africa* (Pretoria University Press 2012) 113.

[9] Bugalo Maripe, 'Development and Balancing of Interests in Environmental Law: The Case of Botswana' in Michael Faure & Willmien Plessis (eds) *The Balancing of Interests in Environmental Law in Africa* (Pretoria University Press 2012) 63.

[10] Oliver Fuo and Sama Semie, 'Cameroon's Environmental Framework Law and the Balancing of Interests in Socio-Economic Development' in Michael Faure & Willmien Plessis (eds) *The Balancing of Interests in Environmental Law in Africa* (2012) 75.

[11] Kihangi Kennedy, 'The environmental Law Framework of the Democratic Republic of the Congo and the Balancing of Interests' in Michael Faure & Willmien Plessis (eds) *The Balancing of Interests in Environmental Law in Africa* (Pretoria University Press 2012) 95.

[12] United Nations Environment Programme (note 4 as above).

2 Recognition and Coverage of Environmental Rights in Ethiopia

What constitutes environmental rights may vary from one jurisdiction to another and from time to time. The minimum bundle of rights common for many jurisdictions and that has been developed in international instruments is the right to live in a clean and healthy environment (also called the right to quality environment), the right to information, the right to public participation in environmental matters, and the right to access to environmental justice.[13] 'Other "new" rights, including rights for climate refugees, have arisen over recent years due to the acceleration of economic globalization and the associated environmental destruction and social disruption.'[14] The first four rights are explicitly covered in the Ethiopian jurisdiction. Like other African countries, Ethiopian legislation on environmental rights is influenced by the international trends, particularly by the principle of sustainable development.[15] It is believed that the Constitution of the 1995 Federal Democratic Republic of Ethiopia (FDRE), hereafter the 1995 FDRE Constitution or Constitution, was informed about the environmental rights by the 1992 conference on the Environment and Development.

The right to a clean and healthy environment gets a high regard from the 1995 FDRE Constitution of Ethiopia. In the first place, in its Article 43 (1), the Constitution enshrines the right to sustainable development for its citizens. Moreover, Article 44 (1) of the same Constitution provides 'All persons have the right to a clean and healthy environment'. What is more, in Article 92 (1), the Constitution provides the environmental objectives of the country where it obliges federal and regional governments to strive to ensure that Ethiopians live in a healthy and clean environment. When it was adopted on 21 August 1995, the FDRE

[13] Dinah Shelton and Alexandre Kiss, *Judicial Handbook of Environmental Law* (United Nations Environmental Program 2005) xix.

[14] Friends of the Earth International, 'Our Environment, Our Rights: Standing up for People and the Planet' (Issue 106 August 2004) https://www.foei.org/wp-content/uploads/2014/07/our-environment-our-rights.pdf accessed 03 July 2019.

[15] James Krueger, Aman Gebru, and Inku Asnake, 'Environmental Permitting in Ethiopia: No Restraint on "Unstoppable Growth?"' 1(1) *Haramaya Law Review* (2012)73.

Constitution was already one of about thirty-two countries in the world with the constitutionally protected right to a healthy environment.[16] In addition to the Constitution,[17] the 1997 Environmental Policy of Ethiopia aims to 'improve and enhance the health and quality of life of all Ethiopians and to promote sustainable social and economic development through the sound management and use of natural, human-made and cultural resources and the environment as a whole'.[18] Under its 'The Key Guiding Principles', the Policy in the first place mentions that 'Every person has the right to live in a healthy environment'. Moreover, the labor law[19] and the civil servants' law provide for the right to a clean and healthy environment.[20]

The environmental law regime of Ethiopia also acknowledges the right to information, the right to access to justice, and the right to public participation in environmental matters. Concerning the right to information, Ethiopia is one of the countries which guarantee the right to information at the constitutional level in addition to recognition given at the level of subsidiary laws. The environmental pollution control proclamation and its council of ministers regulation also recognize the right of citizens to information,[21] including the right to information about the environment.[22]

Access to justice is also a constitutional right in Ethiopia. According to the Constitution, 'everyone' has the right to bring to court 'a justiciable matter'[23] which includes matters relating to breach of environmental rights. In addition, specific provisions on the environmental standing exist on laws regulating environmental rights.[24] Based on the existing

[16] United Nations Environment Program (note 4 as above).

[17] The 1995 FDRE Constitution (Proclamation 1 of 1995) under sec 44.

[18] The 1997 Environment Policy under sec 2.3(a), hereafter the Policy.

[19] Labor Law Proclamation 377 of 2003 under secs 12(4) and 92.

[20] Federal Civil Servants Proclamation 515 of 2007 under sec 48.

[21] Prevention of Industrial Pollution Council of Ministers Regulations 159 of 2008 under sec 11.

[22] Sec 29(2).

[23] Sec 37(1).

[24] Environmental Pollution Control Proclamation 300 of 2002 under sec 11(1) says, 'everyone' has the right to environmental standing without necessarily asserting a vested interest. Moreover, Prevention of Industrial Pollution Council of Ministers Regulations 159 of 2008 under sec 27 lays down the legal regime on the environmental standing.

legal provisions, it is possible to conclude that there is a legal intention of liberalizing environmental standing in Ethiopia. Besides the State which has an inherent right,[25] individuals and non-governmental organizations have the right to an environmental standing without necessarily demonstrating a vested interest in a case involving breach of environmental rights.[26]

Finally, concerning the right to public participation, citizens are entitled not only to information about the environment but also to participation in decision-making on environmental issues. The Constitution includes public participation as a right,[27] which also should include participation in environmental issues. The Policy[28] and the environmental impact assessment law[29] also incorporate this right.

There are also attempts by the government of Ethiopia to incorporate environmental issues that directly or indirectly affect environmental rights into development programs of the country. In 2011, the country developed Climate-Resilient Green Economy (CRGE) strategy, the first of its kind for the whole of Africa.[30] The CRGE strategy builds on the Growth and Transformation Plan (GTP).[31] In GTP, the government aspired for Ethiopia to reach middle- income country by 2025. The first

---

[25] Rose Mwebaza, Philip Njuguna Mwanika & Wondowossen Sintayehu, 'Environmental Crimes in Ethiopia' (July 2009 Situation Report) 10, available at http://globalinitiative.net/wp-content/uploads/2017/01/ethz-environmental-crime-ethiopia.pdf accessed 03 March 2017.

[26] The only or one of few court cases is the following: According to the Constitution, 'any person' or 'everyone', including NGOs and environmentalists, have the right to environmental standing. However, the practicality of the rights given by the FDRE Constitution and Proclamation 300/2002 are being tested in court even though infrequently. A famous case for NGOs is *Action Professionals' Association for the People (APAP)* vs. *Environmental Protection Authority (EPA)*. This was the case where APAP sued EPA for the latter's failure to stop industries (leather industries) from polluting rivers and causing suffering to residents. First APAP requested for an administrative remedy at EPA, but the latter replied it is trying its best but could not stop since there was no then an ambient environmental standard. APAP took the case to First Instance Court which ruled that EPA cannot be sued for it is not the polluter. APAP took the case again to Federal High Court which gave similar decision: sec 11(2) of the Proclamation cannot apply to EPA since it is not polluter.

[27] Secs 43(4), 89(6), and (7).

[28] Sec 4.2.

[29] Environmental Impact Assessment Proclamation 299 of 2002 under sec 15.

[30] Susannah Fisher, 'Low Carbon Resilient Development in the Least Developed Countries'. (IIED Issue Paper 2013) http://pubs.iied.org/pdfs/10049IIED.pdf. Accessed 12 August 2019.

[31] Federal Democratic Republic of Ethiopia, 'Growth and Transformation Plan (GTP) 2010/11–2014/15'. (Ministry of Finance and Economic Development 2010).

GTP period was from 2010/11 to 2014/15. The CRGE strategy is also integrated into the latest development program of the country, which is the second Growth and Transformation Plan set for 2015–2020 (GTP II).[32]

Ethiopia is also party to several regional and international agreements that relate to the environment and the human rights. Needless to mention, the African Union Charter on Human and Peoples' Rights, the country is a signatory state to the Revised African Convention on the Conservation of Nature and Natural Resources that was adopted in 2017. It has also ratified the Bamako Convention on the Ban of the Import into Africa and the Control of Trans boundary Movement and Management of Hazardous Wastes within Africa that entered into force in 1998. Ethiopia also represented Africa to spearhead international negotiations that directly or indirectly affect environmental rights. For instance, it played an important role in shaping the African Union position on climate change.[33]

3 Legislative Challenges

As discussed in the foregoing section, the 1995 Constitution, 1997 Environment Policy and some framework laws of Ethiopia recognize and cover environmental rights that are state-of-the-art under the regional and international environmental law regimes. However, the country faces several challenges in its effort to implement the environmental rights. A chapter cannot address if one recounts all challenges. A few previous studies discussed factors which are lack of capacity and political commitment for enforcement by the government, lack of environmental rights awareness by the people, the state actors and even the judiciary, delayed justice system when people want to take their rights to justice organs, lack of environmental accountability by economic and political actors,

[32] Federal Democratic Republic of Ethiopia, 'Growth and Transformation Plan (GTP II) 2015/16–2019/20'. (Ministry of Finance and Economic Development 2016).

[33] Mahlet Eyasu, 'Ethiopia: Taking Climate Change Issues Seriously' (Climate Action Network International 15 June 2011) http://www.climatenetwork.org/can-blog/ethiopia-taking-climate-change-issues-seriously 23 November 2019.

and emphasis on economic growth at the expense of environmental rights.[34] This section focuses only on factors related to inadequate legislative and institutional practices to execute the flamboyant recognition of environmental rights by the supreme law of the land of the country, which is the Constitution, and other a few framework laws.

Lack of Timely Legislative Making or Revision

Timely lawmaking is essential to realize environmental rights. Even though many African countries embraced environmental rights in time and formally into their constitutions, policies, or proclamations, they seem to learn to legislate framework laws, which refer to more specific laws than constitutions and policies, for several environmental challenges long after they are overdue. Even when they exist, they are not adequate to address multifaceted and emerging aspects of environmental problems that affect the implementation of environmental rights. For the Ethiopian case, it is possible to illustrate this by taking an example of the lacuna of laws governing pollution caused by production, import, or use of chemicals.

At least 84% of about hundred million populations in Ethiopia live in rural areas based on agriculture. Yet, along with the growth of investment and economy, particularly from 2006 to 2016, industries have grown in the country significantly for various business purposes. In what is called Agriculture Development Led Industrialization (ADLI) strategy in effect since 1993, agriculture is expected to grow and lead industrialization and then gradually relinquish itself on behalf of the latter.[35] All development programs since ADLI tried to execute this strategy. GTP II (2015–2020),

[34] Desalegn Amsalu, 'Environmental Rights in Ethiopia: Shifting from Theory to Practical Realization' (2018)9 *Journal of Sustainable Development Law and Policy* 2, 48–66. https://www.ajol.info/index.php/jsdlp/article/view/181258/170652 accessed 20 June 2019; Mwebaza (as in 25 above).

[35] Lulit Mitik, Solomon Lemma, & Befekadu Behuta, 'Public Spending, ADLI, and Alternative Scenarios for Ethiopia: A Dynamic CGE Framework Analysis' (8th PEP General Meeting June 2010).

the current overarching development program in Ethiopia is also developed along this line.[36]

A consequence of the growth of industries is a growth in chemical manufacturing, import, and use. Investment in local manufacturing industries is increasing. For instance, the Chemical and Construction Inputs Industry Development Institute indicated that twenty-two chemical and construction input industries have newly entered production in 2015/16.[37] In 2017/18, Ethiopia's chemical and manufacturing material inputs production showed nearly 18% growth.[38] Absorption from abroad through import and consumption of chemicals in various sectors of the economy also grew in response to the demands of economic growth. Moreover, the country was criticized for having the largest accumulations of obsolete pesticides in the whole of Africa. It was estimated that there were 402 stores at 250 sites containing 1500 tons of outdated pesticides.[39]

There is no doubt about the tremendous benefit of chemicals. Everything in the world is made of them whether organically occurring or artificially synthesized by man. Particularly, some chemicals are created to cause harm for some purpose. But, if not used properly they have also a devastating impact on human health and the environment. Among others, there is also increasing evidence as to the effect of chemicals on our environment too. The priority concerns of their effects on the environment include air pollution, pollution of surface and groundwater, and soil pollution cumulatively compromising efforts toward the realization of environmental rights. UNEP has coordinated different conventions to address the growing challenge of chemicals on our health and environment. Ethiopia is a party to some important international instruments.[40]

---

[36] Federal Democratic Republic of Ethiopia (as in 31 above).

[37] Haftu Gebreegziabher, 'Ethiopia: Chemical, Construction Input Industries Begin Production' *The Ethiopian Herald* (Addis Ababa, 29 July 2016).

[38] Chemical and Construction Inputs Industry Development Institute, 'The Institute Holds Consultation with Investors and Stakeholders in the Sector' (in Amharic). https://www.facebook.com/Chemical-and-Construction-Inputs-Industry-Development-Institute-280553259018222/, accessed October 9, 2019.

[39] MoARD, 'Official Report of the Africa Stockpiles program' (MoARD 2007).

[40] Ethiopia acceded 'Basel Convention on the Control of Transboundary Movements of Hazardous Wastes and Their Disposal' in 2000, and 'Rotterdam Convention on the Prior Informed Consent Procedure for Certain Hazardous Chemicals and Pesticides in International Trade' in 2003.

Though some may appear too remote, we have the most evidential and notable effects of chemicals on the environment in Ethiopia. Residents witness rivers of Addis Ababa, for example, which are extremely polluted because of the effluents of chemical substances from industries. Some toxic chemicals find their way from landfill waste sites into our ground and surface water causing at least health ailments. Chemical fertilizers and nutrients run-off from farms and gardens cause the buildup of toxic algae in rivers, making them uninhabitable to aquatic organisms and unpleasant for humans. Chemicals also enter the air as emissions. One can feel a choking air of our capital city as the result of increasing chemical discharge into the atmosphere, as will be discussed more below.

Despite the growing hazards, the lawmaking for the safe management of chemicals is inadequate or only retroactive. Lawmaking is not strengthened as actual or anticipated chemical production and use grow. As many scholars have argued,[41] interest to grow in an economy far greater overweighs the interest to protect the environment and promote environmental rights in many African countries. Yet, this is not to suggest that laws regulating chemicals affecting the environment are no longer actively being made in Ethiopia. Indeed, new laws are made to regulate environmental problems. An example of this is a recent Proclamation No. 1090/2018 to provide for hazardous waste management and disposal control.[42] In its Proclamation No. 513/2007, Ethiopia also banned either manufacturing or import of easily decomposable plastic bags with a thickness of less than 0.03 mm or any indecomposable plastic bag.[43] But, taken all together, there are not adequate framework laws for immense actual and potential detriments of chemicals against the human rights to the environment. The dissemination and subsequent enforcement of these legal instruments by bodies responsible for their implementation are also weak.

Besides, there is a lack of timely revision of instruments that already exist. For example, it is more than twenty years (1997–2019) since the

[41] See for example Amsalu (as in 34 above) and Maripe (as in 9 above).

[42] Proclamation No. 1090 of 2018 'Hazardous Waste Management and Disposal Control Proclamation'.

[43] Proclamation Number 513 of 2007 'Solid Waste Management Proclamation'.

Environmental Policy of Ethiopia has been in place. However, it does not consider strategies of the country adopted long after the policy came into effect. For example, Ethiopia's CRGE strategy of 2011 was not envisioned in the policy. Strategies come before policy, which is not quite normal in the procedure of legislative drafting. Taking note of this, a revised policy is only in draft.

Finally, hitherto points in this section attest to the fact that legislative practice in Ethiopia, as is presumably the case in many developing countries, is reactive, rather than proactive. Most of the laws and the policy seem to be made in an attempt to solve matters which already have arisen. In proactive legislation, however, legislation is made before the problems become detrimental at least to a serious level. In principle, it is possible to follow both proactive and reactive approaches, but the choice must be made based on the nature of environmental issues. It is possible to develop policies and laws in response to certain problems, but not after the problems cause damage. However, in addition to the fact that it is not comprehensive, the environmental framework law in Ethiopia is not dynamic. The state of the environment can be perceived both as something which is not too often changing on the one hand and as something which is dynamic on the other, depending on which environmental issues we talk about. Even several years may be too short a time to measure climate change, for example. On the other hand, there are also emerging environmental issues that need to be addressed by emerging principles and mechanisms. However, the Ethiopian policy and legal regime is not as dynamic as the environmental problems require, which is so far a common characteristic to developing countries albeit at various levels.

Lack of Standards

In addition to the absence of framework laws, a lack of standards is another problem. Requirements that specify the maximum amount of pollutant(s) an industry be allowed to have in its surroundings should be given by the laws and then by specific standards. Most industrialized countries adopt environmental policies that involve some combination of emissions controls and ambient environmental quality standards.[44] But, Ethiopia's environmental legal framework lacks a number of standards governing emission of any type.

The pollution control proclamation can be mentioned as an example. There are two legal mechanisms to enforce environmental problems of pollution in Ethiopia.[45] The first one is administrative and the second one court proceedings. The administrative mechanisms include requesting a remedy from any local or national governmental administrative entity. This is an out-of-court mechanism of ensuring legal rights to a healthy environment. On the other hand, actions can also be taken to court if administrative remedies are not possible. In the latter case again, the court decision to one's damage as the result of a lack of a healthy environment can be of civil or criminal remedy. Civil remedies include various options such as compensation for damage and injunction.[46] Criminal remedies on their own include fine, imprisonment, or both.[47]

However, the pollution control proclamation is not detailed enough for self-execution. When an industry pollutes the environment, there is no regulation and directives on how to make the environmental valuation. When, for example, a leather industry pollutes the environment by discharging waste without proper treatment, there must be an environmental valuation conducted in order to determine the nature and extent of damage to the environment and human health. However, there is no valuation mechanism for environmental pollution in Ethiopia today. What is being done is just penalize the polluter based on a fixed sum set on the proclamation, that is, a maximum of 10,000 birr for natural persons and a maximum of 20,000 for juridical persons.[48] Even more, what is being done today is focus on correction: just to write a letter for a

---

[44] Stephen Hamilton and Till Requate, 'Emissions Standards and Ambient Environmental Quality Standards in Stochastic Receiving Media' (2010) *JEL Classification*: D62; Q38; Q50. Available at: https://arefiles.ucdavis.edu/uploads/filer_public/2014/03/27/hamilton-emissions-standards.pdf accessed 28 September 2016.

[45] Proclamation 300 of 2000.

[46] See for example Article 8(2) and Article 10 of Proclamation 300/2000 (Environmental Pollution Control Proclamation) and Article 10 of Regulations No. 159/2008 (Prevention of Industrial Pollution Regulations).

[47] See for example, Article 519 of the 2004 Criminal Code of Ethiopia.

[48] Article 12 (1(a)) and 1(b) of Proclamation 300/2002.

company to stop polluting the environment. Even what is set in the proclamation (10,000 birr) is not implemented. According to an expert informant, 'no one has ever incurred monetary penalty arising from the law, no industry has ever been closed for pollution even though they continue polluting the environment even after they were issued a "warning" letter'.

According to a 2012 Vehicular-Smogless Air for Ethiopia (V-SAFE) report on air pollution in Addis Ababa caused by gasoline-powered vehicles, approximately 50% of the city's vehicles produced about 90% of the hydrocarbon (HC) and carbon monoxide (CO) emissions. Furthermore, the same report showed that the city had eleven times, on average, higher concentrations of particulate matter (PM) levels in the city's ambient air than comparable USA or World Health Organization (WHO) standards.[49] The study recommended immediate intervention by the government since the problem has serious health implications. Vehicles are presumably the largest source of air pollution in Addis Ababa with its annual vehicle population growth rate of about 10%.[50] The 2011 Transport Policy of Addis Ababa mentions increasing air pollution as a major environmental problem from the transport sector such as from 'decrepit vehicles'.[51] The policy also identifies 'Absence of legal limits and effective control over air and noise pollution'[52] as an environmental issue in the city and the policy mentions several measures to be taken to address the problem such as favoring mass transport. Ethiopia has 831,000 cars in 2017, of which at least 76% were found in Addis Ababa.[53] Indeed, in Addis Ababa, cars are the main source of air pollution since the majority

[49] Robert Benjaminson, Derebew Shenkute, Garrett Torgerson, Getaneh Gebre, and Rosemary Gallavan, 'The Effect of gasoline-fueled Vehicle Exhaust Emissions on Air pollution in Addis Ababa, Ethiopia' (V-SAFE 2012) file:///C:/Users/3020/Downloads/The%20Effect%20of%20Gasoline-Fueled.pdf accessed 05 July 2019.

[50] World Bank, 'World Development Indicators' *The World Bank* (2015), http://documents.worldbank.org/cuarted/en/795941468338533334/world-development-indicators-2015 accessed 12 March 2015.

[51] See Addis Ababa City Government Transport Policy issued in 2011 at 8.

[52] Ibid. at 14.

[53] https://www.2merkato.com/news/alerts/5294-ethiopia-has-more-than-831000-vehicles-on-its-streets accessed 5 July 2019.

of cars are very old.[54] Yet, there is no standard for pollution of vehicular emission.

The Environmental Protection Authority (EPA), now Environment, Forest and Climate Change Commission, hereafter EPA, functions as an autonomous public institution of the Federal Government.[55] EPA has an Environmental Council (EC), which oversees EPA activities as well as the activities of sectorial agencies and environmental units with respect to environmental management. Even though EC existed since 1995, it became inactive until 2007. Moreover, it has not yet approved any directive or guideline prepared by the EPA.[56] It has drafted only general and nonbinding and draft guidelines published by the EPA in 2000 and 2003.[57] There are some draft regulations and standards, but they remained only in draft since they have never been approved and entered into force.

4 Institutional Factors

Lack of Inter-sectoral Integration

Ethiopia's Environmental Protection Authority (EPA) was established in 1995 under the then Ministry of Natural Resources Development and Environmental Protection. It was re-established in 2002 as an independent institution with a mandate to regulate and monitor all activities in the country that have an impact on the environment, including its responsibility for climate change policies in Ethiopia.[58] EPA became under the Ministry of Environment, Forest and Climate Change

[54] Ibid. at 153.

[55] See proclamation No. 295/2002.

[56] Mellesse Damtie and Mesfin Bayou, 'Overview of Environmental Impact Assessment in Ethiopia: Gaps and Challenges', (MELCA Mahiber 2008). Available at http://www.melca-ethiopia.org/EIA.html, accessed 5 September 2017.

[57] Dominik Ruffeis, Willibald Loiskandl, Seleshi Bekele Awulachew and Eline Boelee, 'Evaluation of the Environmental Policy and Impact Assessment Process in Ethiopia' 28(1) *Impact Assessment and Project Appraisal* (2010) 29–40.

[58] Proclamation No. 295 of 2002 A Proclamation to Provide for the Establishment of Environmental Protection Organs.

(MEFCC) which was established in 2015.[59] In 2018, MEFCC was downsized from the ministerial level to a commission.[60]

Ethiopia's regional states also have a lower level of environmental management institutions. In 2002, the establishment of independent environmental agencies at the regional level was codified. The regional organs responsible for the environment were originally accountable for the implementation of national and regional environmental policies. The federal environmental organ has assisted its regional counterparts in developing their regional-level environmental conservation strategies, through the provision of equipment, training, and environmental education.[61]

In 2002, Ethiopia introduced further environmental decentralization to the *woreda* level, the second lowest-level administrative hierarchy in Ethiopia. The official goal of this devolution was bringing the government closer to the people and strengthening its responsiveness to local needs.[62] In terms of environmental management, the decentralization approach has shifted responsibilities for environmental service delivery and policy implementation from the regional level to the *woreda*. This allowed the regional environmental agencies to re-focus their efforts on developing overarching regional environmental legislation while maintaining only a supervisory role to woreda-level environmental management. In addition to environmental organs vertically placed at federal, regional, and *woreda* levels, there are also sectoral specific ministries and agencies, such as for forest, land, and water.

Integration is an important environmental law principle and environmental rights cannot be implemented without institutions directly or indirectly dealing with the environment harmoniously functioning together. Noting the importance of integration among or between the government or non-government organizations concerned with the

[59] Proclamation No. 916 of 2015 'A Proclamation to Provide for the Definition of Powers and Duties of the Executive Organs of the Federal Democratic Republic of Ethiopia'.

[60] Proclamation No. 1097 of 2018 'Definition of Powers and Duties of the Executive Organs of the Federal Democratic Republic of Ethiopia Proclamation'.

[61] Ministry of Finance and Economic Development (MOFED), 'Ethiopia: Sustainable Development and Poverty Reduction Program' (Addis Ababa 2002).

[62] Ibid.

environment, the principle of integration has been included in the Ethiopian legal framework as well as in the international instruments.[63] Ethiopia's Environment Policy provides for a key guiding principle of integration.[64] The principle of integration is also included in the international environmental law instruments.[65]

There are attempts in the integration of relevant sectors as far as the environment is concerned in Ethiopia. For example, the Ethiopian Water Sector Policy and Ethiopian Water Sector Strategy are good examples. The Water Policy provides specific policy directions for environmental and water resource protection and conservation, use, and management of technology and engineering in the sector.[66] Moreover, Commercial Registration and Business Licensing proclamation provides an industry's license that can be suspended if it has failed to maintain the standards of health and sanitary conditions, environmental protection, safety measures, and the quality of its products or services or has done any other illegal activities in connection with commercial activity or that it failed to observe the conditions under which the business license is issued, as confirmed by the concerned government institution or the appropriate authority.[67]

However, there are still striking gaps in the integration effort from the institutional set-up to the legal and then the practice. As studies by other scholars also corroborate, 'although the [federal] EPA is accountable to the Prime Minister's office, there is no linkage, no desk is installed and no representative is assigned to deal with environmental issues. The EPA is not a member of the Ministerial Council and only reports to the Parliament.'[68] A similar opinion was forwarded by a high-level expert

[63] For a study of environmental non-governmental organizations in Ethiopia, see Matthew Cheever, Katie Graichen, Daniel Homeier, Jillian Howell, Olivia Kefauver, and Tom Kimball, 'Environmental Policy Review: Key Issues in Ethiopia'. (Colby College Environmental Policy Group 2011). http://web.colby.edu/eastafricaupdate/files/2012/01/Environmental-Policy-Review-2011_color_small.pdf accessed 12 November 2012.

[64] Article 2.3(p) and (s) of the Policy.

[65] Principle 4 of the Rio Declaration and Principle 13 of the Stockholm declaration also incorporate this principle.

[66] Ethiopia Water Sector Policy 2001.

[67] Proclamation 686/2010 Commercial Licensing and Registration Proclamation.

[68] Ruffeis (as in 56 above).

informant regarding Addis Ababa Environment Protection Agency (AAEPA) who said the agency has no representation in the executive council of Addis Ababa city.

Pursuant to the principle of integration, any licensing agency has to inform the authority prior to issuing an investment license. In Ethiopia, the responsible organ is Ethiopian Investment Agency now Ethiopian Investment Commission (EIC). Interviews with high-level experts of the AAEPA indicated three obstacles related to this: no functional linkage exists between the AAEPA and EIC. The reasons are diverse. The most crucial aspect in this context is the status of the specific laws. The EIA law is now very dated and has not been adopted since the introduction of the proclamation in 2002. However, the investment proclamation[69] is very dynamic and the current investment proclamation does not include a requirement for environmental impact assessment. This proclamation authorizes EIC can issue the investment license without any requirement for an EIA. Based on the principle that newer laws are superior to older ones, the Investment Proclamation prevails over the EIA proclamation. Therefore the EPA does not have any legal means to enforce the law. In addition to this, no mechanism or rule exists which administers the abidance of the law. According to an informant to this research at AAEPA, the authority knows not more than 5000 industries that have cleared of environmental procedures. However, the list of industries at trade and industries office is at least 15,000. This means that the AAEPA does not know the environmental status of more than 10,000 industries.

Lack of Technological Infrastructure

Modern technology has become a total phenomenon for mechanisms of environmental law; the defining forces of new environmental management in which efficiency is no longer an option. Without modern technology, institutions working in the environment cannot realize human rights to the environment. Indeed, enforcement mechanisms of environmental law require technological infrastructure.

[69] Investment Proclamation No. 769/2012, Proclamation No. 769/2012, Official Gazette No. 63, 17 September 2012.

However, mechanisms in the Ethiopian legal framework for the environment are plagued by a lack of technological infrastructure. The notable example here is a lack of infrastructure for treatment of hazardous chemicals, and even to detect a hazardous chemical purported to have been legally imported. According to the Basel Convention on the Control of Transboundary Movements of Hazardous Wastes and their Disposal of 1992, which Ethiopia has ratified and entered into force in 2000,[70] forty-nine hazardous chemicals are identified. The convention also lists five methods of treatment. According to expert informants, to manage these chemicals, no infrastructure, such as treatment technology is available. This means that EIA is not supported by emerging technologies.

Technology is changing so fast that knowledge about specifics can quickly become obsolete. However, in Ethiopia, even when they exist, most of the existing technologies possessed by industries or government regulatory agencies are obsolete, and hence nominal or dysfunctional. In an observation during a fieldwork, often words came about the industries, such as the leather industry, have waste treatment technologies which are obsolete, and often standing dysfunctional only temporarily fixed to show authorities who came for an environmental monitoring purpose. Even it is the opinion of many expert informants that industries import old technologies at a low price. As an informant indicated, it looks like some industries import technologies that other developed countries have already alienated for disposal.

The importance of technology to solve or cope with environmental problems is mentioned at different national and international instruments. For example, the Environmental Policy of Ethiopia,[71] as well as the Rio Declaration,[72] notes the importance of adoption and dissemination of technologies that use resources efficiently and support for

[70] The Basel Convention on the Control of Transboundary Movements of Hazardous Wastes and Their Disposal, usually known as the Basel Convention, is an international treaty intended to reduce transboundary movements of waste in general, and specifically to prevent transfer of hazardous waste from developed to less developed countries. The Convention is also intended to minimize the amount and toxicity of wastes generation at the source, and to extend assistance to least developed countries in management of hazardous and other wastes they generate (see the Convention).

[71] At 2(3)(e) and 4(8) (d & g).

[72] Principle 7 of the Rio Declaration.

communities and individuals to use and manage such technologies. However, there is a little commitment to back environmental problems management practices with technological infrastructure in Ethiopia.

Lack of Adequate and Qualified Manpower

An adequate supply of environmental specialists is mandatory in order to effectively implement mechanisms in the environmental legal framework and enhance environmental rights implementation. However, a shortage of environmental manpower is critical in the regulatory agencies. There is a lack of manpower both in quantity, which means the number of experts who can handle a volume of work in a particular area, and quality, which means a lack of manpower who can handle specialized technical environmental problems such as the valuation of hazardous chemicals.

This can be seen by investigating AAEPA in 2017. AAEPA had only three personnel who did the environmental review. Environmental review is fact-checking of whether EIA study documents match the reality.[73] There were over forty private consulting firms who made environmental impact assessment studies. An industry may enter into a contract with any of these consultants in order to do the environmental impact assessment of its project. However, the study made by private consulting firms should be approved by AAEPA experts. AAEPA reviewers make sure if the document presented as an EIA is valid after having checked realities on the ground. But, to do the review, there were only three experts, according to an informant in the AAEPA, and this number is far from adequate. The informant who requested anonymity added:

> Recently, Addis Ababa City Administration issued a directive obliging all future building construction projects to undergo EIA (which was not there before). All of a sudden within two months, EIA documents of over 500 building projects arrived on the table of the three experts at AAEA. How can only three experts undertake a review process of all these projects? I

[73] See Article 9 of Proclamation 299/2002 "Environmental Impact Assessment Proclamation".

> asked for the Ministry of Public Service for an accelerated procedure to hire seven additional personnel. However, months have passed without even a response.[74]

The skill problem of those who are hired is also critical. Informants complain about not only the inadequate number of employees but also technical training given, to upgrade their technical skills and knowledge, to those who are already employed. They often mentioned favoritism in specialized skills training. For example, someone is favored to an education or a short-term skills training abroad not because they are doing a great job or can be the ones who close the technical skills gap in the work, but for reasons of political allegiance. Informants regret that even those whose jobs are not directly related to a required skill training can be sent abroad to favor them because they are in political, not environmental, service. It is not contestable to say that leaders who practice favoritism in the workplace have no chance of effective work. Identifying the skills needed to provide a solution to work, and quantitative and qualitative supply of adequate manpower is essential for the enforcement of mechanisms in the environmental law.

5 Conclusion

In Ethiopia, the environmental rights of the public are not only constitutionally guaranteed but also among the guiding principles underpinning the interpretation of the FDRE Constitution and the designing and implementation of public policies and legislation. The Ethiopian case shows that the country's Constitution of 1995 and the Environment Policy of 1997 and a few proclamations recognize the environmental rights to citizens. Seen from this point, indeed, Ethiopia has a robust guarantee of environmental rights for its people. However, the practical realization of the rights is hindered by several challenges, from what this chapter has presented, by a lack of timely legislative practices as well as poor institutional capacities. While the global recognition and a

[74] Expert informant, Addis Ababa Environment Protection Authority, 12 January 2017.

concurrent recognition by the national legal regimes of environmental rights are long overdue, their implementation, particularly for the developing countries, is lagging. There are different factors for this, but as the case study in this chapter attested, the lawmaking itself is hanged only on the higher-level laws. There needs to be a comprehensive, adequate, and timely framework laws and effective institutions that are capable of implementing them.

References

Bugalo Maripe, 'Development and Balancing of Interests in Environmental Law: The Case of Botswana' in Michael Faure & Willmien Plessis (eds), *The Balancing of Interests in Environmental Law in Africa* (Pretoria University Press 2012).

Chemical and Construction Inputs Industry Development Institute, 'The Institute Holds Consultation with Investors and Stakeholders in the Sector' (in Amharic). https://www.facebook.com/Chemical-and-Construction-Inputs-Industry-Development-Institute-280553259018222/ accessed October 9, 2019.

Desalegn Amsalu, 'Environmental Rights in Ethiopia: Shifting from Theory to Practical Realization' (2018) 9 *Journal of Sustainable Development Law and Policy* 2, 48, 66. https://www.ajol.info/index.php/jsdlp/article/view/181258/170652 accessed 20 June 2019.

Dinah Shelton and Alexandre Kiss, *Judicial Handbook of Environmental Law* (United Nations Environmental Program 2005).

Dominik Ruffeis, Willibald Loiskandl, Seleshi Bekele Awulachew and Eline Boelee, 'Evaluation of the Environmental Policy and Impact Assessment Process in Ethiopia' (2010) 28 *Impact Assessment and Project Appraisal* 1, 29, 40.

Donal Worster, 'The Vulnerable Earth: Towards a Planetary History' in Donal Worster (ed), *The Ends of the Earth* (Cambridge University Press 1989).

Federal Democratic Republic of Ethiopia, 'Growth and Transformation Plan (GTP) 2010/11–2014/15' (Ministry of Finance and Economic Development 2010).

Federal Democratic Republic of Ethiopia, 'Growth and Transformation Plan (GTP II) 2015/16–2019/20' (Ministry of Finance and Economic Development 2016).

Friends of the Earth International, 'Our Environment, Our Rights: Standing up for People and the Planet' (*rights* 08 November 2003) https://www.foei.org/wp-content/uploads/2014/07/our-environment-our-rights.pdf accessed 03 July 2019.
Günther Handl, 'Declaration of the United Nations Conference on the Human Environment (Stockholm Declaration), 1972, and the Rio Declaration on Environment and Development, 1992' (United Nations Audiovisual Library of International Law 2012) http://legal.un.org/avl/pdf/ha/dunche/dunche_e.pdf accessed 5 July 2019.
Haftu Gebreegziabher, 'Ethiopia: Chemical, Construction Input Industries Begin Production' *The Ethiopian Herald* (Addis Ababa, 29 July 2016).
James Krueger, Aman Gebru, and Inku Asnake, 'Environmental Permitting in Ethiopia: No Restraint on "Unstoppable Growth?"' (2012)1 *Haramaya Law Review* 1, 73, 102.
Jan Glazewski, *Environmental Law in South Africa* (Butterworths 2000).
Kihangi Kennedy, 'The environmental Law Framework of the Democratic Republic of the Congo and the Balancing of Interests' in Michael Faure & Willmien Plessis (eds), *The Balancing of Interests in Environmental Law in Africa* (Pretoria University Press 2012).
Lulit Mitik, Solomon Lemma, & Befekadu Behuta, 'Public Spending, ADLI, and Alternative Scenarios for Ethiopia: A Dynamic CGE Framework Analysis' *8th PEP General Meeting* (Poverty and Economic Research Network 2010). http://www.pep-net.org/sites/pep-net.org/files/typo3doc/pdf/files_events/8th-PEPmeeting2010-Dakar/papers/Lulit_Mitik.pdf 20 November 2019.
Mahlet Eyasu, 'Taking Climate Change Issues Seriously' (Climate Action Network International 15 June 2011) http://www.climatenetwork.org/can-blog/ethiopia-taking-climate-change-issues-seriously 23 November 2019.
Matthew Cheever, Katie Graichen, Daniel Homeier, Jillian Howell, Olivia Kefauver, and Tom Kimball, 'Environmental Policy Review: Key Issues in Ethiopia' (Colby College Environmental Policy Group 2011). http://web.colby.edu/eastafricaupdate/files/2012/01/Environmental-Policy-Review-2011_color_small.pdf accessed 12 November 2012.
Mekete Bekele, 'The Scope of Citizens' Environmental Rights Protection under Ethiopian Law' in Michael Faure & Willmien Plessis (eds), *The Balancing of Interests in Environmental Law in Africa* (Pretoria University Press 2012).
Mellesse Damtie and Mesfin Bayou, 'Overview of Environmental Impact Assessment in Ethiopia: Gaps and Challenges' (MELCA Mahiber 2008) http://www.melca-ethiopia.org/EIA.html accessed 5 September 2017.

Michael Faure and Willmien Plessis, 'Introduction' in M Faure & W Plessis (eds), *The Balancing of Interests in Environmental Law in Africa* (Pretoria University Law Press 2012).
Ministry of Finance and Economic Development (MOFED), 'Ethiopia: Sustainable Development and Poverty Reduction Program' (Addis Ababa 2002).
Ministry of Agriculture and Rural Development (MOARD), 'Official Report of the Africa Stockpiles Program' (MoARD 2007).
Oliver Fuo and Sama Semie, 'Cameroon's Environmental Framework Law and the Balancing of Interests in Socio-Economic Development' in Michael Faure & Willmien Plessis (eds), *The Balancing of Interests in Environmental Law in Africa* (Pretoria University Press 2012).
Robert Benjaminson, Derebew Shenkute, Garrett Torgerson, Getaneh Gebre, and Rosemary Gallavan, 'The Effect of gasoline-fueled Vehicle Exhaust Emissions on Air pollution in Addis Ababa, Ethiopia' (V-SAFE 2012) accessed 05 July 2019.
Rose Mwebaza, Philip Njuguna Mwanika & Wondowossen Sintayehu, 'Environmental Crimes in Ethiopia' (July 2009 Situation Report) 10, available at http://globalinitiative.net/wp-content/uploads/2017/01/ethz-environmental-crime-ethiopia.pdf accessed 03 March.
Stephen Hamilton and Till Requate, 'Emissions Standards and Ambient Environmental Quality Standards in Stochastic Receiving Media' (2010) *JEL Classification* D62; Q38; Q50. https://arefiles.ucdavis.edu/uploads/filer_public/2014/03/27/hamilton-emissions-standards.pdf accessed 28 September 2016.
Susannah Fisher, 'Low Carbon Resilient Development in the Least Developed Countries'. (IIED Issue Paper 2013) http://pubs.iied.org/pdfs/10049IIED.pdf. Accessed 12 August 2019. https://wedocs.unep.org/bitstream/handle/20.500.11822/27279/Environmental_rule_of_law.pdf?sequence=1&isAllowed=y, accessed 4 July 2019.
United Nations Environment Programme, 'Environmental Rule of Law: First Global Report' (United Nations Environmental Program 2019).
World Bank, 'World Development Indicators' (The World Bank 2015) http://documents.worldbank.org/cuarted/en/795941468338533334/world-development-indicators-2015 accessed 12 March 2015.

12

Women's Right to Participation in Environmental Decision-Making in Uganda

Charlotte Kabaseke

1 Introduction

In order for the environment to be duly protected, there is need for a robust legal, institutional and administrative framework.[1] In order for effective implementation of the framework to be achieved, there is the need for involvement of all stakeholders including political leaders and, most importantly, the grassroots people, who are the subjects of the law.[2] The grassroots people should not only be involved in ensuring proper implementation of the law concerning the environment, but should be involved in its formulation as well as all decisions concerning the environment. This is commonly known as public participation. Public

[1] Donald Kaniaru and Lal Karukulasiriya, 'Capacity Building in Environmental Law,' in Sun Lin and Karukulasiriya L (eds) *UNEP's New Way Forward: Environmental Law and Sustainable Development* 171–184.

[2] Jeroen van Bekhoven, 'Public Participation as a General Principle in International Environmental Law: Its Current Status and Real Impact' 2016 (11) 2 *National Taiwan University Law Review* 229.

C. Kabaseke (✉)
Wuhan University, Wuhan, China

M. Addaney, A. Oluborode Jegede (eds.), *Human Rights and the Environment under African Union Law*, https://doi.org/10.1007/978-3-030-46523-0_12

participation in environmental decision-making was first alluded to by the Stockholm Declaration on Human Environment (Stockholm)[3] and later the United Nations Charter for Nature.[4] The Rio declaration[5] later reiterated the provisions on public participation. The principle of public participation has a strong connection to human rights and has been argued to have its roots in human rights law.[6] Public participation as a right is a rule of international law regarding environmental management.[7] Public participation has also been linked to the right to non-discrimination based on the notion that all persons should be afforded equal opportunity to participate in decision-making without discrimination on any grounds.[8]

Emphasizing the need for public participation, Pirot, Meynell et al.,[9] and Defra[10] point out the fact that local communities and people come face to face with the local environment and the ecosystems which are located within their specific communities. At community level, women have specifically been noted to directly deal with and heavily rely on natural resources. This is because of their socially constructed gender roles which require them to take care of their households by providing water, food and fuel in form of firewood.[11] As a result, women take care of the environment because in the event that these resources are depleted, they

[3] UN Doc A/CONF.48/14/Rev.1, Ch. I (Principle 19).

[4] Adopted by the U.N. General Assembly in 1982 (paragraph 23).

[5] U.N. Doc. A/CONF.151/26/Rev.1 (Vol. I), annex I (Aug. 12, 1992) (principle 10).

[6] Bekhoven (n 2) 236; William Murray Tabb, 'Public Participation in Environmental Decision Making' in Le Roy C Paddock, Robert L Glicksman, Nicholas S Bryner, *Decision Making in Environmental Law (Vol. II Edward Elgar 2016)* 313–327, 315.

[7] Article 18, Berlin rules on water resources, Report of the Seventy-First Conference of the International Law Association held in Berlin, 16–21 August 2004. https://www.unece.org/fileadmin/DAM/env/water/meetings/legal_board/2010/annexes_groundwater_paper/Annex_IV_Berlin_Rules_on_Water_Resources_ILA.pdf accessed on 05 January, 2019.

[8] Bekhoven (n 2) 236.

[9] Convention on Biodiversity (CBD) Secretariat, 'What is Gender and Biodiversity?' https://www.cbd.int/gender/biodiversity/default.shtml accessed 11 December, 2018.

[10] Danny Elder et al., *Ecosystem Management: lessons from Around the World. A Guide for Development and Conservation Practitioners* (2nd Ed. Switzerland: IUCN 2000) 33 and 43.

[11] Nicky Broeckhoven and Ann Cliquet, 'Gender and ecological restoration: time to connect the dots' 2015 (23) 6 *The Journal of the Society for ecological restoration* 729–736, 730.

are the most affected.[12] In the face of resource depletion, women will have to move longer distances in search of these resources, in order for them to continually provide for and take care of their households. As a result, women have acquired knowledge on how to take care of the environment[13] and it is imperative that they are involved in taking decisions that concern the environment. This is important, seeing as these decisions will not only affect them as people that directly deal with the environment, the knowledge they have acquired through their experience taking care of the environment will go a long way in enhancing environmental protection.[14]

The role and contribution of women in environmental protection and management and the need for their participation in decision-making has been recognized by international environmental as well as human rights law. The Rio Declaration on Environment and Development,[15] the Convention on Biodiversity (CBD)[16] the United Nations Convention on Combating Desertification (UNCCD),[17] UNFCCC[18] through its various Conferences of Parties (COPs), the Paris Agreement[19] all recognize the vital role that women play in environmental conservation and invite states parties to ensure the participation of women in environmental decision-making. Similarly, the International Human Rights instruments like the Convention on the Elimination of All Forms of Discrimination Against Women (CEDAW),[20] the Beijing Platform for action[21] and the Protocol to the African Charter on Human and Peoples Rights[22] emphasize the need to do away with discrimination and promote gender equality and the need to involve women in public government function as well

---

[12] Ibid.

[13] Sandeep Kumar Aditya 'Role of Women in Environmental Conservation' (2016) 4 (4) *International Journal of Political Science and Development* 140–145, 141.

[14] Ibid.

[15] UN Doc. A/CONF.151/26 (vol. I); 31 ILM 874 (1992), Principle 20.

[16] United Nations, *Treaty Series*, vol. 1760, No. 30619, preamble.

[17] 1954 UNTS 3; 33 ILM 1328 (1994), preamble, article 5(d), article 10.

[18] Adopted by the United Nations General Assembly, 20 January 1994, A/RES/48/189.

[19] UN Doc. A/CONF. 177/20 (1995) and A/CONF. 177/20/Add. 1 (1995), point k.

[20] Adopted by the UN General Assembly on December 18, 1979, articles 7 (b) and 14.

[21] UN Doc. A/CONF. 177/20 (1995).

[22] CAB/LEG/66.6 (Sept. 13, 2000) Article 18 (2) (a).

as community activities. They also recognize the need to involve women in taking decisions that concern the environment.

In spite of the existence of robust legal provisions at the international and national levels, the effort of women in environmental conservation, women in Africa have been noted to not be involved in decision-making process especially because of social cultural reasons which view women as non-decision-makers.[23] Accordingly, by way of doctrinal analysis, this chapter seeks to examine the adequacy of environmental law at the international, regional and domestic levels in ensuring women's right to public participation in environmental decision-making and management. In analyzing the international environmental law instruments, the study will make reference to international law, specifically environmental law and human rights law. The study will also analyze the extent to which the law at domestic level in Uganda complies with international and regional provisions on women's right to participation in environmental decision-making and if it does, whether and how it has been implemented. Although Uganda has been noted to still face challenges with the integration of women in decision-making and is placed at the center of the study, other countries with similar challenges can draw lessons from the conclusions drawn.

2 Women's Role in Environmental Conservation and the Need for Their Participation in Environmental Decision-Making

Women have widely been noted to be key contributors to environmental conservation and it is imperative to point out some areas in which women have played a vital role in conservation.[24] Further, women's vulnerabilities in the face of environmental degradation and the need to involve women in environmental decision-making is discussed in this section. Women

[23] Lorena Revelo Aguilar et al., *Women in Environmental Decision making: Case Studies in Liberia, Ecuador and the Philippines* (IUCN, 2015) 6.

[24] Aditya (n 13) 141.

have been noted to be responsible for food production across the globe. They constitute 75% in Sub-Saharan Africa, 65% in Asia and 45% in Latin America in agriculture and food production.[25] Similarly in Machakos, Kenya, women dominate the agricultural sector and have been noted to play a significant role in land and soil conservation.[26] Women depend on forests for medicine, firewood and food for their families and have hence been widely noted to be key players in forest management.[27] Similarly in Tanzania, women have planted trees and, as a result, the time spent on these forest resources has been reduced by 80%.[28] Leone however notes that women have continually been left out of decisions concerning forests. This has been noted to be detrimental to forests because women take better care of forests than men.[29] As already pointed out, this is because they have acquired conservation knowledge through their close reliance on forests and should forests get depleted, they will be more affected.

Women have been noted to be instrumental in the conservation of protected areas.[30] They have been noted to apply traditional knowledge in the conservation of protected areas, specifically wetlands and also in the organization of educative events in the community.[31] Women have been recognized to be wise water and wetland users.[32] For instance, indigenous women have been noted to be key players in conservation especially in the preservation of forests, for instance through growing

[25] IPCC, *Climate Change 2007: Impacts, Adaptation and Vulnerability. Contribution of Working Group II to the Fourth Assessment Report of the Intergovernmental Panel on Climate Change*, M.L. Parry, O.F. Canziani, J.P. Palutikof, P.J. van der Linden and C.E. Hanson, Eds., (Cambridge, UK, Cambridge University Press 2007) 976 pp.

[26] Violet N Gathaara et al., 'Gender, soil and water conservation in Machakos district, Kenya' 2011 (10) *African Crop Science Conference Proceedings* 319–322, 320.

[27] Sophie Grouwels, Kata Wagner and Laura Schweitzer, 'Women are key figures in sustainable forest management' http://www.fao.org/forestry/30285-06ea8b7a9e47c9d1469b982dc67ab8092.pdf accessed on 12 October, 2019.

[28] Ibid.

[29] Marinella Leone, 'Women as decision makers in community forest management: Evidence from Nepal' 2019 *Journal of Development Economics* 1–3.

[30] Liliana Bastian, Molly Gilligan and Barbara Clabots, *Gender and Protected Areas: Exploring National Reporting to the Ramsar Convention and the World Heritage Convention* (IUCN and EGI 2016) 5.

[31] Ibid. 24.

[32] Ibid. 8.

medicinal plants as form of conservation.[33] This, they do not because it arises out of their gender roles, but because forests are key to indigenous communities survival and livelihood. In doing this, the indigenous women apply traditional knowledge. This knowledge should therefore not go untapped.

Although women are not the sole 'repositories' of traditional knowledge, there is a gender difference in the way the knowledge is applied.[34] Further, women's reliance on forests for domestic use as already noted above is a more sustainable way of resource utilization compared with men who utilize the natural resources for commercial purposes.[35] Women's drive for conservation has been witnessed through some women's environmental movements such as the Chipko movement of India[36] and the Green belt movement of Kenya.[37] The Chipko movement was established to protest against government tree and forest logging for commercial purposes. The communities depended on the forests for their sustenance and they derived food, fuel and the forests were vital for soil stability and water purification.[38] In addition, the deforestation had led to the loss of top soils, hence frequent landslides and floods. Similarly, women in Kenya established the green belt movement which was aimed to conserve the environment through forestry in order to avoid tree extinction. This had also been as a result of low food production, drying streams and trekking longer distances in search for firewood as a result of deforestation.[39]

Eco-feminist theory argues that women naturally, by the nature of their motherly instinct, are driven to protect. As a result, women possess the

[33] Leah Duran, 'What does Gender have to do with conservation?' 2018 https://www.conservation.org/blog/what-does-gender-have-to-do-with-conservation- accessed 19 November, 2019.

[34] Bina Agarwal, *Gender and Green Governance: The Political Economy of Women's Presence within and Beyond Community Forestry* (OUP 2010) 40.

[35] Angela Martin, 'Gender in the conservation of protected areas' 2007 https://www.cbd.int/doc/pa/tools/Gender%20in%20the%20conservation%20of%20protected%20areas.pdf accessed 12 October, 2019.

[36] Melissa Petruzzello, 'Chipko movement: Indian Environmental Movement' https://www.britannica.com/topic/Chipko-movement accessed 20 November, 2019.

[37] Urvi C. Gupta, Pradeep U. Verma & Hitesh A. Solanki "Role of Ethnic women in Biodiversity Conservation" 2014 (3) 2 *International Journal of Research and Development in Pharmacy & Life Sciences* 856.

[38] Petruzzello (n 36).

[39] Gupta et al. (n 37) 856–857.

drive to protect the environment, especially for the benefit of the future generations (their descendants).[40] Research has showed that priorities for women and men concerning policies differ. Women's presence influences legal and policy outcomes in the sense that they will table or support legislation that encourages gender equality and empowerment as well as legislature that benefits the whole society, for example legislation on education, family life and health, among others.[41] Although it is not ruled out that men would not influence gender-sensitive legislation, their interests vary from those of women and they may not appreciate the urgency of the legislation.[42] A case in point is the drafting process of the Paris Agreement (2015), at the COP 21 where women only constituted 29% participants at the negotiations.[43] Concerns were raised that it was too male dominated and that there is no wonder that issues of women gender equality in the final agreement were only given preambular recognition, which had no binding legal effect.[44] This does not automatically mean that the presence of women means decisions will be made in their favor, but they will be given due consideration.[45] It has also been noted that not all women like to actively participate. Some women prefer to meet outside the lime light, preferably in smaller women's groups and channel their view through a representative. Others prefer that information is channeled to them through their husbands at village meetings, and vice versa.[46]

The quest for women's participation in Environmental Decision-Making is also grounded in the human rights principles of 'gender equality' and 'non discrimination.'[47] Gender equality and gender equity as

[40] Kay Lehman Schlozman et al., Gender and Citizen Participation: Is There a Different Voice? 1995 (39) 2 *American Journal of Political Science* 267–293, 269.

[41] Agarwal (n 34) 6.

[42] Ibid. 21.

[43] European Capacity Building Initiatives (ecbi), *Pocket Guide to Gender Equality under the UNFCCC (ecbi* 2018) 9. https://wedo.org/wp-content/uploads/2018/04/2018-Edition-of-Pocket-Guide-to-Gender_1.pdf accessed 25 November, 2019.

[44] Speech by Mary Robinson, UN Special Envoy on Climate Change, at COP 21. https://www.theguardian.com/environment/2015/dec/08/cop21-is-too-male-dominated-and-has-male-priorities-says-un-special-envoy accessed 25 September, 2019.

[45] Ibid. 13.

[46] Ibid. 14.

[47] Lere Amusan and Oluwole Olutola, 'Contextualising African Women's Empowerment in Agriculture: Challenges from Climate Change and Mineral Extraction Perspectives' 2017 (18) 4 *Journal of International Women Studies* 117–130, 117.

human rights principles have been advanced as key for fostering social justice. Social justice has been noted to be key for the promotion of natural resource management and sustainability.[48] It has further been noted that gender discrimination fosters resource depletion. This is because women's ownership and access to resources is limited, men being the dominant owners and controllers of natural resources like land.[49] This is disadvantageous when it comes to resources management because whereas women deal more closely with natural resources through activities like agriculture, they do not own the resources, making it harder to take decisions concerning the resources.[50] Impliedly, although women seemingly have constructive interests in the land, it is not enough since they still cannot take decisions that would significantly have effect on the land. This hinders women's involvement in decision-making. Women have been confined to the domestic/private space rather than the public space which is believed to exclusively belong to the men.[51] Gender equality is essential because it affords women and men the same opportunities across all society sectors, decision-making inclusive. This by implication includes environmental decision-making. This way, the aspirations and needs of both men and women are valued.[52]

The Sustainable Development Goals (SDGs) set by the UN in 2014 for the period between 2015 and 2030 were highly focused, bringing into sharp focus gendered inequality and strategizing the overcoming of this phenomenon through a series of strong linkages among each of the goals, especially between gender, education, poverty and inequality.[53] SDG 5 on gender equality is aimed at eliminating all forms of discrimination

---

[48] Annelieke Douma, 'Towards a workable approach to mainstream gender in natural resources management' Both ENDS Working Paper Series, January 2012, 8 https://www.bothends.org/uploaded_files/document/120125_Workable_approach_to_gender_report_FINAL.pdf accessed 17 July, 2019.

[49] Ibid.

[50] Ibid.

[51] Sylvia Tamale, 'Gender trauma in Africa: enhancing women's links to resources' 2004 (48) 1 *Journal of African Law* 50–61 at 51–54, 56.

[52] What is Gender Equality? Gender Equality in Ireland http://genderequality.ie/en/GE/Pages/WhatisGE accessed 5 October, 2018.

[53] David Le Blanc, 'Towards integration at last? The sustainable development goals as a network of targets' 2015 DESA Working paper no. 141 United Nations Department of Economic and Social Affairs 3.

against women everywhere. Of the 17 SGDs, 14 have targets that are aimed to eliminate gender inequalities in all spheres, social, economic, political and cultural.[54] Efforts the world over have been geared toward closing the gender inequality gap. Specifically, focus has been geared toward bringing equal rights to women across socio-economic and environmental domains.[55] Nonetheless, a number of challenges still continue to emerge. A historic understanding of gender roles and their stereotypical acceptance is one of the causes why gender inequality is difficult to overcome, for instance leadership has been considered a 'male activity,' and that may be the reason why literature focuses on men as leaders.[56] Having pointed out the role of women in environmental conservation, their vulnerability in the case of environmental changes and the need for their involvement in decisions concerning the environment, the next section analyzes the legal provisions on the role of women in environmental protection and their right to participation in environmental decision-making. The section also analyzes the effectiveness of these legal provisions.

3 Legal Recognition of Women's Right to Participation in Environmental Decision-Making: Assessing the Adequacy of the Law

International environmental law and International Human Rights law alike have put in place measures to ensure reliable environmental protection and conservation through the promotion of women's right to participation in environmental decision-making.

[54] Refer to SDGS 1, 2, 4 5, 6, 7, 8, 9, 10, 11, 12, 13, 16 and 17. https://www.un.org/sustainabledevelopment/sustainable-development-goals/ accessed 19 November, 2019.

[55] Janet G. Stotsky et al., 'Trends in gender equality and women's advancement' 2016 IMF working paper no.16/21 pg. 4. https://www.imf.org/external/pubs/ft/wp/2016/wp1621.pdf accessed on 30 August, 2019.

[56] Lee G Bolman and Terrence E Deal, *Reframing organizations: Artistry, choice and leadership* (6th edition, Jossey – Bass 2017) 351; See notes 51–54 (on which women can be considered for involvement in the decision-making processes).

International Environmental Law Instruments

UN Framework Convention on Climate Change (UNFCCC) (1992)

Although discussions on gender equality in environment and climate governance are gaining traction and whereas slight improvement has been observed,[57] it has been noted that women continue to be under represented in climate change governance and hence more effort is needed in order to achieve this.[58] The UNFCCC does not mention gender or women in the body of its text. This has been attributed to among other reasons, the fact that the UNFCCC was 'state focused, highly technical and the subject of such pronounced wrangling between states and powerful industrial actors that there was arguably little or no viable opportunity at this stage for any influence beyond the political and the technocratic to be made manifest.'[59] It however recognizes public participation and recognizes the need for women participation through its various COPs. Articles 4 (1) (i), article 5 (c) and article 6 (a) (iii) require states parties to involve the public in the fight against climate change. These provisions clearly indicate that the UNFCCC promotes public participation in the fight against climate change. Although women are not specifically mentioned, they can wholesomely be taken to be in included in the general public.

The UNFCCC has made some progress in the recognition of women's role in the fight against climate change, as well as the need for their participation. In 2009, the UNFCCC put in place the Women and Gender Constituency (WGC). Its main purpose was to ensure that women's

[57] Women were noted to hold six of the most influential positions in environmental negotiation in 2016 as compared to two in 2015 'Gender Equality in Political Decision-making' (EIGE). May 17, 2017 https://eige.europa.eu/rdc/eige-publications/gender-equality-political-decision-making accessed 16 January, 2019.

[58] Ibid.

[59] Karen Morrow, 'Ecofeminism and the Environment: International Law and Climate Change' in Margaret Davies and Vanesa E. Munro (eds), The Ashgate Research Companion to Feminist Legal Theory (Ashgate Publishing Co. 2013) 377–393 at 384.

voices are captured into the UNFCCC.[60] In addition, the COP that met in Marrakesh in 2001 is recognized as the first text on gender equality and climate change. It was aimed to improve the participation of women in party representation in bodies established under the UNFCCC and the Kyoto protocol. Later, the Cancun agreements contained provisions on the need for gender equality and women participation in climate action, having recognized that women are vulnerable to climate change.[61] Decision 23/CP.18 put emphasis on 'promoting gender balance and improving the participation of women in UNFCCC negotiations and in the representation of Parties in bodies established pursuant to the Convention.' Even with these promising provisions, progress on women participation has been noted to still be slight. The UNFCCC COPs have been criticized for being ineffective for lack of implementation at domestic level.[62] The UNFCCC and COP provisions lack enforcement mechanisms. Legal provisions alone without guidelines on how they will be implemented are difficult.

The UNFCCC gender composition report of 2018 revealed that only 37% of the party delegates who participated in the COP 23 held in 2017 were women, noting a 5% increase from the COP 22 (2016) percentage.[63] Further, the report revealed that only 24% of the heads of party delegations were women, noting an 8% decrease from the 32% of the women heads that participated in the COP 22 (2016).[64] The report stated that in order to fulfill the requirement of gender balance as required under decision 18 of COP 23, women representation and gender balance in decision-making under UNFCCC is crucial.[65] COP 23[66] adopted the first ever Gender Action Plan (GAP). The main aim of the gender action

[60] Women and Gender Constituency, 'About Us' http://womengenderclimate.org/about-us/ accessed 31 January, 2019.

[61] Decision 1/CP.16, Part I, para. 7; Part III para. E, FCCC/CP/2010/7/Add.1. Report of the Conference of the Parties on its sixteenth session, held in Cancun from 29 November to 10 December 2010.

[62] Sumudu Atapattu *Human Rights and Climate Change* (2015) 204 & 206.

[63] Ibid. 8.

[64] Ibid.

[65] Ibid.

[66] Decision 3/CP.23 FCCC/CP/2017/11/Add.1 8 February 2018. Report of the Conference of the Parties on its twenty-third session, held in Bonn from 6 to 18 November 2017.

plan is to ensure that gender-related decisions are incorporated into the UNFCCC process. The GAP however notes that gender-responsive climate change policy needs strengthening in the areas of adaptation and mitigation-related activities as well as implementation. The climate change regime has been critiqued for lacking a framework through which gender can be integrated into climate change adaptation and mitigation.[67] This is so, despite the fact that gender is a guiding principle for the formulation of the National Adaptation Plans of Action (NAPAs) under UNFCCC.[68]

Convention on Biological Diversity (CBD) (1992)

The CBD is one of the most important documents that arose from the Rio conference in relation to women and the environment. Biological diversity conservation is the main objective of the CBD. The CBD further strongly advocates for public participation and in article 14 (1) (a) provides for the need to involve the public in EIA processes of proposed projects likely to affect the environment significantly. In its preamble, the CBD 'Recognises the vital role that women play in the conservation and sustainable use of biological diversity and affirms the need for the full participation of women at all levels of policy-making and implementation for biological diversity conservation.'[69] Adopted at the CBD Conference of Parties (COP) in Bonn Germany in 2008, the Gender Action Plan of the CBD, in any development or environmental effort employs gender mainstreaming as the methodology for integrating gender. In the realization of these provisions, in the selection of meeting attendees, research work, policy formulation and decision-making at

[67] Charlotte Kabaseke and Emma Charlene Lubaale, 'Making a case for recognition of women's contribution in the legal framework for environmental conservation (in Uganda)' 2019 (3) *Stellenbosch Law Review* (forthcoming).

[68] Ibid.

[69] Para. 12, preamble.

COPs, the CBD secretariat employs gender considerations.[70] The Beijing platform for action recognized the gender provisions of the CBD. The platform called upon states parties to incorporate into their national legislation, the CBD provisions on the recognition of traditional women's knowledge in conservation, for example traditional medicine.[71]

The 2008 Gender Action Plan further came up with development guidelines to have gender mainstreaming as a guiding principle for the National Biodiversity Strategies and Action Plans (NBSAPs).[72] It is important to note that many countries' NBSAPs have no gender equality as a guiding principle.[73] A study conducted by IUCN on 174 countries revealed that only '14% (24) of the countries identify gender equality as a guiding principle in their most recent NBSAPs.' Further, only 7% (12) of the countries have an objective or goal on women or gender considerations in their NBSAPs.[74] Most relevant to study is the fact that only 37% of the plans indicate women as stakeholders in finding solutions to biodiversity depletion.[75] Despite these progressive provisions on the need to have women involved in conservation, it has been observed that these provisions have not yet been widely embraced and put into practice.[76] The only direct provision on the need for the full participation of women is preambular and even if it were binding, it has been further observed that women continue to be left out of taking of decisions that concern the environment.[77] Implementation of the provision is still a challenge, especially amid deep-rooted cultural practices which discriminate against women.

[70] United Nations Environment Programme (UNEP), *Indigenous women and the Convention on Biodiversity: Work of the Secretariat of the Convention on Biological Diversity with indigenous women* (2000–2005) 2.

[71] Point K, paragraph 253 (c), Beijing Declaration and Platform for Action.

[72] Gender Plan of Action. Item 4.13. UNEP/CBD/COP/DEC/IX/24 9 October 2008, 9th meeting, Bonn, Germany.

[73] Barbara Clabots & Lorena Aguillar, *Gender and biodiversity: Analysis of women and gender equality considerations in National Biodiversity Strategies and Action Plans (NBSAPs)* (IUCN 2017) 18.

[74] Ibid.

[75] Ibid. IV.

[76] IUCN *Gender and Biodiversity* (ICUN 2008) 1 & 2 http://www.wocan.org/system/tdf/iucn_fact_sheet_linking_gender_and_biodiversity.pdf?file=1&type=node&id=202 accessed 17 October, 2019.

[77] Aditya (n 13) 142.

Human Rights Instruments

The Convention on the Elimination of All Forms of Discrimination Against Women (CEDAW) (1975)

The CEDAW guarantees the equal right of women to 'participate in the formulation of government policy and the implementation thereof' and to 'participate in non-governmental organizations and associations concerned with the public and political life of the country.'[78] In interpreting article 7, the CEDAW committee pointed out that participation in this case is not only limited to the areas specified in the article but extends to all public and private areas. This implies that participation in this case extends to environmental decision-making.[79] The CEDAW further provides that 'women living in rural areas have the right to "participate in the elaboration and implementation of development planning at all levels."'[80] The CEDAW committee has further emphasized the need to ensure participation in environmental decision-making, in as regards climate policy.[81] The special rapporteur on the right to health noted that women bear the burden of water collection and they are in charge of waste water disposal for their families. They are however left out of decision-making processes that concern water and the management of water and sanitation.[82]

The CEDAW committee,[83] expressing its concern about the absence of gender recognition under the UNFCCC, took note of the fact that state reports reveal that climate change affects women more than men and recommended that climate change response measures should be gender sensitive, put human rights at the fore front and that climate change poli-

---

[78] CEDAW article 7 (b) and (c).

[79] CEDAW General Recommendation no. 23: Political and Public life, 1997, A/52/38, para 5 http://www.refworld.org.docid/453883a622.html accessed 15 October, 2019.

[80] CEDAW article 14 (2) (a).

[81] CEDAW report sect. III.A.1.

[82] A/62/214 para. 84.

[83] Statement of the CEDAW committee on Gender and Climate Change, CEDAW 44th session, New York, 20th July–7th August 2009unfccc.int/resource/docs/2009/smsn/igo/064.pdf. Accessed 02 August, 2018.

cies and program should ensure the participation of women. The committee further observed that older women are specifically vulnerable to climate change and that they should participate in finding measures to reduce their risk during climate change and in finding solutions to climate change mitigation and adaptation.[84] In summary, in mitigating the effects of environmental change and degradation on vulnerable populations and specifically women, it is very imperative to have them involved in seeking the solutions.

The Protocol to the African Charter on Human and Peoples' Rights on the Rights of Women in Africa (2003)

The protocol seeks to promote, protect and realize the rights of women in Africa.[85] In article 9, the protocol recognizes the right of women to equal participation in political and decision-making processes. Article 9 (1) provides that 'States parties should take specific positive action to promote participative governance and the equal participation of women in the political life of their countries through affirmative action, enabling national legislation.' Further article 9 (2) provides that states parties should 'ensure increased and effective representation and participation of women at all levels of decision-making.' At the African regional level, more standards for achieving participation in environmental decision-making have been and are still to be developed. Again, the case of *SERAC v Nigeria (2001)*,[86] in interpreting the right to a satisfactory environment favorable to their development, the African Commission on Human and Peoples' Rights pointed out that in order for the right to be realized, people or the public has to be afforded the opportunity to take part in developing decisions that affect their communities. Further, that 'Government compliance with the spirit of Articles 16 (right to health) and 24 (right to a satisfactory environment) of the African Charter must also include … providing meaningful opportunities for individuals to be heard and to participate in developing decisions affecting their

[84] CEDAW committee general recommendation No. 27 (2010) paras. 25 and 35.

[85] Preamble.

[86] Comm. No. 155/96 (2001).

communities.'[87] It is clear from the above provisions that the right to public participation has laid out points to participation in all areas of governance including environmental affairs.

Similarly, the provisions on non-discrimination encompass the fact that every individual must participate in governance of the affairs concerning them without discrimination. Women therefore must equally participate in the governance of the affairs concerning them, without discrimination. The above international human rights legal provisions afford women the opportunity to equally participate in decision-making together with the men. It has however been noted that the wide practice of patriarchy is still a very big hindrance to the realization of this equality, with women still being viewed as inferior to the men.[88] This makes implementation of the protocol difficult. Further, only 40 African countries have so far ratified this protocol,[89] meaning that women in 16 countries are still not protected under the ambit of the protocol.

4 Women's Right to Participation in Environmental Decision-Making in Uganda: Policy and Practice

Uganda is a party to all the International Conventions outlined above and has taken steps in advancing the rights to women. Uganda is however a dualist state and in order to have international law enforced in Uganda, it has to first be ratified by parliament.[90]

[87] Ibid.

[88] Amusan and Olutola (n 47) 1.

[89] African Union, 'High Level Consultation on the Ratification of the Maputo Protocol' https://au.int/en/newsevents/20180129/high-level-consultation-ratification-maputo-protocol accessed 26 November, 2019.

[90] Manisuli Ssenyonjo, 'Domestic protection and promotion of human rights under the 1995 Ugandan Constitution' 2002 (20) 4 *Netherlands Quarterly of Human Rights* 445–484 at 481.

The Constitution of the Republic of Uganda (1995)

Uganda's Constitution has been advanced as being one of the most gender sensitive in Africa.[91] The Constitution is the *grundnorm* of the land. It sets the pace upon which all the other laws are based. The Constitution of Uganda also has binding force on authorities and peoples of Uganda.[92] In its National Objectives and Directive Principles of State policy, Uganda is committed to the promotion, protection and guarantee of a democratic society and human rights, including gender balance as well as the representation on marginalized groups.[93] This objective perfectly fits within the meaning of interpretation of the right to participate because the rights can only be enjoyed in a democratic society. Objective II (i) more specifically provides that citizens should be given an opportunity to actively participate in their governance at all levels. This objective is more direct to public participation although not in direct reference to women. Objective VI on the other hand directly provides for the direct representation of women on all organs. This is an indicator that women need to be represented on environmental decision-making organs. Objective XV requires that the role of women in society be recognized. Although not legally binding in themselves as their name suggests, the national objectives can be given some effect and consideration whenever it is fair to do so as long as their true and original meaning is not deviated from.[94] This way, they can contribute to the furtherance of the jurisprudence on human rights.[95]

The Constitution in adherence to its international Human Rights Obligations under the Universal Declaration of Human Rights (UDHR) and the International Covenant on Civil and Political Rights (ICCPR) recognizes the right to equal protection of the law and

---

[91] See, Charlotte Kabaseke, 'Justice for Rape Victims in Uganda: Definitional Crises and Errors of Justice in Prosecuting Rape cases' in Michael Addaney, *Women and Minority Rights Law in Africa: Reimagining Equality and Addressing Discrimination* (Eleven Publishing International 2019) 78.

[92] Article 2(1).

[93] 1995 Constitution, objectives II, III, V & VI.

[94] Ssenyonjo (n 90) 456.

[95] Ibid.

non-discrimination.[96] Article 21 (1) of the Ugandan Constitution provides that all persons are equal before and under the law in all spheres of political, economic, social and cultural life and in every other respect and shall enjoy equal protection of the law. Subsection (2) provides that without prejudice to clause (1) of this article, a person shall not be discriminated against on the ground of sex, race, color, ethnic origin, tribe, birth, creed or religion, or social or economic standing, political opinion or disability. This provision is in tandem with article 26 of the ICCPR which provides for equality before the law and prohibits discrimination on any ground. Further, article 33 (1) provides that women shall be accorded full and equal dignity of the person with men. This is in line with the UDHR provision in article 2 which provides that all human beings are equal in dignity and in rights. Hence men and women should be treated equally and without discrimination in all spheres of life. It has however been noted that despite the existence of these robust provisions, majority of African countries (Uganda being no exception), discrimination against women continues to exist in practice.[97] Article 33 (6) prohibits laws, customs and traditions that are against the welfare and interest of women. The CEDAW committee however noted that despite the existence of this constitutional provision, there continue to exist laws and traditional practices which still discriminate against women in Uganda.[98] This state of affairs has been rubber stamped by different concluding observations of various states which have continued to reveal that deep-rooted cultural beliefs and practices, stereotypes, customs and patriarchal beliefs have continuously infringed on the human rights of women.[99] This is so, despite the robust legal provisions that exist in favor of gender equality

[96] Article 21.

[97] Manisuli Ssenjonjo, 'Culture and the Human Rights of Women in Africa: Between Light and Shadow' 2007 (51) 1 *Journal of African Law* 39–67 at 41, 49, 66.

[98] CEDAW Committee, Concluding Observations: Uganda A/57/38 (23 August 2002), para. 154.

[99] CEDAW *Concluding Observations: Mali* CEDAW /C/MLI/CO/5 (3 February 2006) at para 17; *Egypt* A/56/38 (2 February 2001) at para 325; Cameroon A/55/38 (26 June 2000) at paras 53–54; Democratic *Republic of Congo* A/55/38 (1 February 2000) at paras 230–32; *Guinea* A/56/38 (31 July 2001) at paras 122 and 138; *United Republic of Tanzania* A/53/38/ Revl (6 July 1998) at paras 229–30; CESCR *Concluding* Observations: *Zambia* E/C12/1/Add106 (23 June 2005) at para 10, noting that "the persistence of customs and traditions harmful to women" is one of the factors impeding the implementation of the International Covenant on Economic and Social Cultural Rights (ICESCR).

and non-discrimination against women. Reform must go beyond legal reform and serious education and sensitization needs to be practiced.[100]

The Constitution further protects citizens' right to participate in the affairs of their government.[101] This provision has already been settled in this article to include participation in affairs concerning the environment and specifically environmental decision-making. As long as gender inequality and discrimination still persist in practice, however, women will not be duly integrated into the environmental decision-making arena. The Constitution further mandates the State 'to take affirmative action in favour of marginalized groups, on the basis of gender or other reasons created by tradition or custom. This is in a bid to redress the imbalances against them.'[102] This could be interpreted to mean that affirmative action will among other achievements enable women to take part in various environmental decision-making processes. This provision is in line with article 2 (b) of the CEDAW which requires states to put in place appropriate measures, legislative, or otherwise, in order to do away with discrimination. In a bid to advance affirmative action in favor of marginalized groups, Uganda had in place an equal opportunities commission.[103]

Nonetheless, gender parity generally and specifically in the environmental governance sector is still a challenge. In Uganda, out of the 30 Cabinet ministers, only 10 are women and out of the 50 state ministers, only 13 are women.[104] Further only 157 (34.7%) of the 452 parliamentarians are women.[105] Although the minister for Energy and Mineral Development as well the one of state ministers for Water and Environment are women, the Uganda Parliamentary Committee on Climate Change has only 80 members being women, out of the 215 members. Of these,

---

[100] Ssenyonjo (n 97) 65.

[101] Article 38.

[102] Article 32 (1).

[103] Equal Opportunities Commission Act (2007) (Uganda) Section 2.

[104] Cabinet Ministers and Ministers of State (Uganda) https://www.parliament.go.ug/page/cabinet-members-and-ministers-state accessed 26 November, 2019.

[105] Composition of the 10th Parliament of Uganda as at 12th April, 2018. https://www.parliament.go.ug/sites/default/files/COMPOSITION%20OF%20THE%2010TH%20PARLIAMENT%20AS%20AT%2012%20APRIL%202018.pdf accessed 26, November, 2019.

only 3 out of the 9 executive members are women.[106] The Uganda Parliamentary Committee on climate change on Oil and Gas on the other hand, has no female executives.[107] Therefore, whereas Uganda's Constitution has promising provisions, there is still need to put in place all necessary measures, temporary or otherwise, to ensure that the culture, traditions or customs that hinder women's participation are done away with.

The National Environment Act (NEA) (2019)

The NEA[108] is the major law specific to environmental protection in Uganda. It repealed the National Environmental Act of 1995. It deals with a wide range of environmental issues and its main objective among others is to manage the environment for sustainable development.[109] The Act introduces an environmental protection unit which is responsible for the enforcement of the provisions of the Act.[110] The Act also recognizes the continuation of the National Environment Management Authority (NEMA) as established by the repealed NEA (1995). The main function of the NEMA is to monitor environmental matters.[111] In Section 5 (2) (a), the NEA provides for the participation of the people of Uganda in the management of the environment. It is however important to point out that the recognition for the need to have the people participate does not in itself create a right to participate.[112] There would therefore be no legal basis for enforcing the right to participation in environmental

[106] Parliamentary Forum on Climate Change (PFCC) (Uganda) https://www.parliament.go.ug/page/parliamentary-forum-climate-change-pfcc accessed on November 26, 2019.

[107] Parliamentary Forum on Oil and Gas (PFOG) (Uganda) https://www.parliament.go.ug/page/parliamentary-forum-oil-and-gas accessed 26 November, 2019.

[108] Act no. 5 of 2019 (Uganda).

[109] Long title to the NEA (2019).

[110] Act no. 5 of 2019 (Uganda) Section 25.

[111] NEA (2019) Sections 8 and 9; National Environmental Management Authority (NEMA) *State of the Environment Report for Uganda 2014* (2014) 8–9.

[112] Rose Mwebaza, 'Improving Environmental Procedural Rights in Uganda' in Marianela Cedeno Bonilla et al., *Environmental Law in Developing Countries* IUCN Environmental Policy and Law Paper No. 43. Vol. II (IUCN 2004) 50.

decision-making, under the NEA.[113] It has further been observed that participation in environmental affairs is by invitation of a few as well as civil society organizations. The grassroots people are rarely consulted and if at all they are, it is usually in the final stages of projects and they are usually given few days within which to submit written feedback.[114] This continues to be a challenge especially because the information to the grassroots people is usually pasted on local government notice boards, which are located very far away from the grassroots peoples' residences within very long and usually walking distances. As a result, this information rarely gets to the grassroots people.

In addition, not many of the typical grassroots people in Uganda are literate. Another reality is that the public is hardly ever consulted before issuing licenses for big projects. A case in point having been the grant of a license to a company (Kakira sugar factory) to grow sugar canes in Mabira forest, leaving forest-dependent communities, especially the women with no alternatives.[115] In addition, section 5 does not specifically provide for women's participation and neither does it recognize the role women play in environmental management. This is so, despite the fact that as already pointed out in the previous section, the Constitution calls for the recognition of the role women play in environmental management. Ironically, the NEA recognizes Uganda's obligation under International Environmental Law.[116] The NEA only recognizes the need for the NEMA board to be constituted by one-third of women.[117] Currently, the board is constituted by only 4 women out of the 11 members.[118] It should however be noted that the Minister is the appointing authority of the board members, who have to be approved by the cabinet.[119] The problem with this is that it cannot be ascertained for sure

[113] Ibid.

[114] Ibid.

[115] Ibid.

[116] Section 9 (2) (r) and (s) recognizes the binding effect of the Environmental Conventions Uganda has ratified.

[117] Section 12 (4).

[118] NEMA Board of Directors, National Environment Management Authority (Uganda) https://www.nema.go.ug/aboutus/nema-board-directors accessed 26 November, 2019.

[119] Section 12 (2).

whether the women board members will influence decisions according to the needs of the community in general and much less according to the needs of women. The only other time the NEA mentions gender is under the principles of national environment management.[120] The Act also emphasizes the need for 'gender responsive and sustainable use of the environment and natural resources.'[121] The Act does not interpret these provisions and they have been left to the widest interpretation. Hopefully, they can be interpreted to mean that the needs and contributions of all society member, men and women inclusive, need to be given due consideration.

Clearly the NEA does not specifically recognize the need to have women participate in environmental decision-making. This is particularly disturbing, especially seeing as the NEA is the major law on environmental protection in Uganda. This position is continually attributed to the dominant cultural discrimination against women which continues to leave the decision-making spaces to men.[122] Similarly, the other environmental-related laws do not mention the role of women in environmental conservation and much less the need to have women involved in environmental decision-making. These laws include the Forestry and Tree planting Act, 2003 and the Water Act, Cap. 152. In addition to the challenges outlined above, it has been noted that the institutional framework for the enforcement of public participation in Uganda is non-existent and communities and grassroots are unaware about their participatory rights or opportunities for that matter. Educating the people would also be an effort in realizing the right to participation. Having noted the role women play in natural resource protection and conservation, it is imperative that the need for their participation in natural resource management be legally recognized. This can be a good starting point for the enforcement of their international obligations as recognized right to participation in environmental decision-making.

---

[120] NEA Section 5(2)(b).

[121] NEA Section 5 (2)(b).

[122] Government of Uganda (GOU) *Environment and natural resources sub-sector gender mainstreaming strategy 2016–2021* (2016) 2.

5 Conclusion

This chapter analyzes the adequacy of environment and human rights legal framework in ensuring the realization of women's right to participation in environmental decision-making. The chapter examined the extent to which these legal provisions have been implemented and enforced in Uganda. This chapter establishes that women contribute to environmental protection and conservation through their experience dealing with natural resources. Women are also more vulnerable to environmental changes. It is therefore imperative that they be involved in making decisions that concern the environment, without discrimination. Whereas Uganda has a Constitution that greatly advances gender equality and non-discrimination against women, the provisions that specifically concern the need for women's involvement in environmental decision-making are not yet fully developed. Similarly, the NEA, a recently enacted law, which is the major law on environmental protection in Uganda already needs to be reformed. This is in order to capture the current trends on the role of women in environment protection and the need to have them participate in environmental decision-making. Whereas cultural beliefs and practices can be largely blamed for the non-involvement of women in decision-making, it is important to first have the provision well established in law. This is as crucial for the environmental law as it is for the Constitution. This will also be in response to Uganda's international legal obligations. After this, then the discussion on how to break the cultural barriers in order to enforce the legal provisions can begin, through education and sensitization.

References

Aditya, S. K., 'Role of Women in Environmental Conservation' 2016 (4) 4, *International Journal of Political Science and Development* 140–145.

African Union, 'High Level Consultation on the Ratification of the Maputo Protocol' https://au.int/en/newsevents/20180129/high-level-consultation-ratification-maputo-protocol accessed 26 November, 2019.

Agarwal, B., *Gender and Green Governance: The Political Economy of Women's Presence within and Beyond Community Forestry* (OUP 2010) 40.

Aguilar, L. R., Gilligan, M., Luna, M. A., Owren, C., Prebble, M. and Westerman, K., *Women in Environmental Decision making: Case Studies in Liberia, Ecuador and the Philippines* (IUCN, 2015).

Amusan, L. and Olutola, O., 'Contextualising African Women's Empowerment in Agriculture: Challenges from Climate Change and Mineral Extraction Perspectives' 2017 (18) 4 *Journal of International Women Studies* 117–130.

Atapattu, S., *Human Rights and Climate Change* (2015) 204 & 206.

Bastian, L., Gilligan, M., and Clabots, B., *Gender and Protected Areas: Exploring National Reporting to the Ramsar Convention and the World Heritage Convention* (IUCN and EGI 2016).

Bekhoven, J., 'Public Participation as a General Principle in International Environmental Law: Its Current Status and Real Impact' 2016 (11) 2 *National Taiwan University Law Review* 229.

Berlin rules on water resources, Report of the Seventy-First Conference of the International Law Association held in Berlin, 16–21 August 2004. https://www.unece.org/fileadmin/DAM/env/water/meetings/legal_board/2010/annexes_groundwater_paper/Annex_IV_Berlin_Rules_on_Water_Resources_ILA.pdf accessed 05 January, 2019.

Bingeman, K., 'Women's Participation in Forest Management Decisions in the Upper Kullu Valley, Himachal Pradesh, India' 2001 (21)2 article 10 *Himalayan Research Bulletin* 56–57.

Bolman, L.G., and Deal, T. E., *Reframing organizations: Artistry, choice and leadership* (6th edition. Jossey – Bass 2017).

Broeckhoven, N., and Cliquet, A., 'Gender and ecological restoration: time to connect the dots' 2015 (23) 6 *The Journal of the Society for ecological restoration* 729–736 at 730.

Clabots, B., & Aguillar, A., *Gender and biodiversity: Analysis of women and gender equality considerations in National Biodiversity Strategies and Action Plans (NBSAPs)* (IUCN 2017).

Convention on Biodiversity (CBD) Secretariat, 'What is Gender and Biodiversity?' https://www.cbd.int/gender/biodiversity/default.shtml accessed 11 December, 2018.

Douma, A., 'Towards a workable approach to mainstream gender in natural resources management' Both ENDS Working Paper Series, January 2012 pg. 8 https://www.bothends.org/uploaded_files/document/120125_Workable_approach_to_gender_report_FINAL.pdf accessed 17 July, 2019.

Duran, L., 'What does Gender have to do with conservation?' 2018 https://www.conservation.org/blog/what-does-gender-have-to-do-with-conservation- accessed 19 November, 2019.

Easton, P., and Ronald, M., *Seeds of Life: Women and Agricultural Biodiversity in Africa* (World Bank 2000) https://openknowledge.worldbank.org/bitstream/handle/10986/10815/multi0page.pdf?sequence=1&isAllowed=y accessed 06 July, 2019.

Elder, D., Pirot, J. Y., Meynell, P. J., and Pirot, J. Y., *Ecosystem Management: lessons from Around the World. A Guide for Development and Conservation Practitioners* (2nd Ed. Switzerland: IUCN 2000).

European Capacity Building Initiatives (ecbi), *Pocket Guide to Gender Equality under the UNFCCC (ecbi to Gender Equality under the UNFCCC (ecbi* 2018)) https://wedo.org/wp-content/uploads/2018/04/2018-Edition-of-Pocket-Guide-to-Gender_1.pdf accessed 25 November, 2019.

Gathaara, V.N., Gachene, C.K.K., Ngugi, J. N., Thuranira, E. G. and Baaru, M. W., 'Gender, soil and water conservation in Machakos district, Kenya' 2011 (10) *African Crop Science Conference Proceedings* 319–322.

Gender Equality in Political Decision-making (EIGE), May 17, 2017. https://eige.europa.eu/rdc/eige-publications/gender-equality-political-decision-making accessed 16 January, 2019.

Government of Uganda (GOU) *Environment and natural resources sub-sector gender mainstreaming strategy 2016–2021* (2016) 2.

Grouwels, S., Wagner, K., Schweitzer, L., 'Women are key figures in sustainable forest management' http://www.fao.org/forestry/30285-06ea8b7a9e47c9d1469b982dc67ab8092.pdf accessed 12 October 2019.

Gupta, U. C., Verma, P. U., and Solanki, H. A, 'Role of Ethnic women in Biodiversity Conservation' 2014 (3) 2 *International Journal of Research and Development in Pharmacy & Life Sciences* 856.

International Union for Conservation of Nature (IUCN),' IUCN Background Document on 'Gender inequality - A constraint to economic growth, biodiversity protection, poverty reduction and sustainable development: Linkages between MDG 1, 3, and 7' pg. 1 and 2 https://www.cbd.int/gender/doc/iucn-unga-luncheon-en.pdf accessed 13 September, 2018.

IPCC, *Climate Change 2007: Impacts, Adaptation and Vulnerability. Contribution of Working Group II to the Fourth Assessment Report of the Intergovernmental Panel on Climate Change*, M.L. Parry, O.F. Canziani, J.P. Palutikof, P.J. van der Linden and C.E. Hanson, Eds., (Cambridge, UK: Cambridge University Press 2007)

IUCN *Gender and Biodiversity* (ICUN 2008) 1 & 2 http://www.wocan.org/system/tdf/iucn_fact_sheet_linking_gender_and_biodiversity.pdf?file=1&type=node&id=202 accessed 15 October, 2019.

Kabaseke, C., 'Justice for Rape Victims in Uganda: Definitional Crises and Errors of Justice in Prosecuting Rape cases' in Addaney, M., *Women and Minority Rights Law in Africa: Reimagining Equality and Addressing Discrimination* (Eleven Publishing International 2019).

Kabaseke, C., and Lubaale, E. C., 'Making a case for recognition of women's contribution in the legal framework for environmental conservation (in Uganda)' 2019 (3) *Stellenbosch Law Review* (forthcoming).

Kaniaru, D. and Karukulasiriya, L., 'Capacity Building in Environmental Law,' in Lin, S and Karukulasiriya L (eds) *UNEP's New Way Forward: Environmental Law and Sustainable Development* 171–184.

Le Blanc, D., 'Towards integration at last? The sustainable development goals as a network of targets' 2015 DESA Working paper no. 141 United Nations Department of Economic and Social Affairs 3.

Legal Resources Foundation v. *Zambia*, Communication No. 211/98, 14th Activity Report, 2000–2001, Annex V. para. 63.

Leone, M., 'Women as decision makers in community forest management: Evidence from Nepal' 2019 *Journal of Development Economics* 1–3.

Martin, A., 'Gender in the conservation of protected areas' 2007 https://www.cbd.int/doc/pa/tools/Gender%20in%20the%20conservation%20of%20protected%20areas.pdf accessed 12 October, 2019.

Morrow, K., 'Ecofeminism and the Environment: International Law and Climate Change' in Davies, M. and Munro V E (eds), The Ashgate Research Companion to Feminist Legal Theory (Ashgate Publishing Co. 2013) 377–393.

Mwebaza, R., 'Improving Environmental Procedural Rights in Uganda' in Bonilla C. M., Fernandez E F; Jamaiel, S., Mwebaza, R., & Zhandayeva, *D., Environmental Law in Developing Countries* (IUCN Environmental Law Programme 2004) 50.

National Environmental Management Authority (NEMA) *State of the Environment Report for Uganda 2014* (Kampala: NEMA 2014) 36.

National Environmental Management Authority, 'Board of Directors, National Environment Management Authority (Uganda)' https://www.nema.go.ug/aboutus/nema-board-directors accessed 26 November, 2019.

Parliament of Uganda, 'Cabinet Ministers and Ministers of State (Uganda)' https://www.parliament.go.ug/page/cabinet-members-and-ministers-state accessed 26 November, 2019a.

Parliament of Uganda, 'Composition of the 10th Parliament of Uganda as at 12th April, 2018' https://www.parliament.go.ug/sites/default/files/COMPOSITION%20OF%20THE%2010TH%20PARLIAMENT%20AS%20AT%2012%20APRIL%202018.pdf accessed 26 November, 2019b.

Parliament of Uganda, 'Parliamentary Forum on Climate Change (PFCC) (Uganda)' https://www.parliament.go.ug/page/parliamentary-forum-climate-change-pfcc accessed 26 November, 2019c.

Parliament of Uganda, 'Parliamentary Forum on Oil and Gas (PFOG) (Uganda)' https://www.parliament.go.ug/page/parliamentary-forum-oil-and-gas accessed 26 November, 2019d.

Petruzzello, M., 'Chipko movement: Indian Environmental Movement' https://www.britannica.com/topic/Chipko-movement accessed 20 November, 2019.

Regwell, C., 'Implementation and Enforcement: National Implementation' in Daniel Bodansky, Jutta Brunnée and Ellen Hey, *The Oxford Handbook of International Environmental Law* (OUP, 2008).

Ssenyonjo, M., 'Culture and the Human Rights of Women in Africa: Between Light and Shadow' 2007 (51) 1 *Journal of African Law* 39–67.

Ssenyonjo, M., 'Domestic protection and promotion of human rights under the 1995 Ugandan Constitution' 2002 (20) 4 *Netherlands Quarterly of Human Rights* 445–484.

Statement of the CEDAW committee on Gender and Climate Change, CEDAW 44th session, New York, 20th July–7th August 2009. Available at unfccc.int/resource/docs/2009/smsn/igo/064.pdf, accessed August 25, 2019.

Secretariat of the Convention on Biodiversity *Guidelines for Mainstreaming Gender into National Biodiversity Strategies and Action Plans* 19.

Schlozman, K. L., Burns, N., Verba, S., and Donahue, J., 'Gender and Citizen Participation: Is There a Different Voice?' 1995 (39) 2 *American Journal of Political Science* 267–293.

Speech by Mary Robinson, UN Special Envoy on Climate Change, at COP 21. https://www.theguardian.com/environment/2015/dec/08/cop21-is-too-male-dominated-and-has-male-priorities-says-un-special-envoy accessed 25 November, 2019.

Stotsky, J. G., Shibuya, S., Kolovich, L., and Kebhaj, S., 'Trends in gender equality and women's advancement' 2016 IMF working paper no.16/21 https://www.imf.org/external/pubs/ft/wp/2016/wp1621.pdf accessed 20 August, 2019.

UNESCO, 'Women and Girls Education: Facts and Figures' 2014 http://www.unesco.org/new/en/unesco/events/prizes-and-celebrations/celebrations/international-days/international-womens-day-2014/women-ed-facts-and-figure/ accessed 26 November, 2019.

Tamale, S., 'Gender trauma in Africa: enhancing women's links to resources' 2004 Journal of African Law 48 (1) 50–61.

The state of the world's girls 2013. In Double Jeopardy: Adolescent Girls and Disasters (2013) http://www.thejournal.ie/women-and-children-more-at-risk-at-times-of-disaster-1124615-Oct2013/ accessed 09 January, 2018.

United Nations Environment Program (UNEP) *Indigenous women and the Convention on Biodiversity: Work of the Secretariat of the Convention on Biological Diversity with indigenous women* (2000–2005) 2.

United Nations Environment Programme, 'Gender Equality and the Environment: A guide to UNEP's Work' 2017 pg. 4–5. http://www.unpei.org/sites/default/files/publications/Gender_equality_and_the_environment_Guide_to_UNEPs_work-2016%20%281%29.pdf accessed 14 September, 2018.

United Nations Framework Convention on Climate Change 'National Adaptation Plans: Technical guidelines for the national adaptation plan process' at https://unfccc.int/files/adaptation/cancun_adaptation_framework/application/pdf/naptechguidelines_eng_high__res.pdf accessed 09 November, 2018).

What is Gender Equality? Gender Equality in Ireland http://genderequality.ie/en/GE/Pages/WhatisGE accessed 5 October 2018.

Women and Gender Constituency, About Us, http://womengenderclimate.org/about-us/ accessed 31 January, 2019.

World Health Organization (WHO), *Gender, Climate Change and Health* (WHO 2014) 11 https://apps.who.int/iris/bitstream/handle/10665/144781/9789241508186_eng.pdf accessed 20 November, 2019.

13

Connecting Human Rights and the Environment in Cameroon: Successes, Limitations and Prospects

Jean-Claude N. Ashukem

1 Introduction

Generally, one of the crucial issues in any regulatory regime is the provision of adequate and sufficient laws that could ensure respect, protection and fulfillment of people's fundamental rights, including the environmental right. Although complete laws may be adequate, not all complete laws are adequate. This chapter makes use of the former argument, because 'adequate' (laws) relates to something that is enough in quantity or good enough in quality, for a particular purpose.[1] In an environmental

This chapter is an adaption of a 2019 publication by The Comparative and International Law Journal of Southern Africa. The research of this chapter was done during the author's postdoctoral tenure at the University of the Free State.

[1] AS Hornby, *Oxford Advanced Learner's Dictionary: International Student's Edition* (8th edn, OUP 2012) 17.

J.-C. N. Ashukem (✉)
Faculty of Law, University of the Free State, Bloemfontein, South Africa

M. Addaney, A. Oluborode Jegede (eds.), *Human Rights and the Environment under African Union Law*, https://doi.org/10.1007/978-3-030-46523-0_13

context, adequate and sufficient laws would entail laws that have the potential to drive the practical realization of people's environmental rights. This is because adequate and complete laws do not only provide the catalyst for effective environmental protection measures but also highlight state commitments toward environmental protection. Conversely, incomplete and limitative laws apparently become inadequate, insufficient and ineffective in enhancing respect and protection of fundamental human rights generally. Cameroon's environmental legal framework, discussed below, suffers from a plethora of incomplete, inadequate and limitative provisions. These limitations serve as significant challenges for the practical realization of the right to a healthy environment, as stipulated under the *Constitution of the Republic of Cameroon*, 1996 (the Constitution) and the holistic protection of the natural environment. It is this incompleteness and limitations that the author considers inadequate, ineffective and insufficient for environmental protection.

The present chapter critically evaluates Cameroon's legal framework on environmental protection to assess the potential successes, limitations and prospects for the use of human rights norms, principles and standards to ensure environmental protection. It argues that because the legal framework contains several limitative and incomplete provisions, it is insufficient, inadequate and ineffective for protecting the environment, when analyzed through a human rights lens. The discussion begins (in Sect. 2) with a general exposition of, and contextual background on, the link between human rights and the environment and how the environmental right could be addressed through and by means of human rights claims and arguments. Section 3 discusses the connection between human rights and the environment in Cameroon through a thorough examination of Cameroon's legal framework, to ascertain the extent to which the right to a healthy environment is (through a human right lens) used, applied and protected in the country. Section 4 critically assesses this legal framework to provide clarity on its inadequacy, insufficiency and ineffectiveness in strengthening environmental governance in the country. The last part is a conclusion.

2 The Environment as a Human Right: Contextual Background

Commentators have asked why environmental protection should be treated as a human rights issue.[2] Any attempt to answer this question entails an understanding of the link between human rights and the environment. Generally, an environmental right relates to both the right and obligation people have toward the environment. The need to protect the environment through human right claims, as conceived in international and African regional instruments, holds some truth.[3] An environmental right comprises one of those cross-cutting issues that require full attention and protection. This is premised on the fact that the environment plays an increasingly important role in the lives of people, and a highly deleterious environment could significantly impact on peoples' rights, including a wide range of socio-cultural rights. Thus, the right to a healthy environment is apparently a necessary precondition for the enjoyment of other fundamental rights.[4]

The United Nations Conference on the Human Environment (Stockholm Conference)[5] of 1972 clearly provides the bases for, and the need to protect the environment through, human rights claims. Paragraph 1 of the Declaration guarantees both a right to an environmental quality

---

[2] Alan Boyle 'Human Rights and the Environment: Where Next' (2012) 23 EJIL 613.

[3] See for example Principle 1 of the *Stockholm Conference on the Human Environment* of 1972 and art 24 of the *African Charter on Human and Peoples' Rights* of 1981.

[4] See the dissenting opinion of Justice Weeramantry in the case of *Gabcikovo-Nagymaros Project* (*Hungary v Slovakia*) [1997] ICJ Rep at 92. Also see Sumudu Atapattu, 'The Right to a Healthy Life or the Right to Die Polluted? The Emergence of a Human Right to a Healthy Environment under International Law' (2002–2003) 16 TELJ 65; Phillipe Sands, *Principles of International Environmental Law* (2nd edn, CUP 2003); Patricia Birnie, Alan Boyle and Catherine Redgwell, *International Law and the Environment* (3rd edn, OUP 2008); Dinah Shelton, 'Human Rights, Environmental Rights and the Right to Environment' (1992) 28 SJIL 103; Fatma Ksetini, Special Rapporteur for Human Rights and the Environment, *Final Report to Commission on Human Rights, Sub-Commission on Prevention of Discrimination and Protection of Minorities: Human Rights and the Environment,* UN Doc E/CN.4/Sub.2/1994/9, 6 July 1994; Donald Anton and Dinah Shelton, *Environmental Protection and Human Rights* (CUP 2011); Wolfgang Sachs, 'Environment and Human Rights' (2004) 47(1) *Development* 42; Dinah Shelton 'Human Rights and the Environment: Substantive Rights' in Malgosia Fitrmaurice, David M. Ong and Panos Merkouris (eds) *Research Handbook on International Environmental Law* (Edward Elgar Publishing 2010) 265.

[5] Stockholm Declaration 1972 ILM 11:1416.

and the obligation to protect and improve the environment for the benefit of present and future generations. This parallels the values, purport and goal of the recently adopted United Nations Framework Principle on Human Rights and the Environment.[6] In this Report, the former Special Rapporteur on Human Rights and the Environment—John Knox—outlined 16 principles that essentially relate to human rights and the environment and are principally based on existing global human rights systems. These principles motivate the rationale for using human rights obligations as applicable to states, to ensure environmental protection.[7] Speaking of a right to a healthy environment presupposes that the environment is and should be considered a collective right, or that since the environment forms part of people's basic human rights, individuals are as of right entitled to certain rights in relation to, or that are linked to, the environment.[8]

Consequently, environmental protection has conceivably been promoted to the realm of human rights and it is possible to cross-pollinate human rights norms, standards and principles to the environmental realm in order to proactively ensure environmental protection. The right to a healthy environment represents and epitomizes a human right in itself, for which there is an inextricable and formidable relationship between the human right and the environment. As discussed below, this relationship is a very fascinating, uneasy and interesting one, since the use of human rights norms and standards is crucially relevant to provide a holistic approach that serves to ensure environmental protection and other related rights-based entitlements, such as the rights to food, life, human dignity and sanitation among others. These rights are linked to the environment. Commentators have indicated that the rights to human dignity and to environmental quality have much in common,

[6] The Framework Principle on Human Rights and the Environment by Resolution 28/11 https://www.ohchr.org/EN/Issues/Environment/SREnvironment/Pages/FrameworkPrinciplesReport.aspx accessed 3 March 2019.

[7] For details see Principles 1–16 of the Framework on Human Rights and the Environment https://www.ohchr.org/EN/Issues/Environment/SREnvironment/Pages/FrameworkPrinciplesReport.aspx accessed 3 March 2019.

[8] Also see J.G. Merrills 'Environmental Protection and Human Rights: Conceptual Aspects' in Alan Boyle and Michael Anderson (eds) *Human Rights Approaches to Environmental Protection* (OUP 2003) 25; Boyle (n 2).

particularly as they have evolved either coincidentally or independently as drivers in modern constitutionalism.[9]

Human rights, including the environmental right, are by their very nature indivisible, interdependent and interrelated, and complementary.[10] This interconnectivity requires the protection of not only the environmental right per se, but also all other rights that are linked to the environment.[11] As Grear and Kotzé[12] suggest, the fact that human rights are legion and compelling means that human rights-based arguments and claim-making practices are appropriate and importantly useful to play a crucial role in the protection of not only human environmental interests, but also the interest of the environment, as various human rights considerations are inextricably linked to the environment. Human rights have been perceived to have a particularly potent level of symbolic and rhetorical appeal since they are often understood to be urgent ethical demands and not merely putative legal claims,[13] which could also be useful in an environmental context. In this regard, it is evidently unsurprising that environmental claims and interests turn to and have human rights languages and arguments such as those exemplified by the rights-based approach.

Taken together, it emerges hypothetically that the rights to the highest attainable standard of health and to an adequate standard of living are largely dependent to a certain degree on the standard of environmental quality. It follows that environmental degradation, including pollution, deforestation or desertification, could significantly impact negatively on the environment. This constitutes a situation that has been viewed and

---

[9] Erin Daly and James R. May 'Bridging Constitutional Dignity and Environmental Rights Jurisprudence' (2016) 7(2) JHRE 218, 219.

[10] Also see Michael R. Anderson "Human Rights Approaches to Environmental Protection: An overview" in Alan Boyle and Michael Anderson (eds) *Human Rights Approaches to Environmental Protection* (OUP 2003) 3.

[11] Also see the World Health Organisation Report of 2008 "Human Rights-Based Approach to Health and the Environment. Report of a Regional Seminar Bangkok, Thailand, 20–21 August 2007 at 1.

[12] Anna Grear and Louis J. Kotzé 'An Invitation to Fellow Epistemic Travellers – Towards Future Worlds in Waiting: Human Rights and the Environment in the Twenty-First Century' in Anna Grear and Louis J. Kotzé (eds) *Research Handbook on Human Rights and the Environment* (Edward Elgar 2005) 1.

[13] Ibid.

perceived as comprising instances of the violation of some of the rights that are inextricably linked to the environment.[14] Thus, the creation of a reliable, and presumably more effective, system of environmental governance such as those embedded in the rights-based approach could be instrumental in enhancing not only the environmental and human rights of people, but also the well-being of both present and future generations, including indigenous or economically marginalized groups whose livelihoods are dependent on natural resources.[15] It has been indicated that the legal protection of human rights appears to be an effective and necessary means of environmental protection generally, and conservation specifically.[16]

Consequently, it is apposite that there is and certainly should be a right to a satisfactory environment, for which legal means such as those expressed through a rights-based framework should arguably be used to ensure and enforce the protection of the environment, as well as other related rights of people.[17] It is not surprising that many of the rights guaranteed under international human rights and regional human rights laws incorporate an environmental dimension[18] and the need to ensure environmental protection through either substantive or procedural means or both.

Substantively, the formidable relationship between the right to a healthy environment, as it is in some texts, and the enjoyment of human right is recognized in international and regional laws, including soft and hard laws as well as multilateral environmental treaties, with a concomitant obligation on state parties to undertake measures to ensure and protect people's environmental rights at the domestic level. The protection of

[14] Bridget Lewis, "Environmental Rights or a Right to the Environment? Exploring the Nexus between Human Rights and Environmental Protection" (2012) 8(1) MJICEL 36; Anderson (n 10). Also see Committee on Economic, Social and Cultural Rights, *General Comment 12: The Right to Adequate Food*, UN Doc E/C.12/1999/5 (12 May 1999); Committee on Economic, Social and Cultural Rights, *General Comment 15: The Right to Water*, UN Doc E/C.12/2002/11 (26 November 2002); Jan Glazewski *Environmental Law in South Africa* 2edn (LexisNexis, 2005) 76.

[15] Anderson (n 10). Also see Principle 1 of the Stockholm Conference on Human Environment of 1972.

[16] Ibid.

[17] Ibid.

[18] Lewis (n 14).

the environment through human right-based approaches and its rapid growth as an analytical perspective and a mode for normative enquiry and regulatory intervention has long been recognized since the Stockholm Declaration and the United Nations Conference on Environment and Development (Rio Conference) with specific reference to the human right to the environment. Principle 1 of the Stockholm Conference recognizes that adequate protection of the environment is essential as a vital component to human well-being and the enjoyment of other basic human rights. These include the right to life itself and encompass the right of every person to live in an environment that is adequate to his or her health or well-being.

Furthermore, there is a duty to protect and improve the environment for the benefit of present and future generations. In the concluding declaration of the Conference, participating states specifically and unequivocally refer to the fundamental rights to freedom, equality and adequate conditions of life in an environment of a quality that would permit a life of dignity and well-being. Resolution 45/94 of the UN General Assembly further explored and strengthened the link between human rights and environmental protection by stating that all individuals are entitled to live in an environment that is adequate and conducive for their health and well-being, which could be promoted and achieved only if the environment is protected, for example, by and through a rights-based approach. On the other hand, Principle 1 of the Rio Conference provides that 'human beings are entitled to a healthy and productive life in harmony with nature'. Regrettably, neither the International Covenant on Civil and Political Rights nor the International Covenant on Economic Social and Cultural Rights makes provision for an environmental right.

Nonetheless, in the dissenting opinion of the Vice-President of the International Court of Justice in the case of *Gabcikovo-Nagymaros Project—Hungary v Slovakia*,[19] Justice Weeramantry clearly underlined the inevitable nexus between human rights and environmental protection and stated that:

[19] *Gabcikovo-Nagymaros Project (Hungary v Slovakia)* [1997] ICJ.

> The protection of the environment is … a vital part of contemporary human rights doctrine, for it is *sine qua non* for numerous human rights such as the right to health and the right to life itself. It is scarcely necessary to elaborate on this, as damage to the environment can impair and undermine all the human rights spoken of in the *Universal Declaration on Human Rights* and in other human rights instruments.[20]

It has been suggested that a poorly protected environment could also have an indirect impact on the enjoyment of a host of human rights,[21] to the extent that it may either impact on an individual's or community's capacity to realize their human rights, or impede the ability of a government to protect the rights of its citizens. For example, the competing demand for natural resources could apparently result in a state prioritizing immediate human rights concerns to the detriment of the broader environment.[22] It is against this background that the notion of sustainable development and, recently, the Sustainable Development Goals (SDGs) find anchor in the proposition for the protection of both human rights and environmental concerns through the merger of economic, social and environmental needs within the context of development activities. Thus, there exists a significantly important relationship between the environment and human rights, on which basis people are entitled to a justifiable claim to a right to a healthy environment as a human right that deserves similar protection like other fundamental human rights.

At the African regional level, the African Charter on Human and Peoples Rights of 1981 (the African Charter) is highly instructive and provides a collective human right to the environment. The Charter has been considered to have the most explicit normative provision of an environmental right in any binding (regional) human right instrument.[23] Article 24 of the African Charter provides that 'all people shall have the right to a general satisfactory environment favourable to their

[20] At para 92.

[21] Lewis (n 14) 39.

[22] Ibid.

[23] Werner Scholtz 'Human Rights and the Environment in the African Union Context' in Anna Grear and Louis J Kotzé (eds) *Research Handbook on Human Rights and the Environment* (Edward Elgar, 2015) 405.

development'. The term 'people' in Article 24 is used in the collective sense and the African Commission in the well-known decision of *The Social and Economic Rights Action Centre (SERAC) and the Centre for Economic and Social Rights (CESR) v Nigeria*,[24] (SERAC's case) failed to dissect its conceptual or definitional element.[25] However, the language of Article 24 is distinctly clear and brings to light the inextricable relationship between human rights and the environment, if one considers that the right to development serves to provide a point of departure for the development of territorial and extraterritorial dimensions of states' human rights obligations in relation to environmental degradation and pollution. It has been submitted that the inclusion of the reference to 'favourable environment' in Article 24 of the African Charter should be viewed as an attempt to recognize the important link between the environmental right in Article 24 and the right to development in Article 22 within the context of sustainable development.[26] The scope and content of Article 24 was reiterated by the African Commission in the *SERAC's* case where the government of Nigeria was found guilty for contravening Article 24 of the African Charter through its involvement in oil production that led to the violation of the environmental rights of the Ogoni people. Although the Commission did not elaborate on the implication of the linkage between the right to a satisfactory environment and favorable development, it affirmed the ecological element of sustainable development within the broader, substantive meaning of Article 24.[27]

Apart from the recognition granted to the substantive aspect of the environmental right, procedural human rights, such as the rights of access to (environmental) information, public participation and access to justice, are also worthy of consideration in the continuous quest to promote, ensure and enhance environmental governance. They provide potential

---

[24] *The Social and Economic Rights Action Centre (SERAC) and the Centre for Economic and Social Rights (CESR) v Nigeria*, Communication No 155/96 www.cesr.org/downloada/AfricanCommissionDecision.pdf accessed 15 May 2019.

[25] In other cases such as DRC v Burundi, Rwanda and Uganda, Communication 227/99 (2004), paras 87 and 95, and Sudan Human Rights Organisation v Sudan, Communication 297/05 (2009) para 20, the African Commission had the opportunity to develop an interpretation of the term 'people'.

[26] Scholtz (n 23) 407.

[27] SERAC (n 24) para 52.

possibilities of 'greening' existing procedural human rights by assigning to them specific complementary roles in relation to substantive environmental rights.[28] These rights are enshrined in relevant international and regional human and environmental law instruments. The Aarhus *Convention on Access to Information, Public Participation in Decision-making and Access to Justice in Environmental Matters* of 1998, and the *Bali Guidelines for the Development of National Legislation on Access to Information, Public Participation and Access to Justice in Environmental Matters* of 2010 as well as the revised African Convention on the Protection of Nature and Natural Resources of 2003, clearly provide international standards for the use of procedural human rights and are evidently encompassing and instructive, and fully embrace the need to promote and ensure environmental governance by means of procedural human rights frameworks.[29] For space constraints, it is beyond the scope of this chapter to delve into an analysis of these rights in these instruments.

3 Environmental Human Rights Protection in Cameroon

Prior to providing an understanding of the extent of environmental human rights protection in Cameroon, it is necessary and important to first gain an understanding of the nature and substantive normative

[28] Ulrich Beyerlin 'Aligning International Environmental Governance with the 'Aarhus Principles' and Participatory Human Rights' in Anna Grear and Louis J Kotzé (eds) *Research Handbook on Human Rights and the Environment* (Edward Elgar, 2015) 340; Karen Morrow 'Sustainability, Environmental Citizenship Rights and the Ongoing Challenges of Reshaping Supranational Environmental Governance' in Anna Grear and Louis J Kotzé (eds) *Research Handbook on Human Rights and the Environment* (Edward Elgar 2015) 212.

[29] For a detailed understanding of the relevant provisions of these rights and their use and application in promoting environmental governance, see Alan Boyle "Human Rights and the Environment: Where Next" in Ben Boer (ed) *Environmental Law Dimensions of Human Rights* (OUP 2015) 213–218. Also see Jean-Claude N Ashukem "A Rights-Based Approach to Foreign Agro-Investment Governance in Cameroon, Uganda and South Africa (North-West University, 2016) 158–161; S Kravchenko, "The Aarhus Convention and Innovation in Compliance with MEAs' (2007) CJIELP; Jonas Ebbesson "The Notion of Public Participation in International Environmental Law (1997) *Yale Book of International Environmental Law*; Jonas Ebbesson "The EU and the Aarhus Convention: Access to Information, Public Participation in Decision-Making and Access to Justice in Environmental Matters' http://www.europarl.europa.eu/RegData/etudes/BRIE/2016/571357/IPOL_BRI(2016)571357_EN.pdf accessed 3 March 2019.

content of environmental rights as a human right. At the outset, it is important to enquire as to the extent to which human rights approaches provide for environmental interests in Cameroon. The environmental right, like every other right, creates a reciprocal duty, and in this context, it is necessary to ascertain what obligations the environmental right creates, and upon whom they rest. This section attempts to answer these questions by discussing the architecture of Cameroon's human rights framework—substantive and procedural—for environmental protection, the scope of application of the environmental rights, and the potential success, limitations and prospects of the legal framework.

The Environmental Rights

The *Constitution of the Republic of Cameroon*, 1996 (the Constitution) provides the basis for the legal protection of human rights, including the right to a healthy environment. Regrettably, fundamental rights are provided for only in the Preamble, which is part and parcel of the Constitution.[30] In terms of the Preamble, "every person shall have a right to a healthy environment." The protection of the environment shall be the duty of every citizen. The State shall ensure the protection and improvement of the environment.[31] Clearly, the Preamble provides for a distinctly restrictive environmental right, with an accompanying right and two obligations. There is the right of everyone to a healthy environment. This right is reinforced with an explicit obligation on citizens in conjunction although an implicit obligation on the state to protect the environment. While the Preamble bestows a clear-cut obligation on citizens to ensure the protection of the environment, the obligation on the state could be an abstract notion, as the state is only required to ensure the protection and improvement of the environment. The Preamble further provides for incidental rights, including the right to work, the right to life, physical and moral integrity, the right to development, and the right to property, which also have an environmental dimension and are therefore worthy of consideration.

---

[30] See Article 45 of the Constitution.

[31] See the Preamble of the Constitution. Also see s 9(e)(ii) of Law No 96/12.

The foregoing suggests that these substantive rights, could be used in conjunction with the environmental right to promote and ensure the spirit, purport and object of the constitutional right to a healthy environment in Cameroon. As mentioned above, rights do not exist in isolation and it will be erroneous in an environmental context, as in any other, to suggest that the substantive environmental right is the sole component in the arsenal of the human rights-based approach to environmental protection. The interconnectivity of these rights, and particularly so in an environmental context, is derived from the broadly defined environment that is espoused in Law No 96/12 of 5 August 1996.[32] This gives rise to the concomitant perception that an environment that is of an acceptable quality could be perfectly conducive to human health, and consequently is a necessary prerequisite to the realization of and enjoyment of inter alia the right to life, water, food, dignity and health. The right to work suggests, for example, that people have the right to live and work in an environment that is conducive and not detrimental to, or have any negative impact on, the protection of their physical and mental health. Relatedly, the right to development entails that attention must be given to environmental protection when pursuing development-related projects. In other words, a balance needs to be struck between the right to development and the right to the environment, to the extent that the right to development does not trample on the environment or other related rights of people. Logically, the reciprocal respect for and the protection of these rights generally, and the environmental right specifically, could have the potential to enhance and ensure environmental protection.

Law No 96/12 on Environmental Management is Cameroon's main environmental framework legislation and is intended to give effect to the constitutional right to a healthy environment through measures that ensure and promote the protection and management of the environment.[33] Like the Constitution, this law obliges every citizen to safeguard

[32] Law No 96/12 of August 5, 1996, relating to Environmental Management in Cameroon (Law No 96/12). S 1 (k) defines the environment as: all the natural or artificial elements and bio-geochemical balances they participate in, as well as the economic, social and cultural factors which are conducive to the existence, transformation and development of the environment, living organisms and human activities.

[33] See s 1 of Law No 96/12.

the environment and contribute to its protection.[34] Under the law, the environment is a national common heritage, and as such, it is crucially important to ensure its protection in order to ensure the rational management of the country's resources for the general interest of all Cameroonians.[35] Environmental protection is also necessary in order to ensure the conservation and enhancement of the country's cultural and architectural heritage, which constitute an integral part of the state's national policy of environmental protection and development.[36] The Law provides for relevant environmental management principles that are crucially vital for ensuring and promoting the constitutional right to a healthy environment.[37] In furtherance of the protection of the environmental right, the law prohibits pollution of any kind, to the extent that it does not adversely affect the environment. The law obliges the state to establish quality norms for air, water and soil, as well as any other norms that could be necessary to safeguard human health and the environment.[38]

Apart from the above, the National Environmental Management Plan (NEMP) serves as Cameroon's environmental policy. Its objective is to develop policies, strategies and actions for environmental protection and for the rational management of natural resources that contribute to biodiversity conservation and the sustainable exploitation of its components, public participation, benefit sharing and sustainable development. This policy is given full effect by Law No 96/12 of 5 August 1996 on Environmental Management.[39] Law No 96/12 established the Interministerial Committee on the Environment and a National Consultative Commission on the Environment and Sustainable Development to assist the state in the formulation, coordination and implementation and

---

[34] See s 9(e) (ii) of Law No 96/12.

[35] Ss 2(1)–2(2) of Law No 96/12.

[36] See s 39(1–2) of Law No 96/12.

[37] See s 9(a–f). These principles include: the principle of precaution, the principle of preventive action and correction, the polluter pay principle, the principle of liability, the principle of participation and the principle of substitution.

[38] S 10(1) (i).

[39] See ss 10 and 13 of the Law No 96/12.

monitoring of environmental policies.[40] In terms of the Law, the duties, organization and functioning of this body shall be prescribed by a decree. Regrettably, this decree is still awaiting publication. Under the Law, the competent administration in charge of the environment is required to ensure the inclusion of environmental concerns in all sectors, including economic, energy and land, as well as in other plans and programs, and to ensure that Cameroon's international environmental commitments align with and are introduced in national environmental laws, regulations and policies.[41] However, this does not seem to be the case, as will be demonstrated below. Regrettably, the NEMP does not make any (satisfactory) reference to the constitutional guarantee of a right to a healthy environment. By contrast, the Green Paper on Environmental Policy in South Africa, 1996, is more detailed, with specific goals, objectives and visions on environmental rights, than is the corresponding policy of Cameroon.

Procedural Rights

Procedural rights, such as the rights to access to information, public participation and access to justice, including broad locus standi, also form part of the broader rights-based approach in Cameroon. Moreover, their effective respect and implementation could be useful in enhancing environmental protection.

Access to Information

The Constitution does not explicitly provide for the right to access to information.[42] However, the Preamble provides for the right to freedom of expression, which has been considered to be supportive of the right of access to information, since one needs information in order to express an

[40] S 10(2) of Law No 96/12.

[41] S 14(1–2) of Law No 96/12.

[42] Ashukem (n 29) 442; Jean-Claude N Ashukem 'Access to Environmental Information in the Context of Development Activities in the Legal Framework of Cameroon' (2017) 50 VRU 442.

informed opinion.[43] Logically, the right of access to information could be inferred from the right to freedom of expression. Accordingly, the government is obliged to respect people's right to freedom of expression to the extent that it facilitates and permits ordinary citizens to make informed opinions on matters that affect them. In the environmental context, this relates to the making of informed decisions on matters of environmental quality that could guarantee one's good health, as well as on measures to augment environmental protection. The Constitution affirms the country's commitment to upholding ratified international human rights and environmental treaties and conventions—both soft and hard law instruments.[44] Consequently, one could conveniently rely on, for example, the African Charter to assert one's right of access to information and increasingly to environmental information.[45]

Law No 96/12 provides for the right of access to environmental information. It guarantees everyone the right to be informed of the negative effects of harmful activities on their health and the environment.[46] It has been indicated that when read jointly with the constitutional right to a healthy environment, section 7 is highly supportive of the right of access to environmental information.[47] However, the conditions and procedures for this right are still to be provided for by the pending enabling Decree.[48] Nonetheless, the state is also obliged to collect, publish and disseminate information on environmental protection and management to the public.[49] Section 6 requires private sectors to inform the public of the possible impacts that their activities could make on the environment, and on people's environmental right, through the publication of environmental management plans and programs. Collectively, therefore, the state and

[43] George Devenish, *A Commentary on the South African Bill of Rights*, (Durban 1991) 439.

[44] Some of these instruments are the *African Charter on Human and Peoples' Rights* of 1981, the *Rio Declaration on Environment and Development* of 1992; Agenda 21; and the *United Nations Framework Convention on Climate Change* of 1992.

[45] Ashukem (n 29) 442; Ashukem (n 41) 442; Carlyn Hambuda and Rachel Kagoiya (eds.), *Freedom of Information and Women's Rights in Africa. A Collection of Case Studies from Cameroon, Ghana, Kenya, South Africa and Zambia* (UNESCO 2009) pp. 18–19.

[46] S 7(1) of Law No 96/12. Also see s 9(e).

[47] Ashukem (n 29) 443.

[48] See s 7(2) of Law No 96/12.

[49] S 10(1)(vii) of Law No 96/12.

private entities have a duty to inform the public of the potential deleterious impacts of their environment-related activities. Information of this nature, it is submitted, may not only help to promote the implementation of environmental laws and policies, but could also greatly serve to enhance governance for sustainability in the country.[50]

Law No 2003 on Biotechnology[51] also provides for the right to gain access to information. It requires the competent national administration to facilitate public sensitization and education on the use of genetically modified organisms (GMOs), and particularly to ensure that anyone who is involved in the use of modern biotechnology undertakes to sensitize and educate the public on the possible risks and benefits associated with the use of GMOs.[52] Because GMOs are used, inter alia, to enhance the yields of food crops, their use could cause pollution and ecological degradation which could impact negatively on the environment. It is only fair and reasonable that the public should be sensitized to the possible risks that are associated with their use, as this could enable concerned individuals to proactively fulfill their constitutional environmental right. However, section 12 of the Law limits the effectiveness of this right regarding a state security concern. Unfortunately, the Law does not define or state clearly what 'state security concern' means for purposes of the limitation of the right to access to information, and it has been question how this operates as a limitation.[53]

Public Participation

The Constitution does not contain any provision on the right to public participation. Like access to information, the right to public participation can only be inferred from the Preamble, which places an obligation on the state to ensure that land and environmental resources are harnessed in a way that would be beneficial to all Cameroonians. The fact

---

[50] Ashukem (n 41) 444.

[51] Law No 2003/006 of 21 April 2003 lays down Safety Regulation Governing Modern Biotechnology in Cameroon (Law No 2003/006).

[52] See ss 35 and 12(3) of Law No 2003/006.

[53] Ashukem (n 41) 448

that the Constitution requires citizens to ensure the protection of the environment suggests that an effective discharge of this responsibility would entail a situation whereby concerned individuals must be allowed to be involved and effectively participate in environmental decision-making processes. The fact that the Preamble alludes to Cameroon's commitments to international and African regional law instruments, which provide, inter alia, for public participation in environmental matters, suggests that it would be appropriate, as it is legally accepted, for concerned individuals to rely on the provisions of some these instruments (as in the *SERAC's* case noted above) to demand for greater inclusivity and participation in environmental matters, with the intention of safeguarding their right to a healthy environment.

Section 9 of Law No 96/12 provides for the right to public participation in environmental decision-making and explicitly requires that environmental decisions should be made only after full consultation with and the participation of concerned individuals and other relevant actors.[54] The state is also obliged to facilitate, encourage and allow for public participation in environmental management through mechanisms that allow and promote free access to information; to create a consultative mechanism to allow the public to form an opinion; to glean public opinion from public representatives serving on consultative organs on matters relating to the environment; to establish mechanisms that ensure the dissemination of environmental information and to establish mechanisms relating to the sensitization, training, research and education of local communities on the environment and environmental issues.[55] The law makes provision for environmental and social impact assessment regulation[56] that provides for and requires the participation of concerned individuals and other interested and affected persons in decision-making processes as a necessary measure to ensure environmental protection.

It is important to note that decree No. 2013/0171/PM to lay down the conditions for conducting environment and social impact assessment

---

[54] S 9(e)(i–iii) of Law No 96/12.

[55] See ss 72(i–iv) of Law No 96/12.

[56] Decree No. 2013/0171/PM of February 14, 2013, lays down rules for conducting Environmental and Social Impact Studies (EIA Regulations) in Cameroon. The regulation was adopted under Article 9. Also see s 17 of Law No 96/12.

(ESIA) replaced decree No 2005/0577/PM to lay down the procedures for carrying out an environmental impact assessment (EIA). The change from EIA to ESIA caused a simultaneous change in appellation specific to Cameroon.[57] The 2013 decree introduced strategic environmental and social assessment (SEA) and environmental impact statement (EIS) as tools for effective environmental assessment.

Other environmental sectoral legislation that incorporate aspects of public participation include: the 1976 Ordinance on the Management of State Lands,[58] the 1994 Forestry, Fisheries and Wildlife Law[59] and the 2003 Biotechnology Law. Collectively, the participatory approaches enshrined in these laws are relevant for being used toward achieving the fulfillment of the constitutional right to a healthy environment. Article 12 of the 1976 Ordinance provides that the composition of the Land Consultative Board (LCB) includes a government representative in the form of a prefect (a Senior Divisional Officer), a chief and two village elders, and that all decisions on land-based activities must be made in accordance with the participation of all members of the Board. Bearing in mind the broad definition of 'environment' in Law No 96/12, noted above, the composition of the LCB in an environmental context is relevant for two reasons. Firstly, the fact that land-based projects could have a negative impact on people's environmental and other rights suggests that it is only reasonable for a proposed land-based activity to be monitored in this manner. The fact that the environment is a common heritage[60] means that its rational management and protection through participatory governance could serve to accommodate both the interest of all Cameroonians and the need to ensure the conservation and protection of the environment for the benefit of both present and future

---

[57] Christopher Tamasang "Environmental impact assessment under Cameroonian law" in Oliver C. Rupple and Emmanuel D. Kam Yogo (eds) 1 st edn Environmental Law and Policy in Cameroon: Towards Making Africa the Tree of Life (NOMOS), 2018 280.

[58] Ordinance No 76/166 of 27 April 1976 lays down the Management of State Lands in Cameroon (Ordinance 76/166).

[59] Law No 94/01 of 20 January 1994 lays down Forestry, Fisheries and Wildlife Regulations (Law No 94/01).

[60] S 2(1) of Law No 96/12.

generations. Secondly, the inclusion of village representatives in land-based decision-making arguably provides a necessary and meaningful platform for them to be able to make informed decisions on any activity that might impact on their environmental and other rights. The reason is that local communities live close to natural resources and are better informed of the changing dynamics of these resources, to the extent that their inputs during decision-making processes could significantly help to boost the protection of the broadly defined environment.

The 1994 Forestry Law provides a framework for the integrated and sustainable use of forests, wildlife and fisheries resources that is achievable only if local communities are encouraged to participate in forestry decision-making processes.[61] In Cameroon, the government strives to promote, encourage and facilitate an inclusive system of public participation through legal and policy initiatives with regard to the governance of natural resources. The government has created two approaches (of participatory governance) aimed at enhancing sustainable forest management, as required by the Preamble to the 1994 Forest Law.[62] The first approach (under the law) is the creation of community forest tenure. Under this approach, the state is required to facilitate and ensure the active involvement and full participation of all relevant stakeholders, and specifically local communities, in decision-making in order to increase their living standards while concomitantly protecting the ecological integrity of the forest to ensure the sustainable management of forestry resources for the benefit of present and future generations. This approach has led to the creation of over 147 community forest tenure systems in the country, with the aim of encouraging local community participation within this context. A second approach employed by the government to facilitate and promote a participatory approach in the forestry sector is the model forest approach. In terms of this approach, all interested and affected parties, including the state, local communities, indigenous communities, traditional chiefs and civil society organizations, are required to contribute and share ideas on matters relating to and affecting the sustainable management of the forest. The two approaches evidently highlight the impor-

[61] See s 1 of Law No 94/1.

[62] See s 1 of law No 94/1.

tance of the use of public participation as a crucial mechanism for ensuring and promoting sustainable forest management in Cameroon.

Although the law fails to make any explicit provision for public participation in forestry-related decision-making, the right could implicitly be inferred from the provision of Section 23, which requires logging companies to ensure that local communities are given a chance to become involved and participate in forestry management plans (FMPs), prior to their approval by the relevant authority.[63] The intention is that any involvement of local communities in FMPs could be relevant and necessary to contribute toward achieving the ultimate aim of the law, namely to ensure the sustainable use and exploitation of forest resources. Law No 2003/006 reinforces public participation in environmental matters with respect to GMOs. It obliges the competent national authority to facilitate active and effective participation of local communities and other interested and affected parties in decision-making processes.[64]

Access to Justice

Regrettably, the Constitution does not guarantee the right to gain access to justice. The Preamble of the Constitution only refers to the right to a fair hearing before an impartial tribunal or forum. It has been argued that because a right to a fair hearing can be exercised only if one has a right to have access to court,[65] the right to a fair hearing in the Cameroonian context could be conceived as a right to gain access to justice. Consequently, the right to a fair hearing relates to the core right to have access to justice, and because the two are directly related, they could be used where appropriate to enforce claims for the violation of human rights, including the environmental right. The Preamble incorporates the notion of shared responsibility between the government and the people in the effort to foster and enhance environmental protection through the mechanisms of public interest litigation. Locus standi could be interpreted in two ways, one of which is the legal competence to act and to appear in a court of

[63] S 23 of Law No 94/12.

[64] S 35 of Law No 2003/006.

[65] Ashukem (n 29) 243.

law, as either a claimant or a respondent, in which case factors such as legal capacity, mental capacity and age must be considered and are crucially important. It may also relate to a person's competence to present himself or herself as a party in a court as a result of a particular interest in a case under scrutiny which is of interest to him or her.[66] In an environmental context, locus standi relates essentially to the latter interpretation. Law No 96/12 reinforces the constitutional provision on legal standing and provides that:

> Association regularly declared or recognised as publicly useful and exercising their statutory activities in the field of environmental protection may only contribute to the actions of public and semi-public environmental institutions following an authorisation issued in keeping with the terms and conditions laid down by special instruments. Authorised grassroots communities and association contributing to all actions of public and semi-public institutions working for environmental protection may exercise the right of the plaintiff with regard to facts constituting a breach to the provisions of this law and causing direct and indirect harm to the common good they are intended to defend.[67]

Apparently, this right grants local communities and authorized associations the mandate to protect the environment by exercising this right against public and private bodies relating to their activities on the environment and the resulting impacts, direct or indirect, that could cause environment pollution and degradation. In the case of *FEDEV v China Road and Bridge Corporation*,[68] the Bamenda Court of First Instance recognized the right of Foundation for Environment and Development (FEDEV), an NGO with the mission to protect the environment, to have legal standing in the matter. Being the first case to deal with the right, it clearly sets a precedent on the right and the need for future courts decisions to recognize and enforce the right to locus standi of individuals and groups to protect the environment.

---

[66] W Bray 'Locus Standi in Environmental Law' (1999) 22/1 CILJSA 34.

[67] S 8(1) and (2) of Law No 96/12.

[68] [2009] Unreported decision No CFIB/004M/09

4 Limitations

Notwithstanding the provision of some substantive and procedural (environmental) rights aspects in Cameroon's legal framework, as canvased above, these provisions are limited in several respects that render the legal framework ineffective, inadequate and insufficient, and they have been characterized by enormous challenges. These limitations are substantive and procedural, and they could pose significant challenges for the practical realization of environmental rights in Cameroon.

Substantive Limitations

The right to a healthy environment in Cameroon suffers a major substantive limitation. The right is restrictively distinctive, and does not refer to people's health or well-being; nor does it make a link between environmental protection and the health and well-being of people. In this regard, it is unclear what a healthy environment means within the broader context of the right to a healthy environment.[69] For this reason, it might be interesting to know whether a healthy environment relates to protecting the environment from all toxic and dangerous substances and activities that have the potential to cause pollution and ecological degradation that impacts negatively on the environment, or whether a healthy environment is understood as not being excluded from other socio-economic rights which could be negatively affected, should the environment deteriorate.[70] Given this context, it is evident that one cannot rely on the constitutional provision where claims for health-related hazards are associated with an environmental activity that occurs outside the geographical sphere where a person lives. The constitutional right is merely cosmetic and highly illusory, and lacks adequacy and thorough and carefully formulated content relating to its scope of application, enforcement and limitation. It

[69] Jean-Claude N Ashukem 'Exploring the Effectiveness of the Rights-Based Approach to Environmental Governance in Cameroon: What Could be Learned from South Africa' (2019) 50(1) CILJSA 64–65.

[70] Ibid. 65.

has been suggested that both the Constitution and Law No 96/12 on Environmental Management obfuscate the concept of the right to a healthy environment, since they are completely unclear on the meaning of a healthy environment,[71] as would enable the use of human rights norms and standards to guarantee a healthy environment. In other words, it is uncertain what can be achieved by legal means toward achieving the protection of the environment. In this regard, it is apposite to note that until the law is makes clear what constitutes a healthy environment and its link with human rights, the use of human rights approaches to ensure a healthy environment might not be properly achieved and properly implemented, and advocated for. Perhaps the absence of guidance as to what constitutes a healthy environment in Cameroon could account for the futility of its constitutional adjudication.

Relatedly, the Constitution intentionally excludes the state from any obligation to protect the environment. The position is reiterated in Law No 96/12, which bestows a duty and responsibility on citizens and corporations to protect and contribute toward achieving the improvement of the environment. It is strange that the state, which is a duty-bearer of rights under international law, does not have a corresponding obligation to protect the environment. Instead, there is only a vague promise that the state shall ensure and improve the environment, without stating exactly how the state will do this.[72] This vagueness explains the sluggishness of the state in translating its international commitments at the domestic level, as exemplified by the enactment of incomplete environmental laws.[73] It has been suggested that the environmental rights provided for under Cameroon's legal framework constitute a mere declaration of state intent, rather than an obligation on the state to respect, protect and fulfill people's environmental rights.[74] Perhaps the exclusion

[71] Ibid. 45; 64–67.

[72] Ashukem (n 67) 68.

[73] Ibid. n 66. See for example the 21 years and still counting years of the incomplete of section 7 (2) of Law No 96/12 relating to the conditions and procedures of access to information. Also see the glaring avoiding of explicit provisions for the public participation in the 1994 Forestry, Fisheries and Wildlife Law to enable concerned citizens and individuals to effectively engage and participate in forestry-related decision-making processes.

[74] Ibid.

of any obligation on the state is predicated on the belief that the state itself, as a creature of laws, is among the most powerful social artifacts of humanity and, somewhat convolutedly, is equally the primary creator of law. However, without any corresponding obligation and responsibility on the state, any claim to pursue a rights-based approach to environmental protection in Cameroon could be a mere declaration of intent, as the state often undertakes or authorizes activities that negatively impact on the environment. It was inappropriate for the President to have authorized the development of a palm oil project by Herakles Farms in four biodiversity hotspots,[75] for example. The project has had negative impacts on the environment and on the country's rich biodiversity.[76]

The environmental right also lacks a clear link with the overarching notion of sustainable development and clarity on what the right to future generations in relation to the environment might be, notwithstanding its definition in law No 96/12. The fact that the Constitution was adopted in 1996, after the Brundtland Report and several decades after the Stockholm Conference, means that it was reasonably expected for the government to include the all-encompassing relevant notion of sustainable development within the broader ambit of the environmental right, to the extent that it reflected the government's international commitments.[77] However, this is not the case, and evidently the environmental right does not apply to future generations and it is unclear how their environmental right could be protected. Similar to the Constitution and Law No 96/12 on Environmental Management, the NEMP makes no substantive and meaningful contribution to the overarching notion of the environmental right. On this basis, the substantive bases of the right

[75] These are the Bayang-Mbo Wildlife Sanctuary, Korup National Park, Rumpi Hills Forest Reserve and the Bakossi National Park.

[76] For details on this, see Greenpeace "Herakles Farms in Cameroon: A showcase of bad palm oil production" https://www.greenpeace.org/usa/wp-content/uploads/legacy/Global/usa/planet3/PDFs/HeraklesCrimeFile.pdf accessed 5 April 2019; "Palm Oil Plantation Cuts Core from Cameroon's Biodiversity" (2012) https://www.oaklandinstitute.org/palm-oil-plantation-cuts-core-cameroon%E2%80%99s-biodiversity accessed 5 April 2019; "Cameroon's Biodiversity Hotspot in Grave Danger as Palm Oil Conglomerate Quits Sustainability Group" (2012) http://wwf.panda.org/?206114/Cameroon-biodiversity-hotspot-in-grave-danger-as-palm-oil-conglomerate-quits-sustainability-group accessed 5 April 2019.

[77] Also see s 14(2) of Law No 96/12.

to a healthy environment are inadequate and could be challenge in courts of law, where and when the interpretation of its substantive bases are called to question. Perhaps this inadequacy stems from the fact that environmental law is still in its infancy in Cameroon and has not gained traction among academics, researchers and lawyers. To be sure, there is no environmental law programme, either at undergraduate or postgraduate levels in Cameroon, as there are in other countries.

Procedural Limitations

The effective exercise of the right to access to information, and specifically environmental information, remains highly questionable, particularly seeing that there is no national legislation to give effect to the right. The situation has been exacerbated by the fact that Law No 96/12 on Environmental Management has failed completely to provide for the requisite conditions and procedures of access, and only subjects them to a presidential decree, which is still to be promulgated.[78] It has been argued that, without the requisite conditions and procedures of access, the right to access to information, including environmental information, is only illusory and elusive, to the extent that it hinders the possibility of properly enhancing environmental protection in Cameroon.[79]

It has been suggested that the practical realization of the right to public participation has been proven to be more a myth than a reality.[80] It has been indicated that rules, processes and procedures for public participation in Cameroon are not aligned with governance practices that allow for the views of local communities to be taken into account, and that instead, they are harassed and intimidated.[81] The effective participation

[78] See s 7(2) of Law No 96/12.

[79] Ashukem (n 29) 449; 435–450.

[80] Jean-Claude N Ashukem "Public Participation in Environmental Decision-Making in Cameroon: Myth or Reality? in Kameri-Mbote and others (eds) *Law/Environment/Africa* (NOMOS 2018) 357–373.

[81] Ibid. 365. Also see Ashukem 'Included or excluded? An Analysis of the Application of the Free, Prior and Informed Consent Principle in Land Grabbing Cases in Cameroon' (2016) 29 PELJ 20; Freudenthal E, Lomax T and Venant M "The BioPalm Oil Palm Project" in Colchester M and Chao S (eds) *Conflict or Consent: The Oil Palm Sector at a Crossroad* (2013, Forest Peoples Programme) 350; 345.

of local communities and other interested and affected parties during Environmental and Social Impact Assessment processes is a major hurdle,[82] and it is unclear whether the ESIA is meant to facilitate and support public participation, or merely to facilitate internationally financed projects.[83] Without public participation, it is difficult to actually challenge administrative decisions on environmental matters. The 1994 Forestry Law has been criticized for failing to have an explicit provision for public participation.[84] Yet, concession-based forest management[85] is common and continues to play a major role in the forestry sector, as opposed to the envisaged community forest management and the forest model management meant to promote the effective and full participation of local communities in the management of forestry activities. The recent initiation and participation in the EU Voluntary Partnership Agreement (VPA)[86] by the government was widely believed to cement the country's EU timber market, and not to emphatically promote and facilitate public participation, given that there was no participation by forest communities and indigenous people in the VPA negotiations.[87] It is reported that stakeholders were not consulted prior to the decision made by the government to enter into the VPA process, except for informal consultations on how to define legality, traceability and control systems during and in the pre-negotiation phase.[88] External stakeholders find it difficult to have a say on policy-making processes, inter alia. Unsurprisingly, local com-

---

[82] Ibid. Wodschow and others (2016:7).

[83] For more details on the flaws in the system of public participation in Cameroon, see Ashukem (n 78).

[84] Ashukem (n 78) 366.

[85] Concession-based forest management is the allocation of a portion of a forest to a foreign large-scale timber company for the purpose of harvesting timber produce. Also see Alain Karsenty 'Overview of Industrial Forest Concessions and Concession-Based Industry in Central and West Africa and Consideration of Alternatives' (2007) 1–45; Dieudonne Alemagi and Robert A. Kozak, 'Illegal Logging in Cameroon- Causes and the Path Forward' (2010) 12(8) FPE 554–56.

[86] See http://eeas.europa.eu/archives/delegations/cameroon/documents/eu_cameroon/note_information_apv_cameroun_en.pdf accessed 13 May 2019.

[87] Astrid Wodschow, Iben Nathan and Paolo Omar Cerutti 'Participation, Public Policy-Making, and Legitimacy in the EU Voluntary Partnership Agreement Process - The Cameroon Cas' (2016) 63 FPE 5; Sophia Carodenuta 'Local Participation from VPA to REDD+ in Cameroon' 2014 *ETFRN News* 55, 119.

[88] Ibid., Wodschow et al.

munities, indigenous people and individuals, including small-scale chainsaw operators and traders, were not represented and did not participate in the meetings and decisions of the VPA, despite it being crucially important in terms of the legality of the country's forestry sector.

The Ministry of Forestry (MINFOF) decides who should attend and participate at the VPA negotiation process, and this approach has excluded important participants, such as local communities who are often affected by the impact of forestry-related activities.[89] The question arises as to why local communities and/or their representatives are systematically excluded from decision-making processes regarding forestry-related activities. The answer seems to rely on the very objective and purpose of Law No 94/1, which, in addition to promoting sustainable forest management, encourages an increase in tax revenue from the forest sector through the commercial exploitation of forest produce to augment and implement radical socio-economic development that favors a forest management decentralization model.[90] This indicates that the law indirectly favors large-scale forest companies and, arguably, provides and facilitates their eminent role in decision-making processes, at the expense of local communities and their representatives. This is in contradiction to the concept of community forest management, which, as indicated above, requires a participatory approach in benefit sharing and decision-making regarding forestry-related activities. Another reason could lie in the fact that the 1994 Forestry law created and instilled the concept of hegemony control over natural resources, including forestry, which prompts the need to disregard the involvement and participation of local communities in decision-making processes, while placing heavy reliance on the role of the state as the guarantor and protector of these resources. The logic here is that environmental decision-making processes in Cameroon follow a top–bottom approach. It would have been appropriate if a bottom–top approach had been designed, adopted and implemented in the governance of natural resources generally, and specifically in the forestry sector, to provide for the meaningful, full and effective consultation and

[89] Ibid. 7–8.

[90] Phil Rene Oyono, Charlotte Kouna and William Mala 'Benefits of Forest in Cameroon: Global Structure, Issues involving Access and Decision-Making Hiccoughs' (2005) 7 FPE 357, 362.

participation of local communities in decision-making, to the extent that their input could significantly inform and shape the final decision, thus promoting sustainable forest management[91] for the benefit of present and future generations.

According to the UN Food and Agricultural Organization, participatory forest management entails processes and mechanisms that enable people (local communities), with a direct stake in forest resources, to become part of decision-making processes in all aspects of forest management, including policy formulation processes.[92] Thus, participatory forest management is vital and a necessary prerequisite for sustainable forest management,[93] particularly as it serves to ensure that the socio-economic, environmental, cultural and ecological values of local communities generally, and specifically those relating to the use and management of community/concession-based forests, are fully considered. This approach would have been appropriate had the state strictly applied and implemented the concept of concession-based forest management envisaged in the 1994 forestry law. Regrettably, this is not the case. On land matters, Ordinance No 76/166 provides for contradictory provisions for public participation in land-related matters, and it is not clear how and the extent to which the public are provided a chance to become involved in these decision-making processes.

The lack of an explicit and broadly formulated provision for the right to access to justice demonstrates the deliberate failure on the part of the state to meet its international human rights obligation to respect, protect and fulfill the fundamental rights and freedoms of its people at the domestic level,[94] including the environmental right.

[91] Dieudonne Alemagi, Reem Hajjar, Zac Tchoundjeu and Robert A. Zozak 'Cameron's Environmental Impact Assessment Decree and Public Participation in Concession-Based Forestry – An Explanatory Assessment of Eight Forest-Dependent Communities' (2013) 6(10) JSD 8, 8.

[92] FAO 'Website on Participatory Forestry' (2012) http://www.fao.org/forestry/participatory/en/ accessed 13 May 2019.

[93] Robert A. Kozak, W.C. Stetic, H. W. Harshaw, T.C. Maness, and S.R.J. Sheppard 'Public Priorities for Sustainable Forest Management in Six Forest-Dependent Communities of British Columbia' (2008) 38(12) CJFR 3071–3084.

[94] Ashukem (n 29) 244.

5 Conclusion

Axiomatically, an inextricable relationship exists between human rights and the environment, for which human rights norms, principles, standards and approaches remain crucially relevant for ensuring environmental protection and national legislation that professes the use of such an approach, is worthy of praise in this current state of global ecological degradation, largely perpetrated by human-induced activities.

Although legal questions are fundamentally relevant and decisive for the forging of an effective legal framework to properly protect the environment, one cannot ignore the fact that Cameroon's legal framework highlights some prospect for using and following a human rights approach to environmental protection. While such prospect exists, it needs to be complemented with laws that are complete and sufficient laws that would adequately and proactively ensure a significant paradigm shift to environmental protection.

As shown above, the current legal framework suffers from both substantive and procedural deficiencies, and arguably falls short of redressing environmental protection by means of human rights claims. These shortcomings illuminate some of the significant challenging tensions in domestic legal relationships between human rights and environmental protection. While the Preamble in the Constitution provides for the right to a healthy environment, it was clearly established that the constitutional guarantee was limited and therefore problematic, since it exonerates the state from any social justice and moral responsibility relating to environmental protection. It was argued that the right therefore remains as highly abstract. Moreover, a certain standard of environmental quality cannot be ensured by the state, as it currently has no obligation to take measures to control pollution and other environmental degradation that might negatively affect human health and private life. Presumably, such an obligation on the state, as there is on citizens, could significantly augment environmental protection and help to promote the rule of law in that context, given the state would be directly accountable for failing to regulate environmental degradation.[95] Without such an obligation, it is impossible to hold the state accountable for its actions or failures to act.

[95] Also see Boyle (n 2) 613.

The right to public participation needs to be explicitly provided for in the 1994 Forestry Law, and it also needs to be properly reformulated and reconceptualized in the ESIA and Ordinance 76/1. The state should urgently enact the enabling decree that provides for conditions and procedures of gaining access to information so as to permit ordinary people and private bodies to effectively protect the environment, and provide clarity on what exactly constitutes a state security concern. It is evident that a review is required of Cameroon's current legal framework which would address the inadequacies, ineffectiveness and insufficiencies and provide improved prospects for effectively connecting human rights and the environment. Without such a rigorous review, it would be practically difficult to effectively use human rights norms, standards, principles and approaches to ensure environmental protection in Cameroon.

Therefore, it is urgent to address these shortcomings in order to properly connect the linkages between human rights and the environment, and in so doing, deliver an environmental friendly legal framework that protects the environmental rights of both present and future generations. Until this is done, the human rights-based approach to environmental protection in Cameroon would remain highly in the abstract, with little prospect for protecting the environment and human rights.

References

Alemagi Dieudonne and Kozak Robert A 'Illegal Logging in Cameroon: Causes and the Path Forward' (2010) 12(8) *Forest Policy and Economics* 554–561

Alemagi Dieudonne, Hajjar Reem, Tchoundjeu Zac and Kozak Robert A. 'Cameron's Environmental Impact Assessment Decree and Public Participation in Concession-Based Forestry -An Explanatory Assessment of Eight Forest-Dependent Communities' (2013) 6(10) *Journal of Sustainable Development* 8–24

Anderson Michael R. "Human Rights Approaches to Environmental Protection: An overview" in Alan Boyle and Michael Anderson (eds) *Human Rights Approaches to Environmental Protection* (OUP 2003) 1–23

Anton DK and Shelton Dinah L *Environmental Protection and Human Rights* (CUP, 2011)

Ashukem JCN "A Rights-Based Approach to Foreign Agro-Investment Governance in Cameroon, Uganda and South Africa" (2016a, NWU) 1–487

Ashukem JCN "Access to Environmental Information in the Context of Development Activities in the Legal Framework of Cameroon" (2017) 50(4) *Verfassung und Recht in Übersee Journal* 435–450

Ashukem JCN "Included or excluded? An Analysis of the Application of the Free, Prior and Informed Consent Principle in Land Grabbing Cases in Cameroon" (2016b) 29/1 *Potchefstroom Electronic Law Journal* 1–29

Ashukem JCN "Public Participation in Environmental Decision-Making in Cameroon: Myth or Reality? in Kameri-Mbote and others (eds) *Law/Environment/Africa* (NOMOS 2019) 349–365

Ashukem Jean-Claude N 'Exploring the Effectiveness of the Rights-Based Approach to Environmental Governance in Cameroon: What Could be Learned from South Africa' 2019 *Comparative and International Law Journal of Southern Africa* 42–74

Atapattu Sumudu 'The Right to a Healthy Life or the Right to Die Polluted? The Emergence of a Human Right to a Healthy Environment under International Law' (2002–2003) 16(1) *Tulane Environmental Law Journal* 65–126

Balaceanu C, Apostol D and Penu D *'Sustainability and Social Justice' (2012) Procedia-Social and Behavioural Sciences* 6778–681

Bali Guidelines for the Development of National Legislation on Access to Information, Public Participation and Access to Justice in Environmental Matters of 2010

Beyerlin Ulrich 'Aligning International Environmental Governance with the 'Aarhus Principles' and Participatory Human Rights' in Grear Anna and Kotzé LJ (eds) *Research Handbook on Human Rights and the Environment* (Edward Elgar, 2015) 333–352

Birnie Patricia, Boyle Alan and Redgwell Catherine *International Law and the Environment* (3rnd edn, OUP 2008)

Boyle Alan 'Human Rights and the Environment: Where Next' (2012) 23(3) *The European Journal of International Law* 613–642

Boyle Alan "Human Rights and the Environment: Where Next" *in Ben Boer (ed) Environmental Law Dimensions of Human Rights* (OUP 2015) 213–218

Bray W 'Locus Standi in Environmental Law' (1999) 50/2/ *The Comparative and International Law Journal of Southern Africa* 33–58

Carodenuta Sophia 'Local Participation from VPA to REDD+ in Cameroon' 2014 *ETFRN News* 55, 118–124

Daly Erin and May James R. 'Bridging Constitutional Dignity and Environmental Rights Jurisprudence' (2016) 7(2) *Journal of Human Rights the Environment* 218–242

Devenish George *A Commentary on the South African Bill of Rights, Durban* (1991)

Ebbesson Jonas "The EU and the Aarhus Convention: Access to Information, Public Participation in Decision-Making and Access to Justice in Environmental Matters' http://www.europarl.europa.eu/RegData/etudes/BRIE/2016/571357/IPOL_BRI.(2016)571357_EN.pdf (accessed 3 March 2019)

Ebbesson Jonas "The Notion of Public Participation in International Environmental Law (1997) *Yale Book of International Environmental Law* 51–97

Freudenthal E, Lomax T and Venant M "The BioPalm Oil Palm Project: A case study in the Department of Ocean, Cameroon" in Colchester M and Chao S (eds) *Conflict or Consent: The Oil Palm Sector at a Crossroad* (2013, Forest Peoples Programme) 337–354

Glazewski Jan *Environmental Law in South Africa 2edn* (LexisNexis, 2005)

Grear Anna and Kotzé Louis J 'An Invitation to Fellow Epistemic Travellers – Towards Future Worlds in Waiting: Human Rights and the Environment in the Twenty-First Century' in Grear Anna and Kotzé Louis J (eds) *Research Handbook on Human Rights and the Environment* (Edward Elgar, 2005) 1–8

Hambuda Carlyn and Kagoiya Rachel (eds) *Freedom of Information and Women's Rights in Africa. A Collection of Case Studies from Cameroon, Ghana, Kenya, South Africa and Zambia* (UNESCO 2009)

Hornby AS, *Oxford Advanced Learner's Dictionary: International Student's Edition* (8th edn, OUP 2012)

Karsenty Alain *'Overview of Industrial Forest Concessions and Concession-Based Industry in Central and West Africa and Consideration of Alternatives'* (2007) 1–45

Kozak Robert A, Stetic WC, Harshaw HW, Maness TC, and Sheppard SRJ 'Public Priorities for Sustainable Forest Management in Six Forest-Dependent Communities of British Columbia' (2008) 38(12) *Canadian Journal of Forest Research* 3071–3084

Kravchenko S 'The Aarhus Convention and Innovation in Compliance with MEAs' (2007) 18(1) *Colorado Journal of International Environmental Law and Policy* 1–50

Ksetini Fatma Special Rapporteur for Human Rights and the Environment, Final Report to Commission on Human Rights, Sub-Commission on

Prevention of Discrimination and Protection of Minorities: Human Rights and the Environment, UN Doc E/CN.4/Sub.2/1994/9, 6 July 1994

Lewis Bridget "Environmental Rights or a Right to the Environment? Exploring the Nexus between Human Rights and Environmental Protection" (2012) 8(1) *Macquarie Journal of International and Comparative Environmental Law* 36–47

Merrills J.G. 'Environmental Protection and Human Rights: Conceptual Aspects' in Alan Boyle and Michael Anderson (eds) *Human Rights Approaches to Environmental Protection* (OUP, 2003) 25–41

Morrow Karen 'Sustainability, Environmental Citizenship Rights and the Ongoing Challenges of Reshaping Supranational Environmental Governance' in Grear Anna and Kotzé LJ (eds) *Research Handbook on Human Rights and the Environment* (Edward Elgar 2015) 200–218

Oyono Phil Rene, Kouna Charlotte and Mala William 'Benefits of Forest in Cameroon: Global Structure, Issues involving Access and Decision-Making Hiccoughs' (2005) 7 *Forest Policy and Economics,* 357–368

Rio Declaration on Environment and Development of 1992

Sachs Wolfgang, 'Environment and Human Rights' (2004) 47(1) Development 42–49

Sands Phillipe *Principles of International Environmental Law* (2nd edn, CUP 2003).

Scholtz Werner 'Human Rights and the Environment in the African Union Context in Grear Anna and Kotzé LJ (eds) *Research Handbook on Human Rights and the Environment* (Edward Elgar, 2015) 401–420

Shelton Dinah 'Human Rights and the Environment: Substantive Rights' in Malgosia Fitrmaurice, David M. Ong and Panos Merkouris (eds) *Research Handbook on International Environmental Law* (Edward Elgar Publishing, 2010) 265–283

Shelton Dinah *'Human Rights, Environmental Rights and the Right to Environment'* (1992) 28(1) Stanford Journal of International Law 103–138

United Nations Framework Convention on Climate Change of 1992.

Wodschow Astrid, Iben Nathan and Cerutti Paolo Omar 'Participation, Public Policy-Making, and Legitimacy in the EU Voluntary Partnership Agreement Process – The Cameroon Case' (2016) 63 *Forest Policy and Economics* 1–10

World Health Organisation Report of 2008 "Human Rights-Based Approach to Health and the Environment" *Report of a Regional Seminar Bangkok, Thailand, 20–21 August 2007*

"Cameroon's Biodiversity Hotspot in Grave Danger as Palm Oil Conglomerate Quits Sustainability Group" (2012) http://wwf.panda.org/?206114/Cameroon-biodiversity-hotspot-in-grave-danger-as-palm-oil-conglomerate-quits-sustainability-group (accessed 5 April 2019).

http://eeas.europa.eu/archives/delegations/cameroon/documents/eu_cameroon/note_information_apv_cameroun_en.pdf (accessed 13 May 2019).

FAO 'GMOs and the Environment' http://www.fao.org/docrep/003/x9602e/x9602e07.htm (accessed 20 March 2019)

FAO 'Website on Participatory Forestry' (2012) http://www.fao.org/forestry/participatory/en/ (accessed 13 May 2019)

Greenpeace "Herakles Farms in Cameroon: A showcase of bad palm oil production" https://www.greenpeace.org/usa/wp-content/uploads/legacy/Global/usa/planet3/PDFs/HeraklesCrimeFile.pdf (accessed 5 April 2019).

"Palm Oil Plantation Cuts Core from Cameroon's Biodiversity" (2012) https://www.oaklandinstitute.org/palm-oil-plantation-cuts-core-cameroon%E2%80%99s-biodiversity (accessed 5 April 2019);

https://www.ohchr.org/EN/Issues/Environment/SREnvironment/Pages/FrameworkPrinciplesReport.aspx accessed 3 March 2019.

14

The Judiciary and Environmental Protection in Zimbabwe

Brewsters Caiphas Soyapi

1 Introduction

Zimbabwe's colonial period was characterized by a development model that was skewed in favor of economic development at the expense of the environment and natural resources.[1] This partly changed after independence in 1980, with Zimbabwe transitioning to becoming the so-called bread basket of Southern Africa.[2] The period was short-lived, though, as the country "turned from a paragon of post-colonial development to the other extreme of being a model of post-colonial misrule and confusion".[3]

[1] Tumai Murombo, 'Balancing Interests through Framework Environmental Legislation in Zimbabwe' in Michael Faure & Willemien Du Plessis (eds), *The balancing of interests in environmental law in Africa* (Pretoria University Law Press 2011) 559.

[2] Lloyd M. Sachikonye 'Whither Zimbabwe? Crisis & Democratization' (2002) 29(91) *Review of African Political Economy* 13, 13; B Maphosa 'Lessons from the 1992 Drought in Zimbabwe: The Quest for Alternative Food Policies (1994) 3(1) *Nordic Journal of African Studies* 53, 53.

[3] Murombo (n 1) 557.

B. C. Soyapi (✉)
North-West University, Potchefstroom, South Africa
e-mail: Brews.soyapi@nwu.ac.za

M. Addaney, A. Oluborode Jegede (eds.), *Human Rights and the Environment under African Union Law*, https://doi.org/10.1007/978-3-030-46523-0_14

Despite this unfortunate turn of events, Zimbabwe still has an extensive natural resource base which consists of mineral extraction, agricultural activities, water resources and wildlife.[4] Estimates suggest that the proportion of Zimbabweans living in rural areas and relying on the environment and this natural resource base for their livelihoods is 70%[5] and Zimbabwe's economy is largely dependent on the environment and this natural resource base.[6]

On a constitutional governance level, the current Constitution (2013 Constitution)[7] came into being through a referendum that was held throughout the country and was approved by roughly 95% of those who participated.[8] This Constitution is the supreme law of the land, requiring all laws and conduct to accord with it.[9] There is a mandate on the state, all its institutions and agencies, and every person and juristic person to "respect, protect, promote and fulfil" the rights in the Bill of Rights.[10] Further, the Bill of Rights itself binds the executive, the legislature and the judiciary.[11] Unlike the previous constitution, the 2013 Constitution

[4] Mutuso Dhliwayo, 'A Critical Examination of the Scope, Content and Extent of Environmental Rights in the Constitution of Zimbabwe' (Masters Thesis, Midlands State University 2016) 2; Anna Brazier *Climate Change in Zimbabwe: Facts for Planners and Decision Makers* (Konrad Adenauer Stiftung, 2015) 8; ZELA *Strategic Interventions and Focus Issues in Zimbabwe's Extractive and Mining Sector* (ZELA 2014) 11.

[5] Dhliwayo (n 4) 3; Brazier (n 4) 126.

[6] Brazier (n 4) 21. Also, Zimbabwe's current economic blueprint: Zimbabwe Agenda for Sustainable Socio-Economic Transformation (ZIMASSET) is predisposed toward the exploitation of natural resources. However, ZIMASSET does acknowledge the bad state of Zimbabwean environmental management, noting that the challenges include pollution, poor waste management, deforestation, land degradation and biodiversity loss. Government of Zimbabwe, *Zimbabwe Agenda for Sustainable Socio-Economic Transformation (Zim Asset)* (Government of Zimbabwe, 2013) 24.

[7] The *Constitution of the of Zimbabwe Amendment Act* (no 20), 2013.

[8] Brian Raftopoulos 'The 2013 Elections in Zimbabwe: The End of an Era' (2013) 39(4) *Journal of Southern African Studies* 971, 974. More than 3,00,000 people participated, which represented almost half of those registered to vote.

[9] See s 2. Traditionally, however, one of the main problems with the Zimbabwean legal structure is that parliament has always been powerful. It amended the post-colonial constitution on many occasions. Murombo has therefore voiced concern over whether the government would respect the 2013 Constitution. Tumai Murombo 'The Utility of Environmental Rights to Sustainable Development in Zimbabwe: A Contribution to the Constitutional Reform Debate' (2011) 11 *African Human Rights Law Journal* 120, 137.

[10] S 44 of the Constitution.

[11] S 45(1) of the Constitution.

has an elaborate Bill of Rights that recognizes environmental rights (used interchangeably here with the right to a healthy environment).[12]

However, it is imperative to note that there is little in the way of scholarship and commentary on environmental rights in Zimbabwe or on the cases in which the right has been an issue (both before and after the 2013 Constitution). Most of the commentaries are in particular areas, including the constitutionalization of the right to a healthy environmental generally[13] or mechanisms for environmental protection.[14] Specifically, and as far as could be established, there is no scholarship that critiques the development (or lack thereof) of environmental rights in relation to the country's governance structures as well as the work of the courts. In other words, in a quest to see how Zimbabwe could potentially contribute to an understanding of environmental rights in a developing country setting and in Africa in particular,[15] this chapter is an attempt to offer some insight into how, within the context of Zimbabwe's governance setting, the courts have adjudicated environmental rights and to review what hindrances they face in doing so. The first part briefly contextualizes three aspects: the separation of powers, judicial independence and rule of law. Thereafter, the discussion considers the constitutional framework for environmental rights. Following this, the discussion turns to the jurisprudence[16] that has dealt with environmental rights and in the last part, the chapter offers some thoughts on what the cases reveal about the courts

[12] A discussion of some of the features of the 2013 Constitution follows below. It should be noted though, that The 2000 Draft Constitution had an environmental right as part of the state's National Objectives, rather than in the Bill of Rights. This was the same with the Kariba Draft Constitution of 2008, which also had an environmental right under National Objectives.

[13] See Murombo (n 9) 120–146; Tinashe Madebwe 'A rights-based approach to environmental protection: The Zimbabwean experience' (2015) *African Human Rights Law Journal* 15, 110–128.

[14] See Tinashe Madebwe 'Carving Out a Greater Role for Civil Litigation as an Environmental Law Enforcement Tool in Zimbabwe's 2013 Constitution' (2015) 11(2) *Law, Environment and Development Journal*, 106–119.

[15] Africa itself has, on a continental level, a provision that requires a balance between environmental protection and development. This is to be found in Article 24 of the *African Charter of Human and Peoples Rights* (1982), I.L.M., 21, 58.

[16] Many of the cases that are discussed in this article were unreported. They were obtained by the author from the organizations which were involved in the cases (the Zimbabwe Environmental Law Association and the Environmental Management Agency). The cases and their summaries are all on record with the author.

and environmental rights in Zimbabwe and how any challenges identified could be addressed.

2 The Setting: Separation of Powers, Judicial Independence and "Environmental" Rule of Law

Writing on issues related to the Zimbabwean constitutional debate, Murombo expresses the opinion that the efficacy of environmental rights is dependent on many variables which could include a country's politics, the absence or presence of the rule of law, and the implementation and enforcement of environmental law.[17] These variables are particularly pertinent in a country such as Zimbabwe, which has garnered international attention over the years for all the wrong reasons, not least its failure to respect human rights or judicial decisions. With this as background, this section briefly alludes to the separation of powers, judicial independence and the "environmental" rule of law as they show in Zimbabwe's 2013 Constitution.

The Separation of Powers

Looking back to the previous Lancaster House Constitution, it is immediately clear that such Constitution was not supportive of an independent judiciary. For instance, the president had the power to remove any judge for "any cause", as long as a recommendation had been received from a tribunal which the president himself would have set up.[18] The lack of independence took a turn for the worse in the 2000s with the land reform program.[19] During this time, war veterans went as far as invading

---

[17] Murombo (n 9) 121.

[18] S 87 of the Constitution. The judicial service commission was also entirely appointed by the president. See s 90.

[19] This is a government policy initiated around the year 2000 as a means to redistribute land which was majority owned and occupied by white farmers. The policy was chaotic and signaled the start of the economic meltdown that Zimbabwe has had over the last few years.

the Supreme Court and preventing the court from hearing a case brought by the Commercial Farmers Union.[20] Also, the police had begun to disregard court orders at will. For instance, without any sound basis, in *Commissioner of Police* versus *Commercial Farmers Union*,[21] the Commissioner of Police challenged a court order that required invaders of farmlands to be evicted. Thereafter, the lack of judicial independence was so severe that a number of highly regarded judges were forced into exile.[22] The 2013 Constitution sought a break from this. It divides governance powers in Zimbabwe between the *trias politica*: executive, the legislature and the judiciary. The first indication of the existence of this separation of powers can be found in the supremacy clause, where it is noted that the Bill of Rights binds the executive, legislative and judicial institutions.[23] The Constitution of Zimbabwe notes that power for the executive,[24] legislature[25] and judiciary[26] is derived from the people, with these organs exercising such powers on behalf of the people. Also, there is a system of checks and balances in Zimbabwe. For example, the Constitution of Zimbabwe requires the state to assist the courts to ensure their independence and proper functioning,[27] while the decisions of the courts bind the state.[28] This confirms that while the hallmark of the *trias politica* is the division of powers and the uniqueness of the functions of each arm of government, there will always be a degree of reliance and dependence among the three arms of the state.

---

[20] Robert Martin 'The rule of law in Zimbabwe' (2006) 95(384) *The Round Table* 239, 250; Human Rights Watch, 'Our Hands Are Tied: Erosion of the Rule of Law in Zimbabwe' 14 <https://www.hrw.org/report/2008/11/08/our-hands-are-tied/erosion-rule-law-zimbabwe> accessed 01 December 2019, 14; Gugulethu Moyo 'Corrupt judges and land rights in Zimbabwe' in Transparency International (ed) *Global Corruption Report 2007* (Cambridge University Press 2007) 36. Some of the war veterans were shouting slogans like "Kill the judges".

[21] (HC 3985/2000).

[22] Human Rights Watch (n 20); Moyo (n 20) 36.

[23] S 2(2) of the Constitution.

[24] See s 88(2) of the Constitution.

[25] See s 117(1) of the Constitution.

[26] See s 162 of the Constitution.

[27] S 164(2)(b) of the Constitution.

[28] S 164(3) of the Constitution.

Judicial Independence

Structurally, the Zimbabwean courts are divided into superior and lower courts. The superior courts are the Constitutional Court, the Supreme Court, the High Court, the Labour Court and the Administrative Court.[29] The lower courts include Magistrates Courts, customary courts and any other courts established by an Act of Parliament.[30] For the first time in the country's history, this Constitution created a Constitutional Court,[31] which sits as the highest court in constitutional matters.[32] The Supreme Court is the highest court in all other matters.[33] As far as hierarchy is concerned, Zimbabwe follows the doctrine of precedents, with the decisions of higher courts binding lower courts.[34]

Currently, judicial independence is safeguarded through a provision that subjects the courts only to the Constitution, "which they must apply impartially, expeditiously and without fear, favour or prejudice".[35] No one and no institution may interfere with the work of the judiciary and the state is required, through legislative and other measures, to "assist and protect the courts to ensure their independence, impartiality, dignity, accessibility and effectiveness".[36] Ironically (considering current political events), of all the countries in southern Africa, Zimbabwe's Constitution has the most elaborate set of guidelines for judicial functions. To start with, the Constitution recognizes that public trust in the judiciary is essential to the work and legitimacy of the judiciary. For this reason, the Constitution requires members of the judiciary to "honour their judicial office as a public trust and … strive to enhance their independence in

---

[29] Ss 162(a)–(e) of the Constitution.

[30] Ss 162(f)–(h) of the Constitution.

[31] S 162(a) of the Constitution.

[32] S 167(1)(a) of the Constitution.

[33] S 169(1) of the Constitution.

[34] Murombo (n 1) 561. Interestingly, during the COPAC constitutional drafting process the drafters briefly considered the idea of creating an environmental court for the country. However, this was left out of the final Constitution, with no official reason or explanation being given. See COPAC *Drafting Instruments: Second All Stakeholders Conference, October 2012* (COPAC 2012) 25.

[35] S 164(1) of the Constitution.

[36] S 164(2)(b) of the Constitution.

order to maintain public confidence in the judicial system".[37] Further, and perhaps owing to a history where the judiciary was very much under the control of the executive, the Constitution of Zimbabwe has specific prohibitions for the judiciary. The judiciary is prohibited from engaging in political activities,[38] holding office in political parties/organizations,[39] and soliciting funds or contributing toward political organizations,[40] and there is a prohibition on judges attending any political meetings.[41] This clearly sets the judiciary at a distance from the other members of the *trias politica,* and it could facilitate their independence.

It is perhaps also important to note that under the 2013 Constitution, the courts are tasked with defending and protecting the Bill of Rights, and when doing so, they are required to, for example, take into account international law and consider foreign law in their adjudication.[42] As a testament to how the judiciary holds a place that is unique among the organs of state, the Constitution also requires the judiciary to continuously learn and to keep abreast of both national and international developments for the purposes of enhancing its knowledge and skills: "Members of the judiciary must take reasonable steps to maintain and enhance their professional knowledge, skills and personal qualities, and in particular must keep themselves abreast of developments in domestic and international law".[43] While it is not clear what "reasonable steps" specifically entails, the existence of such a provision in the Constitution is quite unique.[44] This arguably reveals that the drafters of the Constitution were mindful of the fact that in a highly globalized world, legal development is dependent not only on developments in the domestic context but also on developments in foreign jurisdictions and internationally. It is such provisions which could encourage judges to engage in comparative borrowing.

---

[37] S 165(2) of the Constitution.

[38] S 165(4)(a) of the Constitution.

[39] S 165(4)(b) of the Constitution.

[40] S 165(4)(c) of the Constitution.

[41] S 165(4)(d) of the Constitution.

[42] S 46 of the Constitution.

[43] S 165(7) of the Constitution.

[44] This is something not required of the executive, the legislature or any of the courts in the other southern African countries like South Africa or Namibia.

"Environmental" Rule of Law

Nothing exemplifies the erosion of the rule of law in Zimbabwe more than the land reform that was undertaken in 2000.[45] Arguably, nothing justified disregarding judicial pronouncements on the issue or the intimidation of judges. Thus, as a matter of context, the rule of law in Zimbabwe has generally been weak. For this reason, the rule of law is elaborated comprehensively in the 2013 Constitution. First, the Constitution notes that the rule of law is one of the founding values and principles of the Constitution.[46] This implies that the rule of law must either directly or indirectly inform all actions of the state and all functions of the judiciary. Second, the Constitution connects judicial independence with the rule of law, noting that independence, impartiality and effectiveness are central to the rule of law.[47] This confirms that there is an inseparable link between the separation of powers, judicial independence and the rule of law. Third, the Constitution of Zimbabwe explicitly requires the judiciary to be guided by the rule of law. When exercising judicial authority, the Constitution requires the judiciary to be guided by the principle that "the role of the courts is paramount in safeguarding human rights and freedoms and the rule of law".[48] Accordingly, the rule of law is the foundation upon which courts must measure state actions in, for example, protecting, respecting, promoting and fulfilling human rights, including environmental rights.

In a developing country like Zimbabwe a strong rule of law will be important because this could mean stronger measures for environmental protection. In fact, the last few years have seen rapid development of concepts like global environmental constitutionalism and "environmental" rule of law (EROL), which are considered to be necessities for stronger environmental laws. This will be important in evaluating the extent to which the government of Zimbabwe safeguards or operates within "environmental" rule of law (EROL). Loosely described, global environmental

---

[45] Martin (n 20) 249.

[46] S 3(1)(b).

[47] S 164(2) of the Constitution.

[48] S 165(1)(c) of the Constitution.

constitutionalism is tied to the rise of constitutional environmental rights (or broader environmental protection measures) and the manner in which courts are adjudicating these rights, including processes of transjudicial communication or judicial cross-fertilization. On the other hand, EROL is tied to the idea of the rule of law, as it applies in environmental settings.[49] In other words, courts enforce the rule of (environmental) law as a means to achieve environmental sustainability:

> courts could, for example, use the idea of EROL as a yardstick against which to measure environmental decisions. Similarly, decision makers could be guided by the EROL in legislating, as well as in enforcing, environmental law. This would mean that, where a project requires an EIA to be carried out, EROL would be violated if such a project commences before an EIA has been duly conducted. On this view, EROL facilitates the fulfilment of the environmental rights protection.[50]

Thus, if courts are to better protect the environment and be the vanguard of environmental rights, they have to operate within and enforce EROL.

3 Constitutional Right to a Healthy Environment

Zimbabwe's erstwhile environmental policies have been consistently described as archaic, since they were centered on promoting exploitation rather than conservation.[51] Indeed, before the enactment of the *Environmental Management Act* (EMA) in 2002, Zimbabwe's environ-

---

[49] For a brief reflection of how these issues play out and interact in a developing country setting, see Caiphas B Soyapi, 'A Multijurisdictional Assessment of the Judiciary's Role in Advancing Environmental Protection in Africa' (2019) *Hague Journal on the Rule of Law* <https://doi.org/10.1007/s40803-019-00128-9> accessed 25 November 2019.

[50] Caiphas B Soyapi, 'Environmental Protection in Kenya's Environment and Land Court' (2019) 31 *Journal of Environmental Law* 151, 154.

[51] Murombo (n 9) 138. Such environmental laws were largely natural resource extraction oriented, and they were also discriminatory. Also see Rajah Naome, Dino Rajah and Steven Jerie, 'Challenges in Implementing an Integrated Environmental Management Approach in Zimbabwe' (2012) 3(4) *Journal of Emerging Trends in Economics and Management Sciences* 408, 408.

mental laws were sectoral and fragmented, with at least 18 different statutes (which were further administered by over eight different ministries)[52] regulating environmental matters without any framework law in place and without these laws being embedded in a constitutional rights-based foundation. Such an approach predictably impacted negatively on the efficacy of the environmental protection effort. It stands to reason that if courts are to enforce EROL, a constitution has to have elaborate provisions on environmental rights. Accordingly, with a view to grounding the manner in which environmental protection plays out in Zimbabwe, this section considers the constitutional framework that supports environmental rights.

The Constitutional Framework on Environmental Rights

As far as the inclusion of the right to a healthy environment in the Constitution is concerned, Constitution Parliamentary Committee (COPAC) found that out of 1946 wards in the country, 951 (48.87%) wards supported the idea of a right to a healthy environment.[53] This indicates that a sizeable percentage of the country acknowledged that there is indeed a need for environmental protection through a rights-based approach. Thus, the right to a healthy environment found its way into the 2013 Constitution. The right reads as follows:

1. Every person has the right—
    - (a) to an environment that is not harmful to their health or well-being; and
    - (b) to have the environment protected for the benefit of present and future generations, through reasonable legislative and other measures that—
        - (i) prevent pollution and ecological degradation;

[52] Naome, Rajah and Jerie (n 51) 409.

[53] COPAC *National Statistical Report Version 1: Second All Stakeholders Conference*, October 2012 (COPAC 2012) 189.

(ii) promote conservation; and
(iii) secure ecologically sustainable development and use of natural resources while promoting economic and social development.

2. The State must take reasonable legislative and other measures, within the limits of the resources available to it, to achieve the progressive realization of the rights set out in this section.[54]

As a starting point, the environment envisioned in this provision is one that is not "harmful" to the "health" and "well-being" of people. From this wording, the right is clearly anthropocentric, as the determination of harmful is in relation to "every person". However, what counts as healthy for human beings is subject to contestation, as a healthy environment is undefinable to the satisfaction of all, and a proper determination will probably have to be made on a case-by-case basis.[55] Yet, it is reasonable to assume that "health" within the provision is related to being free from the harmful/unhealthy effects of issues like "pollution, waste disposal, and other sorts of toxic contamination, since the most immediate threats to health and well-being concern contamination of air, water and food".[56] By this reading, a threat to human health—no matter how minor—could trigger the violation of the right to a healthy environment.[57] This would differ from well-being, which would go further and include issues like

[54] S 73 of the Constitution. This provision is similar to s 4(1) of the EMA, Zimbabwe's framework environmental law which preceded the 2013 Constitution. However, unlike the constitutional provision, the EMA provision specifically acknowledges that the enforcement of the right to a healthy environment is dependent on access to information. Such access to information is now also constitutionally guaranteed for both juristic persons and specifically the media under s 62(2) of the Constitution.

[55] Ernst Brandl and Hartwin Bungert, 'Constitutional Entrenchment of Environmental Protection: A Comparative Analysis of Experiences Abroad' (1992) 16(1) *Harvard Environmental Law Review* 1, 82. Also see James May and Erin Daly 2015 *Global Environmental Constitutionalism* (Cambridge University Press 2015) 93–94 for a broader discussion on the difficulties inherent in assigning the adjectives healthy, balanced or quality to the environmental right.

[56] Tim Hayward, 'Constitutional Environmental Rights: a case for Political Analysis' (2000) 48 *Political Studies* 558, 559. Also see James W Nickel, 'The Human Right to a Safe Environment: Philosophical Perspectives on its Scope and Justification' (1993) 18 *Yale Journal of International Law* 281, 284.

[57] This is perhaps why there is framework legislation that builds on from a constitutional right to a healthy environment.

noise, which disturbs comfort and welfare. To be sure, Kotzé and Du Plessis view well-being as having to do with issues that go beyond the environment and health effects associated with the environment in that it covers a person's welfare.[58] They go on to argue that contentment is at the center of well-being, with the result that if environmental harm disturbs such contentment, then the right to well-being is violated.[59] It could be possible then that environmental damage might not be harmful to health yet affect one's well-being; well-being thus conceptualized is broader than health. An example would be odor from a factory that might make one uncomfortable but not sick.[60] The words "to have the environment protected" imply that an obligation is placed on the state to protect the environment. While it is not stated as such in the wording of the right itself, such an obligation would correlate with the state duty to promote, protect, respect and fulfill rights.[61] The phrase "for the benefit of present and future generations" would require that planning is not done on a short-term basis but on a long-term one, in view of the fact that decisions might not only have implications for the present generation but also for generations yet to come, and also that as a developing country, Zimbabwe has to engage in development activities.

Sustainable development is mentioned only once in the Constitution. By requiring development to be ecologically sustainable,[62] the

[58] Louis J Kotzé and Anel Du Plessis "Some brief Observations on Fifteen Years of Environmental Rights Jurisprudence in South Africa" (2010) *Journal of Court Innovation* 157, 167. Also see Du Plessis 2011 *South African Journal on Human Rights* 291 for the view that well-being "is necessary to living meaningfully beyond simply being healthy".

[59] Kotzé and Du Plessis (n 58) 167. Also see Erika De Wet and Anel Du Plessis, 'The Meaning of Certain Substantive Obligations Distilled from International Human Rights Instruments for Constitutional Environmental Rights in South Africa' (2010) 10 *African Human Rights Law Journal* 345, 357, where they note that being uncomfortable could be a symptom of a violation of one's well-being.

[60] See for instance the South African case *Hichange Investments (Pty) Ltd v. Cape Produce Co (Pty) Ltd t/a Pelts Products and Others*, 2004 2 SA 393 (E), where subjectivity in relation to smell was considered to be an aspect that could determine whether well-being has been affected.

[61] See s 44 of the Constitution.

[62] Harding explains that sustainability is an ultimate goal (outcome), while sustainable development is the means to reach that goal. Ecologically sustainable development then denotes development that is 'ecologically' or environmentally friendly, rather than purely economic. See discussion in Rachel Harding, 'Ecologically sustainable development: origins, implementation and challenges' 2006 187 *Desalination* 229, 233.

Constitution creates a benchmark through which development will be measured. For example, if an issue is brought before them, courts will thus have the task of balancing the competing interests of development and environmental protection. This is clearly an unavoidable eventuality for courts because for the most part, African states are in a quest to develop, with the result that a certain level of environmental degradation will have to be experienced if any development is to ensue.[63] In the context of Zimbabwe, and considering the immediate past, it seems that courts have not had the chance to do such balancing, as proven by the lack of any significant jurisprudence on this particular issue (see the discussion below). What is clear, though, is that socio-economic development has for the most part been occurring at the expense of the environment. Much of this can also be tied to the country's "Look East" policy, which has resulted in various neo-colonial-type deals with countries from the East, mostly China.[64]

Zimbabwe's provision of environmental rights also subjects the fulfillment of the rights to the concept of the "progressive realization of rights". This concept is largely understood to be applicable to economic, social and cultural rights and means that fulfilling economic, social and cultural rights obligations "requires economic resources, and the financial constraints faced by many developing countries may make simultaneous and immediate fulfilment of all ESR [economic and social rights] rights obligations impossible".[65] For this reason, it is possible that the right to a healthy environment may well be considered less important than "pure"

[63] United Nations. Economic Commission for Africa (1992–02). *African common position on the African environment and development agenda. UN. ECA Meeting of the Technical Preparatory Committee of the Whole* (13th: 1992, Apr. 10–16: Addis Ababa, Ethiopia), para 56; Tiyanjana Maluwa, 'Environment and Development in Africa: An Overview of Basic Problems of Environmental Law and Policy' 1989 1(4) *African Journal of International and Comparative Law* 650, 651.

[64] The "Look East" policy was initiated by the Zimbabwean government after it had been hit with sanctions by the Western world for human rights violations. For a comment on this and some of the environment-related problems that have been caused by this policy, see generally Caiphas B Soyapi, 'Zimbabwe's 'Look East' Policy: A socio-legal perspective' 2015 30(1) *Southern African Public Law* 176–192.

[65] See generally Sakiko Fukuda-Parr, Terra Lawson-Remer and Susan Randolph, 'Measuring the Progressive Realization of Human Rights Obligations: An Index of Economic and Social Rights Fulfillment' (2008) *Economics Working Papers* 2008, 22 1, 6–7. Also see, Jeanne Luh, Rachel Baum and Jamie Bartram, "Equity in water and sanitation: Developing an index to measure progressive

socio-economic interests and rights such as access to housing. Indeed, past events in the country seem to demonstrate this. For example, there have not been sufficient efforts to tackle the sanitation problems in Zimbabwe, which led to one of the worst cholera outbreaks in the history of the country.[66] This was even before the constitutional provision that allowed for the progressive realization of the right to a healthy environment. Yet, even with the 2013 Constitution in place Zimbabwe has been hit by typhoid at least twice in 2017, with ten reported deaths.[67] The government has been unable to sufficiently respond to these events, despite it is being required by the Constitution to take measures to do so. What is peculiar in this respect is that the same conditions as those which resulted in the 2008 cholera outbreak still persist. This suggests that at a governance level at least, the 2013 Constitution has not yet started to generate benefits.

The above discussion demonstrates that the right to a healthy environment is broadly worded to include a range of aspects relevant for environmental protection. There are, however, some factors in the Zimbabwean context that could render the protection of the right to a healthy environment difficult to realize. At an institutional level, the institutions tasked with enforcing and investigating issues related to the right to a healthy environment have not been very active. For example, the Environmental Management Authority of Zimbabwe (EMA Zimbabwe) lacks both the human and financial resources to implement and operationalize the EMA.[68] An example that clearly demonstrates this is the building of a Chinese mall on a wetland. EMA Zimbabwe protested against this

---

realization of the human right" (2013) 216(6) *International Journal of Hygiene and Environmental Health* 662, 662.

[66] Dhliwayo (n 4) 26.

[67] The Zimbabwean, 'Harare City Council should take blame for continued typhoid outbreak' (21 June 2017) <http://www.thezimbabwean.co/2017/06/harare-city-council-take-blame-continued-typhoid-outbreak/> accessed 12 November 2019 NewsDay, 'Typhoid outbreak hits Mbare' (23 October 2017) <https://www.newsday.co.zw/2017/10/typhoid-outbreak-hits-mbare/> accessed 12 November 2019.

[68] Mar M Monyau and Amarakoon Bandara *Zimbabwe 2017* (Africa Economic Outlook Report 2017) 11; ZELA (n 4) 19.

development, but then bowed under political pressure.[69] It is questionable that a government should constitute a body like EMA Zimbabwe to serve as its overall environmental regulator and advisor and then ignore the advice it provides. In some cases there is bureaucratic and legislative fragmentation between, for example, the ministry responsible for mines and the ministry responsible for the environment.[70] The former ministry believes that environmental issues related to mineral activities must be handled by the minerals industry, a belief that has been streamed into the country's current *Mines and Minerals Amendment Bill*.[71] Such a construction arguably allows for environmentally harmful extractive activities to take precedence over environmental protection.

Also, despite having a specific portfolio dedicated to environmental rights, the Zimbabwe Human Rights Commission[72] is not recorded to have dealt with any environmental issues.[73] As far as could be established, the only reference to the right to a healthy environment is in a report made after the Commission visited flood victims in Tsholotsho, but even here the reference is marginal.[74] Thus, these institutional challenges, along with the country's struggles with corruption, lack of good governance and failure to observe the rule of law stand as serious challenges to the protection, promotion, respect and fulfillment of the right to a healthy environment. In view of these observations, the following section considers the cases in which the right to a healthy environment has been in issue.

---

[69] Soyapi (n 64) 184. Also see ZELA (n 4) 19, where it is noted that EMA Zimbabwe is at times plagued by political interference.

[70] Soyapi (n 64) 184. The ministry of Mines and Mining often clashes with the ministry of Environment, Water and Climate when it comes to issues related to mining and the environment. Dhliwayo (n 4) 57.

[71] See for example s 257B(2)(c), where it is noted that officials from other bodies established by EMA Zimbabwe must be in agreement with the Ministry of Mines and Minerals Development. Worse still, the Bill envisages the Minister of Mines and Minerals Development and his/her experts to determine the best methods for mining on places like riverbeds. See s 257B(2)(a).

[72] In terms of s 243 of the Constitution, some of the powers of the Zimbabwe Human Rights Commission include promoting awareness and respect for human rights, monitoring the observance of human rights and investigating allegations of violations of human rights.

[73] See the official website, http://www.zhrc.org.zw/environmental-rights/

[74] ZHRC 2014 *Report on the Mission Visit to Tsholotshoto* 22.

4 The Courts and the Right to a Healthy Environment

Most of the environment-related cases discussed below were instituted by the Zimbabwe Environmental Law Association (ZELA) in its capacity as an environmental NGO. Zimbabwe does not have a central database that records and provides public reports for all cases, a fact which in itself demonstrates the challenges both the judiciary and the people face in upholding the right to a healthy environment as well as the challenging circumstances under which courts operate in the country.[75] These few environmental judgments were obtained during a research visit to ZELA's offices and to EMA Zimbabwe.

There are two cases which demonstrated the inability of governance institutions to protect the right to a healthy environment. The first was *Manyame Park Residents v. Chitungwiza Municipality* (*Manyame Park Residence*),[76] where ZELA sought an order to compel the Chitungwiza Municipality to construct proper sewage systems. ZELA demonstrated that the municipality was negligently discharging raw sewage into a public stream and a residential area in contravention of the then recently enacted EMA and its right to a healthy environment. The court granted the order, and the municipality admitted to the pollution but pleaded that it did not have the resources to either rehabilitate the contaminated land or build proper sewage structures. Places like Chitungwiza were subsequently hit hard by the 2008–2010 cholera outbreak,[77] which clearly illustrates that government has been unable to apply a precautionary approach and take preventive measures.[78]

[75] The existing database is https://www.zimlii.org/content/judgments. The website is difficult to work with. A search of the database also revealed no case dealing with the right to a healthy environment.

[76] HC 11552, 2003.

[77] Ahmed et al., 'The 2008 Cholera Epidemic in Zimbabwe: Experience of the ICDDR, B Team in the Field' (2011) 5 *Journal of Health and Population Nutrition* 541, 541; ZADHR *Cholera in a time of Health System Collapse: Violations of Health Rights and the Cholera Outbreak* (ZADHR 2009) 4.

[78] The precautionary approach stems from the precautionary principle which is a guiding standard which holds that where there is uncertainty as to the environmental consequences of a proposed development or activity, environmental protection must be prioritized. See generally David Krielbel

Further, in *Dora Community* versus *Mutare City Council* (*Dora Community*),[79] ZELA initiated a case similar to *Manyame Park Residence* when it approached the courts for an order compelling the Mutare City Council to stop depositing semi-treated waste into a river (which the Council controlled), and to monitor, control and prevent the unbridled discharge of untreated industrial waste into the river in violation of the right to a healthy environment in terms of the EMA. ZELA obtained judgment by consent, after the sitting councilors had admitted to the pollution and agreed to put remedial measures in place. However, the Minister of Local Government dissolved the council that had admitted to the pollution and replaced it with a Commission that then refused to cooperate with ZELA or the community. This effectively meant that the court order was impotent. Also, there seems to be no record of a return to court in an attempt to have the judgment enforced. As is probably clear by now, this is the trend in Zimbabwe, where judgments are honored more in the breach than in the observance.

Arguably, it becomes clear through *Manyame Park Residence* and *Dora Community* that for environmental protection to be effective, the courts rely on other organs of state to play their part. In other words, the fulfillment of the right to a healthy environment is entirely reliant on the executive's playing its part. Clearly the courts have not been hesitant to order government environmental agencies to take action, yet the government and its agencies have been found wanting. This points to an uncomfortable disregard of EROL and the respect that should be given to the judiciary. What is the worth of a judicial decision if it does not bring any change for those who have the decision in their favor? What is the use of an environmental remedy if it cannot be effected because the state blatantly disregards the order? These questions all have grim answers, both for governance in general and the right to a healthy environment in particular; arguably more so for the latter, because the right needs to be enforced by the state.

---

et al., 'The Precautionary Principle in Environmental Science' (2001) 109(9) *Environmental Health Perspectives* 871–875, 871.

[79] (HC 1312/05).

The implication of these cases is that with the provision in the Constitution that requires the state to protect the right to a healthy environment subject to the progressive realization of rights,[80] it is possible that the state could argue that a lack of resources prevents it from fulfilling the right. As such, even if courts declare (as they did in these cases) that there are violations of the right to a healthy environment, such declaration is immaterial if it cannot be enforced/operationalized. This then means that the role of the courts in providing effective remedies is stifled due to the absence of EROL.

In another case, the Zimbabwean High Court had the occasion to adjudicate a special plea on a pollution matter related to diamond mining. In *Zimbabwe Environmental Law Association & Others v Anjin Investments (Private) Limited and Marange Resources (Private) Limited and Diamond Mining Company (Private) Limited* (*ZELA* vs. *Anjin and others*),[81] ZELA approached the High Court seeking an order against diamond mining companies that were discharging untreated waste (effluents, raw sewage, metals and chemicals) into the Odzi, Shingwizi and Save rivers.[82] The defendants filed a special plea opposing this. The facts were that the diamond companies dumped the waste into the rivers every second week, which left the rivers "silted, muddied, dirty and loaded with chemical and metal deposits".[83] The court granted the order and ruled that two of the arguments raised by the defendants (that the claim be barred because EMA Zimbabwe had not been cited and that the court lacked jurisdiction because the plaintiffs could have received the relief they sought under the EMA) were of no merit, since EMA Zimbabwe could not issue interdicts or a declaration of rights.[84] The order was granted only in 2015[85] although the case was filed in 2012.

[80] S 73(2) of the Constitution, which reads as follows: "The State must take reasonable legislative and other measures, within the limits of the resources available to it, to achieve the progressive realisation of the rights set out in this section."

[81] (HC 9451/12).

[82] Page 1, *ZELA* vs *Anjin and others*.

[83] Page 2, *ZELA* vs *Anjin and others*.

[84] Page 6, *ZELA* vs *Anjin and others*.

[85] The file name written on the judgment indicates that judgment was passed in 2015: HH 523/15.

To start with, *ZELA* versus *Anjin and others* illuminates the longevity of cases and the difficulties that can be faced in public interest litigation. It took over three years for the issuance of a declaration to the effect that Anjin was indeed contravening environmental laws by discharging effluents in the river. The inordinate length of time it took for just a declaration to be issued in this matter had obvious financial ramifications,[86] which were made worse by the current Zimbabwean economic situation. With the dollarization of the Zimbabwean economy, living costs have further increased and public interest litigation, the backbone of most cases dealing with the right to a healthy environment, suffers. The second issue to be noted about the case is its failure to mention the right to a healthy environment or sustainability in any way. Although directly alluding to the fact that the conduct of the defendants was clearly posing a hazard to the environment and the ecosystem around the rivers, the court merely stated that the conduct violated the EMA,[87] without specifying that it violated the right to a healthy environment. The court failed to interpret what the right could mean or entail, especially given that there were rivers involved, which are public goods. This is disappointing, because the court's hand could ultimately have been strengthened had it relied on the right to a healthy environment. Several questions arise, the answers to which are not immediately apparent: if there is such an elaborate environmental right, why do courts not rely on it? Is it a matter of ignorance? Or are courts intimidated to the extent that they fear reliance on constitutional provisions in the face of threats of victimization by the government? Notably in this regard, as recently as in 2016 former president Robert Mubage openly threatened judges who had made decisions in favor of people who wanted to protest.[88]

In *Marange Development Trust v Zimbabwe Consolidated Diamond Company (Private) Limited and Environmental Management Agency* (*Marange Development Trust case*),[89] ZELA approached the High Court on behalf of the Marange Development Trust seeking an order to halt

---

[86] Murombo (n 1) 576.

[87] Page 2, *ZELA* vs *Anjin and others*.

[88] See New Zimbabwe 2016 http://www.newzimbabwe.com/news-31051-Stop+allowing+protests;+Mugabe+warns+judges/news.aspx

[89] (HC 902/17).

diamond mining activities until an EIA had been obtained from the EMA. The facts as contained in the applicant's founding affidavit were that the first respondent had started conducting mining activities before the issuance of an EIA. The applicant had requested to see the EIA and had failed to get a response from the first respondent.[90] The applicant had also written to the second respondent, EMA Zimbabwe, but the second respondent had failed to respond and to halt the mineral extraction activities that had commenced.[91] In their heads of argument the first respondent argued that the applicant did not have standing to represent the community since it did not physically stay in the community and because it had not produced any authorization letter from the community.[92] Substantively, the first respondent denied that an EIA had not been undertaken, but acceded that the certificate was pending, though it expected it to be issued.[93] The applicants were successful in obtaining an interdict against the first respondent, who was ordered to desist from conducting any mining operation until an EIA certificate had been issued by the EMA Zimbabwe.

Marange Development Trust is the one case where the High Court exercised its powers of review in an attempt to uphold EROL. The High Court did so by ordering mining activities to be halted until an EIA certificate had been issued. However, what is disturbing from a governance point of view is that the first respondent admitted that they had started operations without the EIA certificate as required in terms of the EMA, but seemingly had the assurance that the certificate would be issued. This arguably suggests that corrupt practices were at play. Consistent with most of the declaratory order cases in Zimbabwe, the court went only as far as giving the order. It did not castigate EMA Zimbabwe for having failed to halt operations before the certificate had been issued. This further suggests, as noted above, that the courts could be hesitant to be outspoken against a government that either ignores their

[90] *Marange Development Trust v Zimbabwe Consolidated Diamond Company (Private) Limited and Environmental Management Agency* HC. 9451/12, Para 8–9.

[91] Para 18–19.

[92] Respondent's Heads of Argument, *Marange Development Trust v Zimbabwe Consolidated Diamond Company (Private) Limited and Environmental Management Agency* (HC 902/17), Para 4.1–4.10.

[93] Ibid., Para 4.15.

orders or that victimizes them. Thus, even where courts are willing to uphold EROL, they do so under inhibiting circumstances without the assurance of judicial independence.

In *Augar Investments OU* versus *Minister of Water and Climate and EMA* (*Augar Investments*),[94] the question before the High Court was whether the Minister of Environment, Water and Climate could declare a piece of land to be a wetland or whether the Minister's powers extended only to declaring existing wetlands to being ecologically sensitive areas.[95] The Minister had declared—through a general notice—the applicant's lands to be a wetland, without having allowed the applicants or anyone who was likely to be affected to make representations. The applicants argued that the Minister could not unilaterally declare any land to be a wetland, unless such land fell within the description of a wetland as defined in the EMA.[96] In delivering the judgment, the court noted that the Minister had acted ultra vires by failing to allow for representations, and that the applicants could lodge their appeal with the Administrative Court.[97] The Court also made some *obiter dicta* related to the right to a healthy environment and sustainable development. The judge noted the following regarding the EMA: "the purpose of EMA is to define environmental rights and to set out the principles of environmental management, as well as to provide an enforcement mechanism against recalcitrant offenders".[98] The Court went on to add that the ability of the EMA to nurture and protect the environment could in part depend on the interpretation given to its provisions, since it is relatively new in Zimbabwe.[99] The judge concluded that:

---

[94] HC 1017/14.

[95] Page 1, *Augar Investments.*

[96] S 2 of the EMA defines a wetland as meaning "any area of marsh, fen, peatland or water, whether natural or artificial, permanent or temporary, with water that is static or flowing, fresh, brackish or salt, and [it] includes riparian land adjacent to the wetland".

[97] Page 5, *Augar Investments.*

[98] Page 1, *Augar Investments.*

[99] Although at the time the EMA had been in existence for over 12 years, this could have been a sign to show that the courts had not seen much traffic related to the enforcement of the EMA and its provisions.

> It is hoped that the citizens of Zimbabwe will vigorously pursue and enforce their rights as provided in terms of the Environmental Management Act, lest we be judged and found wanting, by future generations, for failing to play our part in preserving and protecting the environment.[100]

Augar Investments is the only recorded case the author could find where mention is made of sustainability and the rights of future generations. Although alluding to the fact that much of the meaning and impact of the EMA will depend on how it is interpreted, the court did not go into detail on the potential meaning and impact of the right to a healthy environment within the context of the country. What is significant, though, is that the court at least acknowledged the existence of sustainable development and of the rights of future generations, which recognition could open the door for future courts to venture into unpacking these issues.

5 An Appraisal of the Judiciary and Environmental Protection in Zimbabwe

As is evident from the above discussion, there is not much to work with in Zimbabwe when it comes to the right to a healthy environment and the courts. However, a few observations can be made from the above discussion. While the country was provided with the right to a healthy environment for the first time through the EMA in 2002 and through the 2013 Constitution, the courts have had very little to do when it comes to potentially contributing to the development of the right. An investigation of the few cases that exist reveals that there is no single case in which the right to a healthy environment was defined and there is no case in which the courts actively considered the potential contours of the right. The norm is rather that cases rarely get to be argued in court and courts rarely have the opportunity to dissect the law on environmental protection and how this certainly relates to the right to a healthy environment. Of the cases discussed, almost all involve declaratory orders,

---

[100] Page 5, *Augar Investments*.

usually containing nothing but the order itself, with little to no reasoning offered at all. Thus, the judgments are notorious for being brief and to the point.[101] If one needs to appeal the decision, it is difficult to fully identify the grounds for appeal, simply because the cases are not sufficiently detailed to offer grounds for appeal. In the face of such judgments, it is difficult to see how precedent could be set.

Second, "[g]ood political governance is an integral appendage of good environmental governance".[102] That good political governance in Zimbabwe has not been settled for a long time is a known fact, and this has often resulted in governance failures which have had ripple effects on the environment. For example, the lack of enforcement of environmental laws by the government and its agencies ultimately resulted in one of the worst cholera outbreaks to have plagued Southern African in recent times, and it had a severe impact on the health and well-being of the very people that the right to a healthy environment seeks to protect. Owing to the failure to collect waste, provide clean water and repair dysfunctional sewage pipes, many citizens suffered from an extremely contagious cholera outbreak that killed over 4000 people.[103] This had varied and unintended consequences such as the spread of diseases across borders.[104] Such was the case with the cholera outbreak, which spilled into South Africa, Zambia and Malawi. Also, the lack of enforcement can be imputed from the fact that despite mineral extraction activities having been undertaken without environmental authorizations and EIAs in the Marange diamond areas,[105] there is no case law that demonstrates efforts to bring accountability for such actions.[106] There is also no tangible evidence of mining

[101] A search on the official site for Zimbabwean case law revealed nothing on environmental law or the environmental right. See https://www.zimlii.org/content/caseindex.

[102] Murombo (n 1) 590.

[103] Dhliwayo (n 4) 26.

[104] Soyapi 2016 *Potchefstroom Electronic Law Journal* 16.

[105] Dhliwayo (n 4) 20. Also, it is common knowledge that companies like DMC, Anjin and Marange Resources have polluted rivers such a Save and Odzi, but there is seemingly no tangible record of the enforcement of the laws to protect the environment. See generally ZELA (n 4) 19.

[106] Madebwe (n 14) 112. In fact, Madebwe has noted that perpetrators of environmental crimes in Zimbabwe prefer to pay fines rather than to rehabilitate the environment, a situation made worse by the fact that environmental crimes attract fines between $5 and $5000. Thus, Naome, Rajah and Jerie argue that the fixed nature of the fines, irrespective of the environmental damage, is not deterrent enough. (n 51) 412.

licenses having been canceled. Thus, the right to a healthy environment demands a continued facilitation of its realization and as noted earlier, the effectiveness of the judgments that courts make in relation to the right will always be dependent on the other branches of the *trias politica*. For its part, the judiciary in its role of advancing the right to a healthy environment serves as the ultimate defender of the right, but as the Zimbabwean example shows, the courts are often only as effective as the other arms of government are. Thus, there is a need to find ways to foster the relationship between the judiciary and other arms of the *trias politica* in overseeing the fulfillment of the right to a healthy environment, especially when courts have made orders against the executive or legislature.

Third, the fact that there are only a few cases related to the right to a healthy environment could be a result of the decay in the rule of law and judicial independence, which has resulted in passivity among Zimbabweans. There is a marked passivity among Zimbabweans when it comes to the enforcement of their rights, which could be imputed to the events occurring within the country. Obvious examples which have already been noted include graphic and international newsworthy events like the cholera outbreak and river and land pollution in mining towns, which directly violate the right to a healthy environment. One of the reasons could be that citizens are aware of the difficulties that exist when the government is being challenged for not regulating or for violating rights. Past events like the intimidation of judges and the disregard of their judgments bear testimony to this. Another reason could be that citizens are not sufficiently aware of or educated on the potential and meaning of those laws in the country aimed at protecting the environment and specifically of the potential power of the right to a healthy environment itself.[107] Thus, regulation is just one component of the entire environmental governance process, as education is also required to complete the cycle, as it were.[108] To some extent, this is partly being implemented, as EMA Zimbabwe has for the past few years been organizing high school debates related to environmental issues under a "Catch them Young

[107] Madebwe (n 14) 112; Dhliwayo (n 4) 54.
[108] Dhliwayo (n 4) 57.

Programme".[109] The theme/topic for 2017 was "Environmental rights are being given less recognition in development programmes than other rights in Zimbabwe".[110] Such efforts might also be enhanced by the constitutional provision requiring the 2013 Constitution and its provisions to be integrated into school curricula.[111]

In addition, since the role of the courts in advancing the right to a healthy environment could be limited by the fact that courts cannot invent facts and that they generally deal only with what is argued before them, there is a need for litigants, especially NGOs, to craft innovative and persuasive arguments that directly invoke the right to a healthy environment. If there are going to be creative and innovative ways of advancing the right to a healthy environment, such a critical task cannot be left to the judiciary only. It will instead require a more deliberate and collective effort from all arms of government and from civil society if the lofty provisions and objectives of the right to a healthy environment that exist on paper are to be comprehensively, innovatively and effectively translated into practice.

Fourth, Dhliwayo argues that Zimbabwean judges are not activist in nature as their knowledge of environmental law is very limited.[112] The cases discussed above suggest that there is merit in Dhliwayo's argument. For instance, if an EIA is an important and indispensable requirement for the commencement of development works, why would the court in *Marange Development Trust* simply grant an order and not rationalize it through a discussion of the necessity of EIA and how it connects to the right to a healthy environment? Why would the court in *Augar Investments* conclude that the ability of the EMA to protect the environment will be dependent on the interpretation given to it, yet the court itself did not give any meaningful interpretation beyond the statement "the purpose of

[109] The Herald, 'Schools environmental debate competition' (17 May 2017) <http://www.herald.co.zw/schools-environmental-debate-competition/> accessed 22 November 2019.

[110] Ibid.

[111] See s 7(b), which reads as follows: "The State must promote public awareness of this Constitution, in particular by — … (b) requiring this Constitution to be taught in schools and as part of the curricula for the training of members of the security services, the Civil Service and members and employees of public institutions". Of the countries analyzed in this work, no other constitution has such a provision.

[112] Dhliwayo (n 4) 9.

EMA is to define environmental rights and to set out the principles of environmental management"? This could point to a lack of proper appreciation of environmental law and its potential by the courts. It is hoped that the constitutional provision that requires judges to keep themselves abreast of international developments[113] might assist in facilitating judges' obtaining the requisite knowledge of environmental issues.

6 Conclusion

This chapter has evidently demonstrated that there is very little in the way of available information when it comes to the right to a healthy environment, courts and judicial decisions in Zimbabwe. This could be directly linked to the absence of focused jurisprudence related to the right to a healthy environment and to the handful of academic publications on the matter itself. This damning *status quo* means that so far as the quest for the judicial development of a form of African Union law on the environment, or even possibilities for judicial comparative borrowing, Zimbabwe does not have anything to offer. Overall, the state of environmental degradation and the lack of judicial decisions on violations of the right to a healthy environmental could be tied to the fragile political and critical socio-economic challenges the country has faced for a while. For that reason, Zimbabwe's post-independence history offers the most vivid illustration of the deterioration of the (environmental) rule of law, of the difficult circumstances that the courts could find themselves operating under, and the challenges related to their independence.

What the Zimbabwean situation ultimately illustrates is that a country could have provisions on the right to a healthy environment entrenched in its laws, and it could have a structured court system, but the efficacy of these systems is to a large part conditional on political will and respect for the (environmental) rule of law. Yet the future might not be as grim as was depicted in the discussion. The country now also has a new Constitution which seeks to be defined by a state which has a clear separation of powers, recognizes judicial independence and observes the rule

[113] See s 156(7) of the Constitution.

of law. The work of the judiciary could be enhanced by the various provisions on judicial independence and the rule of law. While laws on paper are not the ultimate solution to environmental sustainability, there can be no denying that such laws at least set the foundation and the framework from which action and efforts could stem. For now, the jury is still out on the issue of whether the 2013 Constitution will help foster a better appreciation of environmental protection and give effect to the lofty ideals and objectives of environmental rights.

References

Books and Book Chapters

Anna Brazier, *Climate Change in Zimbabwe: Facts for Planners and Decision Makers* (Konrad Adenauer Stiftung, 2015)

Government of Zimbabwe, *Zimbabwe Agenda for Sustainable Socio-Economic Transformation (Zim Asset* (Government of Zimbabwe, 2013)

Gugulethu Moyo 'Corrupt judges and land rights in Zimbabwe' in Transparency International (ed), *Global Corruption Report 2007* (Cambridge University Press 2007)

James May and Erin Daly, *Global Environmental Constitutionalism* (Cambridge University Press 2015)

Mar M Monyau and Amarakoon Bandara, *Zimbabwe 2017* (Africa Economic Outlook Report 2017)

Tumai Murombo, 'Balancing Interests through Framework Environmental Legislation in Zimbabwe' in Michael Faure & Willemien Du Plessis (eds), *The balancing of interests in environmental law in Africa* (Pretoria University Law Press 2011a)

Articles

Anel Du Plessis, 'South Africa's Constitutional Environmental Right (Generously) Interpreted: What is in it for Poverty?' (2011) 27 *South African Journal on Human Rights* 279

Brian Raftopoulos 'The 2013 Elections in Zimbabwe: The End of an Era' (2013) 39 *Journal of Southern African Studies* 971

B Maphosa 'Lessons from the 1992 Drought in Zimbabwe: The Quest for Alternative Food Policies (1994) 3(1) *Nordic Journal of African Studies* 53

Caiphas B Soyapi, 'A Multijurisdictional Assessment of the Judiciary's Role in Advancing Environmental Protection in Africa' (2019a) *Hague Journal on the Rule of Law* <https://doi.org/10.1007/s40803-019-00128-9> accessed 25 November 2019.

Caiphas B Soyapi, 'Environmental Protection in Kenya's Environment and Land Court' (2019b) 31 *Journal of Environmental Law* 151

Caiphas B Soyapi 'Water security and the right to water in Southern Africa: An overview' (2016) 19 *Potchefstroom Electronic Law Journal* 1

Caiphas B Soyapi, 'Zimbabwe's 'Look East' Policy: A socio-legal perspective' 2015 30 *Southern African Public Law* 176

David Krielbel *et al,* 'The Precautionary Principle in Environmental Science' (2001) 109 *Environmental Health Perspectives* 871

Erika De Wet and Anel Du Plessis, 'The Meaning of Certain Substantive Obligations Distilled from International Human Rights Instruments for Constitutional Environmental Rights in South Africa' (2010) 10 *African Human Rights Law Journal* 345

Ernst Brandl and Hartwin Bungert, 'Constitutional Entrenchment of Environmental Protection: A Comparative Analysis of Experiences Abroad' (1992) 16(1) *Harvard Environmental Law Review* 1

James W Nickel, 'The Human Right to a Safe Environment: Philosophical Perspectives on its Scope and Justification' (1993) 18 *Yale Journal of International Law* 281

Jeanne Luh, Rachel Baum and Jamie Bartram, 'Equity in water and sanitation: Developing an index to measure progressive realization of the human right' (2013) 216 *International Journal of Hygiene and Environmental Health* 662

Lloyd M. Sachikonye 'Whither Zimbabwe? Crisis & Democratization' (2002) 29 *Review of African Political Economy* 13

Louis J Kotzé and Anel Du Plessis 'Some brief Observations on Fifteen Years of Environmental Rights Jurisprudence in South Africa' (2010) *Journal of Court Innovation* 157

Rachel Harding, 'Ecologically sustainable development: origins, implementation and challenges' 2006 187 *Desalination* 229

Rajah Naome, Dino Rajah and Steven Jerie, 'Challenges in Implementing an Integrated Environmental Management Approach in Zimbabwe' (2012) 3 *Journal of Emerging Trends in Economics and Management Sciences* 408

Robert Martin, 'The rule of law in Zimbabwe' (2006) 95 *The Round Table* 239

Sirajuddin Ahmed *et al*, 'The 2008 Cholera Epidemic in Zimbabwe: Experience of the ICDDR, B Team in the Field' (2011) 5 *Journal of Health and Population Nutrition* 541

Tim Hayward, 'Constitutional Environmental Rights: a case for Political Analysis' (2000) 48 *Political Studies* 558

Tinashe Madebwe, 'A rights-based approach to environmental protection: The Zimbabwean experience' (2015a) 15 *African Human Rights Law Journal* 110

Tinashe Madebwe, 'Carving Out a Greater Role for Civil Litigation as an Environmental Law Enforcement Tool in Zimbabwe's 2013 Constitution' (2015b) 11 *Law, Environment and Development Journal* 106

Tiyanjana Maluwa, 'Environment and Development in Africa: An Overview of Basic Problems of Environmental Law and Policy' 1989 1 *African Journal of International and Comparative Law* 650

Tumai Murombo 'The Utility of Environmental Rights to Sustainable Development in Zimbabwe: A Contribution to the Constitutional Reform Debate' (2011b) 11 *African Human Rights Law Journal* 120

Reports and Other Secondary Sources

African common position on the African environment and development agenda. UN. ECA Meeting of the Technical Preparatory Committee of the Whole (13th : 1992, Apr. 10 - 16 : Addis Ababa, Ethiopia)

COPAC *Drafting Instruments: Second All Stakeholders Conference, October 2012* (COPAC 2012a)

COPAC *National Statistical Report Version 1: Second All Stakeholders Conference,* October 2012 (COPAC 2012b)

Sakiko Fukuda-Parr, Terra Lawson-Remer and Susan Randolph, 'Measuring the Progressive Realization of Human Rights Obligations: An Index of Economic and Social Rights Fulfillment' (2008) *Economics Working Papers* 200822

ZADHR *Cholera in a time of Health System Collapse: Violations of Health Rights and the Cholera Outbreak* (ZADHR) 2009)

ZELA *Strategic Interventions and Focus Issues in Zimbabwe's Extractive and Mining Sector* (ZELA 2014)

ZHRC *Report on the Mission Visit to Tsholotshoto* (ZHRC 2014)

Zimbabwe Agenda for Sustainable Socio–Economic Transformation (ZIMASSET)

Thesis

Mutuso Dhliwayo, 'A Critical Examination of the Scope, Content and Extent of Environmental Rights in the Constitution of Zimbabwe' (Masters Thesis, Midlands State University 2016)

Legislation

African Charter of Human and Peoples Rights (1982), I.L.M., 21, 58
Constitution of the of Zimbabwe Amendment Act (no 20), 2013
Environmental Management Act Chapter 20:27 (2002)

Cases

Augar Investments OU vs Minister of Water and Climate and EMA HC 1017/14
Commissioner of Police vs. Commercial Farmers Union (HC 3985/2000)
Dora Community vs Mutare City Council (HC 1312/05
Hichange Investments (Pty) Ltd v. Cape Produce Co (Pty) Ltd t/a Pelts Products and Others, 2004 2 SA 393 (E)
Manyame Park Residents v. Chitungwiza Municipality HC 11552/2003
Marange Development Trust v Zimbabwe Consolidated Diamond Company (Private) Limited and Environmental Management Agency (HC 902/17)
Marange Development Trust v Zimbabwe Consolidated Diamond Company (Private) Limited and Environmental Management Agency HC. 9451/12
Zimbabwe Environmental Law Association & Others v Anjin Investments (Private) Limited and Marange Resources (Private) Limited and Diamond Mining Company (Private) Limited (*ZELA vs Anjin and others*) (HC 9451/12)

Online Sources

Human Rights Watch, 'Our Hands Are Tied: Erosion of the Rule of Law in Zimbabwe' 14 <https://www.hrw.org/report/2008/11/08/our-hands-are-tied/erosion-rule-law-zimbabwe> accessed 01 December 2019
New Zimbabwe 2016 http://www.newzimbabwe.com/news-31051-Stop+allowing+protests;+Mugabe+warns+judges/news.aspx

NewsDay, 'Typhoid outbreak hits Mbare' (23 October 2017) <https://www.newsday.co.zw/2017/10/typhoid-outbreak-hits-mbare/> accessed 12 November 2019

The Herald, 'Schools environmental debate competition' (17 My 2017) <http://www.herald.co.zw/schools-environmental-debate-competition/> accessed 22 November 2019

The Zimbabwean, 'Harare City Council should take blame for continued typhoid outbreak' (21 June 2017) <http://www.thezimbabwean.co/2017/06/harare-city-council-take-blame-continued-typhoid-outbreak/> accessed 12 November 2019

15

Human Rights-based Approach to Environmental Protection: Kenyan, South African and Nigerian Constitutional Architecture and Experience

Collins Odote

1 Introduction

When the community of African states adopted the African Charter on Human and Peoples' Rights in 1981,[1] it became the first treaty to explicitly settle the debate about the relationship between human rights and the environment by providing that "all people shall have the right to a general satisfactory environment favorable to their development."[2] This recognized that environmental rights are human rights and thus part of the responsibility of states to promote and protect like all other human rights. At the heart of the challenge in linking the two disciplines is not

[1] Adopted in Nairobi on June 27, 1981. Available at http://www.humanrights.se/wp-content/uploads/2012/01/African-Charter-on-Human-and-Peoples-Rights.pdf accessed 15th August 2019
[2] Ibid., Article 24.

C. Odote (✉)
University of Nairobi, Nairobi, Kenya

M. Addaney, A. Oluborode Jegede (eds.), *Human Rights and the Environment under African Union Law*, https://doi.org/10.1007/978-3-030-46523-0_15

just the question of the historical stages in their evolution, with human rights having been recognized as a distinct branch of law at a time when environmental law had not emerged in international discourse, but more about the relationship between the environment and human beings. While it is true that human beings require the environment for their sustenance, their reliance on the environment at the same time also leads to the destruction of the same environment. Consequently, it is sometimes necessary to protect the environment from human beings. The controversy that ensues is whether adopting a human rights lens will promote the conservation of the environment or limit the extent of the conservation of the environment. This controversy pitted ecocentrics against anthropocentrics, with the former arguing that the environment is to be conserved for its intrinsic value, while the latter see the environment within the prism of human support.[3] Finding a balance between this two became an international contestation among states for several years. The leadership role of the African continent in making the link between these two contesting approaches provides a useful backdrop against which to assess the appropriate linkages between human rights and the environment.

Discussing the relationship between the environment and human rights must be undertaken within the context of the development of the principle of sustainable development. When the first international conference on the environment was held in Stockholm, Sweden, developed and developing countries could not agree on the importance and focus of an international legal framework on the environment. Efforts to reconcile the divide between the two perspectives led to the adoption of the concept of sustainable development as the organizing concept for environmental law worldwide. At the basis of sustainable development is the principle of integration. Originally conceptualized as a balance between environment and economics, sustainable development is now seen as requiring a triple balance between not just the environment and economics, but also social issues. These linkages were restated in 2015 when the international community developed the Sustainable Development Goals, with the resolution adopting the SDGs providing that "(t)hey are

[3] Suzanne C Gragnon Thompson and Michelle A Barton, 'Ecocentric and Anthropocentric Attitudes Toward the Environment' (1994)14 *Journal of Environmental Philosophy* 149.

integrated and indivisible and balance the three dimensions of sustainable development: the economic, social and environmental."[4] The Declaration underscores the importance of linking human rights and environment in the process of resolving the world's problems, which problem the Declaration sees within the framework of poverty eradication. It mentions human rights several times and recognizes both the key human rights instruments and the main international declaration on the environment, an indicator of the importance of both a human right and an environmental lens in pursuing sustainability.

Against the above background, this chapter explores the extent to which a human rights lens has been adopted in protecting the environment in Africa. It argues that applying a human rights lens will help provide an additional layer for environmental conservation as long as it is not applied as the sole or primary methodology for environmental management, due to the limits and dangers of a purely anthropocentric approach. It does so by comparing the experiences of Kenya, South Africa and Nigeria. Following this introductory section, Sect. 2 explicates the relationship between human rights and the environment and how both branches have influenced the development and implementation of the other at the global and regional levels. This is followed, in Sect. 3, by a discussion of the situation in Africa, a discussion that is grounded on the content and implementation of the provisions of the African Charter on Human and Sect. 4 discusses the experiences of Kenya, South Africa and Nigeria in terms of recognizing the link between human rights and environmental management in their legislative provisions and examples of implementation. Section 5 provides a critique of the shortcomings of a human rights approach to environmental management, while Sect. 6 concludes the chapter.

---

[4] United Nations, *Transforming our World: the 2030 Agenda for Sustainable Development,* A/Res/70/1 https://www.un.org/ga/search/view_doc.asp?symbol=A/RES/70/1&Lang=E accessed 17 July 2019.

2 Conceptualizing the Link Between Human Rights and the Environment

The link between human rights and the environment both internationally and nationally is recent. The two branches of law originally grew separately, each distinct from the other. Nationally, "beginning in the 1970s, public recognition of worldwide environmental degradation and the inadequacy of state responses prompted constitutional changes and recourse to the powerful language of human rights."[5]

At the international level, emergence of international environmental law is traceable to the convening and outcome of the United Nations Conference on the Human Environment, convened by the United Nations General Assembly.[6] The Conference laid the basis for subsequent discussions on sustainable development. Its outcome included three non-binding instruments, a Declaration, an Action Plan and a resolution on institutional and financial arrangements.[7] The Stockholm Declaration stated the link between human rights and the environment. The Declaration recognized that man has a fundamental influence over the environment, indicating that man is "both the creature and moulder of his environment."[8] Further that "(b)oth aspects of man's environment, the natural and the man-made, are essential to his well-being and to the enjoyment of basic human rights-even the right to life itself."[9] Principle one of the Declaration then expressly provided for the link between human rights and the environment, stating that:

> Man has the fundamental right to freedom, equality and adequate conditions of life, in an environment of a quality that permits a life of dignity

---

[5] David R Boyd, *The Environmental Rights Revolution: A Global Study of Constitutions, Human Rights, and the Environment* (UBC Press 2012) 3.

[6] UNGA Res. 2398(XXIII) (1968).

[7] United Nations, *Report of the United Nations Conference on the Human Environment,* A/CONF.48/14. Available at https://www.un.org/ga/search/view_doc.asp?symbol=A/CONF.48/14/REV.1 accessed 19 July 2019.

[8] Stockholm Declaration, Preamble. Ibid.

[9] Ibid.

and well-being, and he bears a solemn responsibility to protect and improve the environment for present and future generations.[10]

Principle 1 of the Stockholm Declaration thus gave the hook for a human rights and environment link. However, subsequent international instruments largely failed to build on this hook, instead either remaining silent or using much weaker language. The main human rights conventions, the International Covenant on Civil and Political Rights and the International Covenant on Economic, Social and Cultural Rights are both silent on environmental rights.[11] The Rio Declaration of 1992 too failed to build on the language of the Stockholm Declaration, instead using more modest language on the issue. It provided that "human beings are at the center of concerns of sustainable development."[12] Although it then states that human beings are "entitled to a healthy and productive life in harmony with nature,"[13] its language is extremely mild largely because at the time of its adoption, there was both uncertainty and debate about the proper place of human rights law in the development of international environmental law.[14]

From 1990, the UN Commission on Human Rights has continued to pass resolutions on the linkages between human rights and the environment, an issue that has been continued by the Human Rights Council.[15] In 2012, it established the position of special rapporteur on the issue of human rights obligations relating to the enjoyment of a safe, clean, healthy and sustainable environment and appointed John Cox as the first rapporteur. His mandate ended in 2018 and he was replaced by David Boyd. The rapporteur's position has been to explore and make a case for

[10] Principle 1, Stockholm Declaration n 7.

[11] Ulrich Beyerlin and Thilo Marauhn, *International Environmental Law* (Hart Publishing, 2011) 393.

[12] Principle 1, Rio Declaration on Environment and Development, 1992. Available at http://www.unesco.org/education/pdf/RIO_E.PDF (Last accessed on 25th June 2019).

[13] Ibid.

[14] Dinah Shelton, 'What Happened in Rio to Human Rights?' (1992) 3(1) Yearbook International Environmental Law 75, 82.

[15] Phillipe Sands, et al. *Principles of International Environmental Law* (4th Edition, Cambridge University Press, 2018) 81.

an international link between human rights protection and environmental management.

While the linkages between human rights and the environment have several perspectives, a human rights approach to environmental management has been justified on the basis that it directly addresses environmental impacts on the life, health, private life and property of individual humans rather than on other states or the environment in general[16]; secures higher standards of environmental quality by obligating states to better control pollution and also promotes rule of law.[17] Even as the justification is made, while the influence of the environment on the enjoyment of human rights is well recognized, the recognition that human beings have a right to the environment, and that this right is over and above what human beings need for their sustenance has been slow to be accepted.[18]

The appointment of the special Rapporteur on Human Rights and the Environment by the United Nations has led to advancement of the knowledge on the link between environment and human rights. In a report to the Human Rights Council that appointed him, the first Special Rapporteur, John Knox, called for international recognition of the human right to a healthy environment and presented a set of framework principles to ensure that states promote enjoyment of a clean and healthy environment.[19] The Framework Principles were included as part of his final report as special rapporteur to the UN Human Rights Council in 2018.[20]

[16] Alan Boyle, 'Human Rights and the Environment: Where Next?" (2012) 23(3) European Journal of International Law 613.

[17] Ibid.

[18] Bridget Lewis, 'Environmental Rights or the Right to Environment: exploring the Nexus Between Human Rights and Environmental Protection' (2012) 8(1) Macquarie Journal of International and Comparative Environmental Law 36, 36.

[19] UN expert calls for global recognition of the right to safe and healthy environment, Geneva, 5 March 2018 https://www.ohchr.org/EN/NewsEvents/Pages/DisplayNews.aspx?NewsID=22755&LangID=E.

[20] United Nations General Assembly, *Report of the Special Rapporteur on the issue of human rights obligations relating to the enjoyment of a safe, clean, healthy and sustainable environment,* 37th Session of the Human Rights Council, 26 February-23 March 2018, A/HRC/37/59 https://www.right-docs.org/doc/a-hrc-37-59/

The special Rapporteur in the 2018 report underscored the interdependence between human rights and the environment,[21] since environmental harm affected the enjoyment of human rights just as much as human rights violations negatively impacted on the environment. The Framework Principles articulated in the report seek to facilitate the implementation of human rights obligations relating to the enjoyment of a safe, clean, healthy and sustainable environment.[22] They are a set of 16 Principles, whose aim is to "reflect the application of existing human rights obligations in the environmental context"[23] and not to create new principles.[24] Their adoption is intended to spur international consciousness and action with a view to eventual recognition and adoption of a global instrument explicitly providing for the human right to a clean and healthy environment.

The 16 Principles provide for both substantive and procedural issues. The first two principles demonstrate the link between human rights and the environment, with Principle 1 providing that "states should ensure a safe, clean, healthy and sustainable environment in order to respect, protect and fulfil human rights"[25] while Principle 2 states that "states should respect, protect and fulfil human rights in order to ensure a safe, clean, healthy and sustainable environment."[26] Realization of one category of right depends on and impacts the other hence the need for promoting and protection of both human rights and environmental protection.

Principles 3 and 4 deal with protection against discrimination and safe working environment respectively, with Principle 3 requiring states to "prohibit discrimination and ensure equal and effective protection against discrimination in relation to the enjoyment of a safe, clean, healthy and sustainable environment"[27] while Principle 4 calls for provision by states of "a safe and enabling environment in which individuals, groups and organs of society that work on human rights or environmental issues can

---

[21] Ibid.

[22] Ibid.

[23] Ibid.

[24] Ibid.

[25] Ibid.

[26] Ibid.

[27] Ibid.

operate free from threats, harassment, intimidation and violence."[28] Related issues of freedom of association, expression and assembly and their link to the environment are dealt with in Principle 5, requiring states to "respect and protect the rights to freedom of expression, association and peaceful assembly in relation to environmental matters."[29]

The other issues that the Principles deal with are procedural rights. The Rio Declaration recognized access to information, public participation, and access to justice as critical elements in the realization of a sound and safe environment. These principles have also been captured in a regional Convention within the European level, the Convention on Access to Information, Public Participation in Decision-Making and Access to Justice in Environmental matters.[30] Principles 6 and 7 deal with access to information issues, while Principle 7 explicitly provides that "(s)tates should provide public access to environmental information by collecting and disseminating information and by providing affordable, effective and timely access to information to any person upon request."[31] Principle 6 supplements this by requiring states to provide for public awareness and education in environmental matters.[32] The second limb of procedural rights is that of public participation. Principle 9 captures this by requiring states to "provide for and facilitate public participation in decision-making related to the environment, and take the views of the public into account in the decision-making process."[33] The last aspect of procedural rights relate to the rights to access courts and other judicial bodies so as to enforce rights. This is captured in Principle 10, which calls on States to "provide for access to effective remedies for violations of human rights and domestic laws relating to the environment."[34]

The next set of Principles address issues of environmental impact assessment, establishment and enforcement of environmental standards, international cooperation, additional measures to protect those most

[28] Ibid.

[29] Ibid.

[30] https://www.unece.org/fileadmin/DAM/env/pp/documents/cep43e.pdf.

[31] UNGA, n20.

[32] Ibid.

[33] Ibid.

[34] Ibid.

vulnerable or at risk from environmental harm, complying with the rights of indigenous peoples and traditional communities.[35] Finally Principle 16 demonstrates that the linkages between human rights and the environment are imperative for the realization of sustainable development, detailing States to "respect, protect and fulfil human rights in the actions they take to address environmental challenges and pursue sustainable development."[36]

The newly appointed special Rapporteur for Human Rights and the Environment, David Boyd, recently restated the interlinkages between Human Rights and the Environment, in an address to the 4th session of the United Nations Assembly in March 2019.[37] He emphasized that recognition of the linkages between human rights challenges and environmental issues would "catalyze stronger action by States and businesses and empower the millions of people worldwide working so hard to achieve the Sustainable Development Goals."[38] He consequently reiterated the need for global recognition of the right to a clean and healthy environment. In doing so he traced the history and key elements of that right, highlighting that:

> *The human right to live in a healthy environment was first mentioned in the Stockholm Declaration in 1972. Over the past fifty years, this right has been defined as including procedural and substantive elements. The procedural rights include access to information, participation in decision-making, and access to justice. The substantive elements include clean air, safe water and adequate sanitation, healthy and sustainably produced food, healthy biodiversity and ecosystems, a non-toxic environment, and a safe climate. Today, 130 States are parties to regional treaties--in Africa, Latin America and the Caribbean, Europe, Asia, and the Middle East-- explicitly incorporating the right to a healthy environment. In more than 110 States, this right enjoys constitutional*

[35] Ibid.

[36] Ibid.

[37] David Boyd, "Statement of the Special Rapporteur on the issue of human rights obligations relating to the enjoyment of a safe, clean, healthy and sustainable environment," 4th Session UNEA Assembly, 12th March 2019. http://web.unep.org/environmentassembly/un-special-rapporteur-human-rights-and-environment

[38] Ibid.

> *protection. It is also included in environmental laws of over 100 States. In total, at least 155 States recognize, in law, the right to a healthy environment.*[39]

The link between the two branches of law, while contested over the years, is increasingly becoming clearer and clearer. Dealing with the obstacles that prevent full recognition of the right to a clean and healthy environment at the international level will help to ensure that the benefits of the conceptual link between environmental rights and human rights are not only recognized but utilized in the quest for delivering sustainable development.

3 The African Context and Realities

Human rights discourse has long grappled with the question of universalism and relativism. This debate has accompanied human rights from the time of the adoption of the Universal Declaration of Human Rights, with some complaining that the Declaration represented a Western conception of human rights.[40] The Universalists held the view that human rights are universal, indivisible and inhere in all human beings irrespective of the diversity in cultures.[41] Cultural relativism on the other hand argues that human rights are culture specific.[42] Despite the existence of this controversy, international human rights law continued to evolve with the adoption of the International Covenant on Civil and Political Rights and the International Covenant on Social, Economic and Cultural Rights in 1976. The controversy though did not die, instead it only intensified.[43] The issue was dealt with at the World Conference on Human Rights held by the United Nations in Vienna, Austria, on 14 to 25 June 1993. The debate about universalism and relativism continued at the conference. In

[39] Ibid.

[40] Istvan Lakatos, 'Thoughts on Universalism versus Cultural Relativism with Special Attention to Women's Rights' (2018) 1 Pécs Journal of International and European Law 6.

[41] Ibid., 9.

[42] Ibid., 14.

[43] Yash Ghai, (2000) 21 'Universalism and Relativism: Human Rights as Framework for Negotiating Interethnic Claims' Cardozo L Rev. 1095, 1095.

the end agreement was reached that human rights are universal, indivisible and interdependent thus ending the debate in 1976 about civil and political rights versus social, economic and cultural rights.[44]

The above was the background within which the African Charter on Human and People's Rights was adopted in Nairobi in 1981. Its provisions should be seen considering the contemporary move in the discourse which acknowledges both the merits of universal human rights projects and the significance of culture in the conceptualization and implementation of human rights.[45] The Adoption of the African Charter sought to demonstrate the linkages that African states wanted between civil and political rights and social and economic rights. In addition, African States were uncomfortable with a focus on the individual only in the then obtaining human rights discourse, holding the position that in Africa, society was largely communitarian. Consequently, it was necessary to balance both the Western culture with the African culture[46] in elaborating a rights regime for the continent. This thinking led to the title of the Convention on Human Rights being named the *African Charter on Human and People's Rights*.[47] It is a recognition that rights are to be enjoyed both by human beings and by peoples collectively. As Professor Ojwang has argued, the development of a rights' regime in Africa had to take into account the contexts and realties obtaining, and those realities were and still are a reality of "abysmal poverty; with its largescale shortages of food, shelter, clothing, education and health; with its unsatisfactory planning of resource use; with its poverty of basic administrative structure; with its uncontrolled pace of population growth; with its numerous political upheavals, cases of turmoil, war and destruction."[48]

---

[44] UNHCR, "Vienna Declaration and Programme of Action," Adopted by the World Conference on Human Rights, Vienna, 25 June 1993.https://www.ohchr.org/EN/ProfessionalInterest/Pages/Vienna.aspx

[45] Lakatos, (n 40) 18.

[46] J.B. Ojwang, *Laying a Basis for Rights: Towards a Jurisprudence of Development,* Inaugural Lecture Delivered Before the University of Nairobi on 9th July 1992, (University of Nairobi Press, 1992) 20.

[47] Adopted June 27, 1981, OAU Doc. CAB/LEG/67/3 rev. 5, 21 I.L.M. 58 (1982). http://hrlibrary.umn.edu/instree/z1afchar.htm.?

[48] Ibid.

Dealing with the above issues resulted in a Charter that sought to provide a human rights architecture that dealt with civil and political rights as well as social and economic rights in the same document. One of the main issues was the place of property and property rights. To deal with poverty required that there was greater security and guarantee of property rights for citizens of Africa. This was particularly important for land, as majority of the population in Africa relied on land for their sustenance.[49] Consequently, the Charter recognizes the right to property.[50] The other critical issue is the place of development in the human rights discourse, in light of the endemic poverty levels in the content. The consequence was the inclusion of the obligation on states to "individually or collectively, to ensure the exercise of the right to development."[51]

The next critical issue is the linkages between sound management of the environment and realization of traditional human rights. The Charter recognized the need for including the environment as part of its catalogue of human rights, hence the provision guaranteeing that "all peoples shall have the right to a general satisfactory environment favourable to their development."[52] This is in recognition of both the abundant natural resource base of the continent and also the huge environmental challenges that Africa faces.[53] It is the rationale for the continent adopting a Convention as early as 1968 to deal with the conservation of its natural resources[54] and revised in 2003.[55] The objectives of the revised Convention are to enhance environmental protection, foster conservation of nature and natural resources and harmonize and coordinate policies in these fields.[56] The continent has over the years focused on environmental issues,

---

[49] AU, ADB and ECA, *Framework and Guidelines on Land Policy in Africa* (Addis Ababa, Ethiopia, 2010. https://www.uneca.org/sites/default/files/PublicationFiles/fg_on_land_policy_eng.pdf

[50] ACHPR, Article 14.

[51] Ibid., Article 22(2).

[52] Ibid., Article 24

[53] UNEP, *Global Environment Outlook (GEO-6), Regional Assessment for Africa (*Nairobi, 2016). http://wedocs.unep.org/bitstream/handle/20.500.11822/7595/GEO_Africa_201611.pdf?sequence=1&isAllowed=y

[54] African Convention for the Conservation of Nature and Natural Resources, 1001 UNTS 3.

[55] https://au.int/sites/default/files/treaties/7782-treaty-0029__revised_african_convention_on_the_conservation_of_nature_and_natural_resources_e.pdf

[56] Ibid., Article II.

developing a body of environmental law for Africa addressing such diverse issues as waste management, energy, illegal trade in flora and fauna; and protection of the marine environment,[57] among others. As a result of this importance of nature not just for its conservation imperatives but also for sustenance, the debate about whether environmental rights are a Human Rights, was positively answered in 1981, when the African Charter was adopted. Article 24 of the Charter consequently included environmental rights within an essentially human rights instrument.

Article 24 was elucidated by the African Commission on Human and People's Rights following a reference to it, in what is commonly referred to as the *Ogoni* Case.[58] The case revolved around complaints by two non-state actors against the Government of Nigeria as a result of oil drilling in Niger Delta.[59] In the communication, the organizations accused the Nigerian Government of contributing both directly and indirectly to the violation of the rights of the Ogoni people.[60] It was alleged that the actions complained about violated Article 24 of the African Charter relating to environment. This was as a result of the "actions of the Nigerian Military Forces, in protecting government interests in the oil venture, and the negligent and environmentally unsound management of the oil exploration in the Niger Delta."[61] The Nigerian Military Government was accused of being directly involved in oil production through the Nigerian National Petroleum Company whose operations were causing both environmental degradation and health problems to the Ogoni people.[62] The Government failed to protect the Ogoni people from the harm caused by the NNPC shell Consortium but instead used its security

[57] Robert Wabunoha, "Environmental Law of Africa," in Patricia Kameri Mbote and Collins Odote (eds), *Blazing the Trail: Professor Charles Okidi's Enduring Legacy in the Development of Environmental Law* (School of Law, University of Nairobi, 2019).

[58] Social and Economic Rights Action Center and the Center for Economic and Social Rights v. Nigeria ('SERAC v. Nigeria – the Ogoniland Case'), AfCHPR, Communication 155/96 (2002).

[59] Ibid.

[60] Morne Van Linde, "Considering the Interpretation and Implementation of Article 24 of the African Charter on Human and Peoples' Rights in Light of the SERAC Communication' (2003) 3 African Human Rights Law Journal 167, 168.

[61] Ibid.

[62] SERAC, (n58).

forces to facilitate the damage.[63] Other complaints against the Government included contaminating air, water and soil, and failure to either permit or conduct studies on actual and potential environmental and health risks of oil production.[64]

The Commission found the Government of Nigeria culpable for violating the rights complained about including the Right to a satisfactory environment as guaranteed under Article 24 of the African Charter. In doing so, the Commission made the point that environmental rights were essential elements of human rights in Africa.[65] Consequently, the Government of Nigeria had a duty to ensure its respect, protection, promotion and fulfillment, for in the Commission's verdict, "there is no right in the African Charter that cannot be made effective."[66] Instead of adhering to its responsibilities under the Charter, the Nigerian Government was "actively involved in the pollution, the contamination of the environment and related health problems of the Ogoni people, by condoning and facilitating the activities of the oil companies through the placing of the state's legal and military powers at the disposal of the oil companies."[67] In addition, "the government also kept the Ogoni communities uninformed about the damages created by the activities of the oil companies. It also failed to produce basic health and environment impact studies, nor asked the oil companies."[68]

The background to the case is tracebale to 1958 when oil drilling started in Niger Delta in Nigeria.[69] The oil exploration was taking place in the delta, which is the largest wetland in the world,[70] justifying the environmental concerns. What the decision in this case demonstrates is that the inclusion of the right to a satisfactory environment is a progressive step and potentially a powerful mechanism in addressing environ-

[63] Ibid.

[64] Ibid.

[65] Ibid.

[66] Ibid.

[67] Fons Cooman, "The Ogoni Case Before the African Commission on Human and Peoples' Rights' (2003) 52(2) The International and Comparative Law Quarterly 749, 754.

[68] Ibid.

[69] Van Linde, (n60), 168.

[70] Ibid.

mental concerns in Africa.[71] It is also part of the continental strides to develop an African human rights system and human rights law, one which is autochthonous and navigates the relationship between the African Union as a supranational organ and sovereign nation states, while also pursuing and promoting African concepts such as those of *ubuntu*. How these play out in practice in protecting the environment while also promoting human rights is explored much more in the next sections of this chapter.

4 Experiences of Select Countries in Africa

While international legal frameworks have been slow in recognizing the nexus between human rights and the environment, several countries in the continent include environmental rights within their national human rights systems. In dealing with environmental issues at the national level, there has been increase in the inclusion of environmental provisions in the Constitution of several African states.[72] This progress continues as demonstrated by the fact that countries that have reviewed their constitutions in recent years, such as Kenya, continue to include the right to a clean and healthy environment, responding to the complaints by Environmental Law Institute that African countries had largely underutilized the power that the Constitution gave for protecting the environment.[73] Instead, there is a growth of environmental constitutionalism. Environmental constitutionalism has tended to focus on the right to a healthy environment and related procedural rights necessary for their effective implementation.[74] This section discusses the experience of three countries, Kenya, South Africa and Nigeria not just with environmental

[71] Van Linde, (n60), 176.

[72] Boyd, (n5), which demonstrates that there has been an increase from several 4 in 1990 to 32 in 2011.

[73] Environmental law Institute and United Nations Environment Programme, Constitutional Environmental Law: Giving Force to Fundamental Principles in Africa (2nd Edition, Washington, 2007), vii.

[74] Carl Bruch and others, 'Natural Resources, Power Sharing, and Peacebuilding in Post conflict Constitutions' in Erin Daly and James R May (eds), Implementing Environmental Constitutionalism: Current Global Challenges (Cambridge University Press, 2018) 100.

constitutionalism but with implementing the linkages between human rights and the environment at the national level.

Kenya: The 2010 Transformation Constitution

Kenya's journey to a new Constitution was marked by the desire to improve the quality of governance. Concerns about human rights violations provided strong impetus for pro-democracy activists who led the agitation for the 2010 Constitution. The 2010 Constitution includes a robust Bill of Rights addressing the criticisms of the former Constitution, which was a Bill of exceptions, for including several claw-back clauses. Instead the 2010 Constitution captures the limitations in only one clause.[75] In addition, it discusses issues of recognition, protection, promotion and enforcement of human rights within its provisions. On recognition, unlike its predecessor that only dealt with civil and political rights, the Bill of Rights in Kenya's Constitution captures all the gamut of rights, earning it the tag of being among the most progressive and transformative Constitutions in the world, at least in terms of content.

The Kenyan Constitution deals with environmental management in several of its provisions,[76] including the preamble, the provisions on the Right to a clean and healthy environment,[77] on socio-economic rights,[78] on state obligations relating to the environment,[79] on access to environmental justice,[80] and on the principle of sustainable development,[81] among others. These provisions were a marked departure from the previous Constitution, which did not have a provision on the environment, with enforcement and implementation only relying on the provisions on the Right to life.[82] A weak constitutional and legal framework was a major

[75] Constitution of Kenya, 2010, article 24.

[76] For a discussion of these, see Collins Odote, 'Kenya: Constitutional Provisions on the Environment' (2012) 1 IUCN Academy of Environmental Journal, E-journal 136.

[77] Constitution of Kenya, 2010, Article 42

[78] Ibid., Article 43

[79] Ibid., Article 69

[80] Ibid., Article 70

[81] Ibid., Article 10.

[82] See High Court case of Peter K. Waweru V. Republic, (2006) eKLR

drawback to the realization of environmental rights in the country.[83] The 2010 Constitution was thus a breath of fresh air in terms of environmental governance, earning it the label of a "green Constitution."[84]

Three issues stand out in Kenya's Constitutional architecture as relates to the discussion on the link between human rights and the environment. First is the recognition of the right to a clean and healthy environment as a substantive right. Although there had been instance before 2010 when Kenyan courts relied on the right to life to address environmental violations[85] the lack of a substantive right dealing with environmental issues limited the use of a human rights lens to govern the environment. By constitutionalizing environmental rights within the Bill of Rights, the state is thus put under a duty to utilize the machinery available for the protection, promotion of human rights generally. In ensuring that the right to a clean and healthy environment is realized within Kenya, the Constitution places the state under specific obligations, including the obligation to ensure sustainable utilization and management of the environment and natural resources[86]; achieving and maintaining a tree cover of at least 10 percent of the land area[87]; protecting and enhancing intellectual property and indigenous knowledge as relates to biodiversity and genetic resources of community[88]; encouraging public participation in environmental management processes[89]; protecting genetic resources and biological diversity[90]; establishing environmental impact assessment, audit and monitoring processes[91]; eliminating processes and activities likely to endanger the environment[92] and utilizing the environment and

[83] Joel Kimutai Bosek, 'Implementing Environmental Rights in Kenya's New Constitutional Order: Prospects and Potential Challenges' (2014) 14 African Human Rights Law Journal 489, 491.

[84] Donald Kaniaru, (2011–2012) 29 'Environmental Courts and Tribunals: The Case of Kenya' Pace Environmental Law Review, 566, 581.

[85] Collins Odote, 'The Role of the Environmental and Land Court in Governing Natural Resources in Kenya' in Patricia K. Mbote, et al., *(eds) Law, Environment Africa* (Nomos, 2019) 340.

[86] Constitution of Kenya, Article 69(1)(a)

[87] Ibid., Article 69(1)(b).

[88] Ibid., Article 69(1)(c)

[89] Ibid., Article 69(1)(d).

[90] Ibid., Article 69(1)(e)

[91] Article 69(1)(f).

[92] Article 69(1)(g).

natural resources for the benefit of the people.[93] The 2010 Constitution by including these elements, not only recognizes the right, but also elaborates on its contents. In addition, under socio-economic rights recognized in the Constitution,[94] there is included the right to "clean and safe water in adequate quantities."[95]

The entire machinery within the country for human rights protection is thus available for the protection and promotion of the right to a clean and healthy environment. For example, even before the adoption of the Constitution in 2010, the country had a national human rights Commission, the Kenya National Commission on Human Rights. This was then given constitutional status in 2010.[96] As part of supporting the protection and promotion of human rights within Kenya, the Commission can investigate or research on any human rights issue and violations within the country. The Commission has in the past carried out investigations on areas relating to environmental rights violations. In July 2005, the Commission undertook a public inquiry into allegations of human rights violations as a result of salt manufacturing being undertaken by several companies in Magarini within the current Kilifi County.[97] Among the complaints raised to the Commission by the local communities as part of human rights violations by the salt mining companies were environmental grievances.[98] The key environmental grievances were mangrove destruction; salinization of freshwater wells and springs; flooding as a result of dyke-building; deforestation; accelerated coastal erosion; loss of biodiversity, evidence being declining fisheries and other marine resources dependent on mangroves; and pollution on both coastal and marine ecosystems and soils, thus affecting their productivity.[99]

[93] Article 69(1)(h).

[94] Constitution of Kenya, Article 43.

[95] Ibid., Article 43(1)(d).

[96] Constitution of Kenya, 2010, Article 59.

[97] For a report of the findings of the inquiry, see KNCHR, *The Malindi Inquiry Report 2006: Report of a Public Inquiry into Allegations of Human Rights Violations in Magarini, Malindi*, July 2006 http://www.knchr.org/Portals/0/Reports/Malindi_Inquiry.pdf

[98] Ibid.

[99] Ibid.

In addressing the environmental complaints arising from salt mining, the Commission reiterated the link between human rights and the environment.[100] Relying on provisions from international human rights instruments and decisions of international tribunals, the Commission pointed out that the link is undeniable.[101] Based on its human rights mandate and lens, the Commission made several short-term and medium-term recommendations in order to "mitigate the adverse environmental impacts to enable measures, actions and processes that will enhance improved environmental performance to be put in place."[102] In 2017, the Commission conducted an audit of the state of the implementation of the findings of the inquiry it had conducted 11 years earlier and concluded that while there were still several challenges, including some ongoing environmental violations, progress had been made in implementing the recommendations of its report.[103] The second issue that arises from the Kenyan experience due to constitutionalization is the mechanisms for ensuring that in cases where there are violations of the right to a clean and healthy environment, then appropriate redress is obtained. This relates to two interconnected issues. First is the locus standi to be able to litigate environmental cases. The second is the appreciation by the judiciary of environmental issues and its performance in dealing with environmental cases. For long, Kenya's jurisprudence on locus standi on environmental cases was restrictive, insisting on proof of either having suffered injury greater that other members of the public or personal proprietary interest in the mater in dispute.[104] As a consequence many public-spirited individuals were unable to use the court to protect the environment. Professor Wangari Maathai, the Nobel laureate from Kenya, is most known for her environmental activism in the country seeking to protect Kenya's forest and other critical environmental resources from destruction and grabbing by politically correct individuals. Her attempts at one time to use

---

[100] Ibid.

[101] Ibid.

[102] Ibid., page 128.

[103] KNCHR, (n97)

[104] Collins Odote, 'Public Interest Litigation and Climate Change-An Example from Kenya', in Oliver C. Ruppel et al., *Climate Change: International Law and Global Governance, Volume 1: Legal Responses and Global Responsibility* (Nomos, 2013) 817.

the court to protect Kenya's biggest recreational park, Uhuru Park, failed due to the argument by the courts that she had no locus standi to bring such a case for determination by the judiciary.[105] The 2010 Constitution has dealt with this issue by allowing any person who feels that the right to a clean and healthy environment is either being, had been or likely to be threatened to seek court redress.[106] In approaching court, the rules of locus standi have been relaxed since one does not need to demonstrate that any person has incurred loss or suffered injury.[107] This is a huge improvement from the pre-2010 Constitutional; position where one had to prove not just loss or injury but the loss or injury had to be associated with the person applying to court for remedies.

As regards the judiciary, there has been established the Environment and Land Court to determine environmental and land cases.[108] The Courts are established as specialized courts as the same status as the High Court. Its existence is important for improved environmental governance[109] as it sought to encourage specialization and thus improved in the quality of jurisprudence and thus environmental justice. While the court has made some strides in delivering on this expectation, its "utility has been hampered by structural and normative challenges."[110] For example, in practice both the Environment and Land Court and the High Court have jurisdiction on enforcement of rights of an environment nature, depending on whether these are couched as environmental rights or human rights. This is because while the constitutional provisions in Article 162 as read with the Environment and Land Court Act is to ensure that constitutional issues relating to the environment are dealt with by the Environment and Land Court, The High Court also has jurisdiction to deal with interpretation and enforcement of all human rights.[111] This jurisdictional challenge has dogged ELC since its

[105] Wangari Maathai v Kenya Times Media Trust, 1 KLR(E&L) 2006, 164–171.

[106] Constitution of Kenya, 2010, Article 70(1).

[107] Ibid., Article 70(3).

[108] Constitution of Kenya, Article 162(2)(b).

[109] Odote, (n 85), 336.

[110] Ibid.

[111] Collins Odote, 'Kenya: The New Environmental and Land Court' (2013) 4 IUCN Academy of Environmental law E-Journal 171, 175–6

establishment[112] and moves beyond the question of which court has the mandate to interpret issues relating to the right to a clean and healthy environment to extend to the entire relationship between the ELC and the High Court. The challenge though has not limited the judiciary's appreciation of the link between human rights and the environment with Kenyan courts using a human rights lens to enforce environmental rights.

The High Court of Kenya, in a case relating to the Construction of the Lamu Port-South Sudan-Ethiopia-Transport Corridor project (LAPSSET), in which local community members in Lamu argued that the project was violating their fundamental human rights, including the right to a clean and healthy environment, the court demonstrated not just the link between traditional human rights and environment but also that the right to a clean and healthy environment is a human right.[113] The Court identified the rights whose violation were in question to include the right to public participation, the right to a clean and healthy environment, the right to livelihood including fishing rights, cultural rights, rights of access to information and access to justice.[114] The court then held that these rights were intertwined with the achievement of one affecting the other rights. In the words of the Court:

> Thus, violation of rights to a clean and healthy environment can easily lead to the violation of other rights in the Bill of Rights such as the right to life. Yet, the determination of violations or threats of violation of any rights in the Bill of Rights undoubtedly falls within the province of this Court.[115]

Consequently, the court proceeded to assess the evidence before it, including a site visit it conducted and reached the conclusion that the right to a clean and healthy environment was at risk of being violated as a result of the project.[116]

In addition to substantive rights, the nexus between human rights and the environment is aided by the procedural rights of access to

[112] Supra, note 85 at page 342

[113] Mohamed Ali Baadi and others v Attorney General & 11 others [2018] eKLR.

[114] Ibid.

[115] Ibid.

[116] Ibid.

information, public participation and access to justice. The Kenyan Constitution and legal framework recognize the importance of these procedural rights giving meaning to the human rights to a clean and healthy environment. Courts have affirmed these and relied on them to enforce rights to a clean and healthy environment.[117]

South Africa

The South African Constitution is among the most progressive of the continent. Adopted as a charter to move the country from the apartheid regime with its human rights violations, it included robust human rights provisions and enforcement mechanisms. In the negotiations for the 1995 South African Constitution proposals were made to include environmental norms.[118] While debate occurred about the extent of the right, the recognition of its importance and elements, including an environmental human right, an obligation on the state to undertake positive environmental action and a duty on citizens to defend it[119] was never in dispute. Consequently, when the South African Constitution was adopted it included environmental provisions in its Bill of Rights. The South African Constitution is celebrated as a leading example of environmental constitutionalism.[120] Section 24 is, in this vein, highlighted as "a remarkably progressive provision, embodying a subjective right to environmental quality for everyone, as well as requiring legislative action to protect the environment for future generations."[121] Section 24 of the South African Constitution provides that:

> Everyone has the right: (a) to an environment that is not harmful to their health or well-being; and (b) to have the environment protected, for the benefit of present and future generations, through reasonable legislative

[117] Ibid. See also, In the Mater of the Mui Coal Basin (2015) eKLR.

[118] Jan Glaweski, 'The Environment, Human Rights and a New South African Constitution' (1991)7 South African Journal of Human Rights 167, 180.

[119] Ibid., 182.

[120] Susana, Borras, 'New Transitions from Human Rights to the Environment to the Rights of Nature' (2016) 5(1) Transnational Environmental Law 113, 126.

[121] Ibid.

> and other measures that: (i) prevent pollution and ecological degradation; (ii) promote conservation; and (iii) secure ecologically sustainable development and use of natural resources while promoting justifiable economic and social development.[122]

The provision has not only been given judicial support in several cases but has also influenced the development of constitutional provisions dealing with environmental rights in other countries across the African continent, like Zimbabwe and Kenya.[123] In its implementation, Government in South Africa is required to take "reasonable legislative and other measures." This responsibility is shared by all the three spheres of Government in South Africa,[124] national, provincial and local. Even though the level of responsibility may differ among the three levels as the South African court affirmed in the case of *Grootboom v Oostenberg Municipality*[125] all levels have some responsibility to ensure that the right is realized in South Africa.

One of the greatest environmental challenges facing the world is climate change. In South Africa the constitutional recognition of the right to a clean and healthy environment has been used to successfully litigate climate change. This is particularly important considering the mention of the linkage between climate change and human rights in the preambular sections of the 2015 Paris Agreement.[126] In the South African case of *Earthlife Africa Johannesburg v Minister of Environmental Affairs*,[127] the case revolved objections by an NGO, Earthlife Africa to the construction of a 1200 MW coal-fired power station in the Limpopo Province due to its potential environmental impacts. The NGO largely went to court complaining of the administrative action of the Department of

---

[122] Section 24, South African Constitution.

[123] Anel du Plessis, 'The Promise of "Well-being" in Section 24 of the Constitution of South Africa' (2018) 34(2) South African Journal of Human Rights 191, 192.

[124] Anel du Plessis, 'Local Environmental Governance and the Role of Local government in the Realization of Section 24 of the South African Constitution' (2010) 21(2) Stellenbosch Law Review 265, 268.

[125] 2000 3 BCLR 277 (CC) paras 82–84.

[126] Paris (France), 12 Dec. 2015, in force 4 Nov. 2016 http://unfccc.int/paris_agreement/items/9485.php (Paris Agreement) accessed on 26 August 2019.

[127] (65662/16) [2017] ZAGPPHC 58.

Environmental Affairs (DEA) to grant the environmental authorization for the power station and a subsequent appeal decision of the Minister. The main argument of the NGO was that National Environment Management Act of South Africa envisaged that a climate change impact assessment be conducted before such authorization could be issued, despite the lack of an express legislative requirement for such an assessment. The NGO argued that their position was supported by among other things the provisions of Section 24 of the South African Constitution on the right to a clean and healthy environment.[128] The case was eventually decided in favor of the NGO, but the court did not comprehensively deal with the issue of Section 24 of the Constitution and its linkages to climate change litigation and adaptation and mitigation support. However, the court made a statement which reinforces the place of section 24 of the South African Constitution. In the court's words:

> Section 24 recognizes the interrelationship between the environment and development. Environmental considerations are balanced with socio-economic considerations through the ideal of sustainable development. This is apparent from section 24(b)(iii) which provides that the environment will be protected by securing ecologically sustainable development and use of natural resources while promoting justifiable economic and social development. Climate change poses a substantial risk to sustainable development in South Africa. The effects of climate change, in the form of rising temperatures, greater water scarcity, and the in-creasing frequency of natural disasters pose substantial risks. Sustainable development is at the same time integrally linked with the principle of intergenerational justice requiring the state to take reasonable measures to protect the environment 'for the benefit of present and future generations' and hence adequate consideration of climate change. Short-term needs must be evaluated and weighed against long-term consequences.[129]

[128] Jacqueline Peel and Harry M Osofky, 'A Rights Turn in Climate Change Litigation?' (2018) 7(1) Transnational environmental Law 37, 59.

[129] Earthlife Africa Johannesburg v. Minister for Environmental Affairs & Others, Case No. 65662/16, Judgment of High Court of South Africa, Gauteng Division, Pretoria (South Africa), 8 Mar. 2017, available at: http://cer.org.za/wp-content/uploads/2017/03/Judgment-Earthlife-Thabametsi-Final-06-03-2017.pdf. (Earthlife Africa Johannesburg) para 82 accessed 20th August 2019

Although Earthlife won the case, the substantive and procedural environmental rights in the Constitution did not receive comprehensive and substantive attention in its arguments.[130] The reliance on environmental impact assessment requirements and provisions in the case, has, however been argued to provide the nexus with the constitutional right to a clean and healthy environment, since, "environmental impact assessment (EIA) as an environmental management tool is one of the ways in which to work towards environmental outcomes such as those envisaged in the Constitution."[131]

Literature assessing Article 24 of the South African Constitution is unanimous that the article captures both negative and positive obligations on the state.[132] In addition, judicial pronouncements have helped to clarify the meaning of the right. Although an assessment of the role of that the Courts have played in explicating environmental rights in South Africa undertaken in 2010 concluded that the courts had played a minimal role,[133] the South African Judiciary has underscored its importance in "upholding the rule of environmental law, and that courts will not hesitate to intervene where questions of sustainability and environmental rights arise."[134]

Nigeria

Nigeria is special within the African continent on matters linkages between human rights and the environment due to the fact that the first case determined by the African Commission on Human and Peoples' Rights on Article 24 of the African Charter on Human and Peoples' Rights, the *Ogoniland* case,[135] originated from Nigeria. The case went to the Commission due to the failure of Nigeria to address the infractions

[130] Du Plessis, (n123), 203.

[131] Ibid.

[132] Louis J. Kotze, And Anel du Plessis, 'Some Brief Observations on Fifteen Years of Environmental Rights Jurisprudence in South Africa' (2010) 3(1) Journal of Court Innovation 157, 165.

[133] Ibid., page 175

[134] Ibid., Page 174.

[135] SERAC, note 58.

complained about. This is despite the constitutionalization of environmental issues in Nigeria. Section 20 of the Nigerian Constitution obligates the State to "protect and improve the environment and safeguard the water, air and land, forest and wildlife of Nigeria."[136] The section though is contained in the chapter of the Constitution that deals with Fundamental Objectives and Directive Principles of State Policy, hence, as some scholars have argued, non-justiciable and unenforceable.[137] While the 1999 Constitution, like previous Constitutions of Nigeria does not specifically mention the right to a clean and healthy environment,[138] the inclusion of section 20 signals environmental consciousness and constitutionalism in the country. This is even though the Constitution specifies that this article is not enforceable in Court.[139]

Despite the above position, environmental rights can be and have been inferred from human rights included in the Nigerian Constitution, especially the constitutional right to life. Courts in Nigeria have used this approach to protect the right to a clean and healthy environment.[140] One of the seminal decisions in Nigeria on the right to a clean and healthy environment as a human right is the case of *Jonah Gbemre v. Shell Petroleum Development Corporation & 2 Others*.[141]

In the above case, Jonah Gbemre filed a case on his behalf and on behalf of Iwherekan Community in Delta State, Nigeria, arguing that the continued flare gas in the course of their oil exploration and production activities in their community was violating their constitutional rights, including the right to a clean and healthy environment. The court held in

[136] Chapter II, Section 20, 1999 Constitution of the Federal Republic of Nigeria.

[137] Theodore Okonkwo, 'Environmental Constitutionalism in the United States and Nigeria: Recognizing Environmental Rights' (2017) 5(7) International Journal of Social Science Studies 66, 69.

[138] Edwin Egede, 'Human Rights and the Environment: Is There a legally Enforceable Right of a Clean and Healthy Environment for the "Peoples" of the Niger Delta under the Framework of the 1999 Constitution of the Federal Republic of Nigeria' (2007) 19(1) Sri Lanka Journal of International Law 51,52.

[139] Section 6(6), Nigerian Constitution,

[140] Jonah Gbemre v. Shell Petroleum Development Company Nigeria Ltd. and Others, Federal High Court of Nigeria, Benin Division, Judgment of 14 November 2005, Suit No. FHC/B/CS/53/05; Oronto Douglas v. Shell Petroleum Development Company Ltd. (1999) 2 N.W.L.R. (Part 591); Gani Fawehinmi v. Abacha (1996) 9 N.W.L.R. (Part 475) 710.

[141] Suit No. FHC/B/CS/153/05, of 2005

favor of the applicants, pointing out that the constitutional right to life and dignity of the applicants included the "right to clean, poison-free, pollution-free and healthy environment."[142] Despite the above case, the assessment is largely that the Nigerian Judiciary has not been as robust on enforcing environmental rights. While one expects the Nigerian Judiciary to be bolder in its enforcement of environmental rights, Nigeria may consider enacting environmental rights expressly as a fundamental right.[143]

5 The Merits and Limits of a Human Rights-based Approach

Despite the international discussions on the necessity of a human right to a healthy environment within international law, its recognition within the African Charter and incorporation in several Constitutions and protection by courts as the discussions of the experiences of Kenya, South Africa and Nigeria demonstrate, there is still lack of unanimity on the content of the right. The scope and utility of the right to a healthy environment remain the subject of ongoing debate.[144] This debate revolved around whether the environment has a right and can be protected for its own sake or whether it is only useful for the benefit it provides to human beings. A human rights-based approach to the environment is viewed by several scholars as perpetuating the notion that the environment and natural resources exist only for the benefit of humans and have no intrinsic worth.[145]

The reality is that this debate is not about to end. There are several sides to the argument, with some contending that the right exists while others arguing against it and as Rodriguez-Rivera aptly states, the human rights approach to the environment is as controversial as other debates

[142] Ibid., para 3.

[143] CT Emejuru, 'Human Rights and Environment: Whither Nigeria?' (2014) 30 Journal of Law, Policy and Globalization 19, 27.

[144] Borras, (n120) 128.

[145] Ibid.

concerning new or emerging human rights like the right to development, the rights of peoples and indigenous rights.[146] The proponents of the right see it as the basis for unleashing the human rights framework in the protection of the environment. This approach is supported by the experience of the three African Countries reviewed in this chapter. An experience which has both Kenya and South Africa including within their Bill of Rights, the right to a clean and healthy environment. This empower human beings to be able to claim protection of the environment and argue that environmental management is a human right and how it is dealt with has implications on the enjoyment of human rights by all citizens in a society. Such an approach will make the process of environmental management linked with the livelihoods and lived realities of human beings. It moves environmental management from an inanimate and abstract discussion to one that is real and that human beings can relate to. It is arguable that the environment in Africa is the basis of and impacts on all human rights in Africa, a continent that has a large natural resource base and whose people rely on the environment and natural resources for their sustenance. In addition, such a link will make the constitutional efforts to protect human rights applicable to environmental management leading to the emergence of environmental constitutionalism in the continent. Courts in Africa are using human rights provisions to resolve environmental disputes and make decisions protecting the environment from degradation.

Some of the advantages of adopting a human rights approach include the fact that a human rights perspective directly addresses environmental impacts on the life, health, private life and property of individual humans rather than on other states or the environment in general,[147] may serve to secure higher standards of environmental quality, based on the obligation of states to take measures to control pollution affecting health and private

[146] Louis, E. Rodriguez-Rivera, 'Is There a Human Rights to Environment Recognized Under International Law? It Depends on the Source' (2001) 12 Colorado Journal of International Environmental Law and Policy 1,1.

[147] Allan Boyle, 'Human Rights and the Environment: Where Next?' (2012) 23(3) The European Journal of International Law 613.

life,[148] helps to promote the rule of law,[149] gives new life to the idea that there is, or should be, in some form, a right to a decent environment.[150] It, therefore, enhances rather than compromises environmental management. On the other hand, taking a human rights approach elevates anthropocentrism as the model for environmental management. The criticism of such an approach is that the environment is seen purely as a tool for satisfying human interests. It ignores the seminal contributions by Christopher Stone[151] who argued for the protection of the environment for its own sake by making a case for granting rights to inanimate objects like lakes, trees and wetlands. The argument requires recognition of the right of the environment as a distinct entity and not through the prism of human beings that would be necessary if we were to adopt a human rights approach. The justification for this argument derives from the fact that at the center of degradation efforts are human beings. Making them central to conservation efforts limits the strategies for environmental conservation.

It is essential to view the environment on its own prism so that when the interest of development and that of the environment are being balanced as required by the concept of sustainable development, this is not done from a human rights lens. This is the greatest limitation to a human rights approach especially in Africa where poverty is endemic. Taking a human rights approach will justify degradation of the environment so as to address the fundamental human rights of individuals, including the right to life and livelihoods which requires exploitation of the environment and natural resources. In such circumstances, environmental degradation will be justified as being a lesser evil in the process of satisfying the more fundamental development needs of citizens and thus helping to protect and guarantee their human rights.

African experience provides useful lessons for dealing with controversy. As this chapter has demonstrated, creating links between human rights and the environment is both important and necessary. It helps to ensure

---

[148] Ibid.

[149] Ibid.

[150] Ibid., at 614.

[151] Christopher, D. Stone, 'Should Trees Have Standing? Toward Legal Rights for Natural Objects?' (1972) 45 Southern California Law Review 450.

that a human rights lens and approach for protection of rights are utilized for environmental conservation. This approach is already evident with success in Kenya, South Africa and Nigeria as in many other parts of the African Continent. Secondly, the phraseology of the rights in national constitutions also imposes a duty on human beings thus recognizing that human beings are also responsible for the degradation of the environment. As a result, in protecting and promoting the right, human beings have responsibilities to undertake measures to ensure that the right is realized. This is over and above the traditional responsibilities placed on states by international human rights instruments. Essentially this requires greater focus not just on rights language but also responsibilities so that human beings are, as per Hohfeldian jural relations and postulates,[152] required not just to seek the protection of the environment as part of their human rights to a clean and healthy environment but also recognize that environmental conservation is a responsibility that all human beings owe not only fellow human beings but also the environment. This will help ensure that a human rights approach also addresses the ecocentric approach.

The African experience with its focus on collective as opposed to individualism demonstrates that it is possible to create a mutually reinforcing and supportive framework between human rights and environmental rights, a link that recognizes that the environment is important both for its own sake and also for the benefit of human beings.

6 Conclusion

By recognizing environmental issues as a human rights issue, we should not advocate for subsuming environmental law into human rights law. Instead, there should be realization that in solving the huge environmental challenges facing the globe, adoption of a multitude of approaches are necessary and human rights can be relied on to support the global quest for environmental sustainability. This requires a move away from a purely

[152] WN Hohfeld, 'Some Fundamental Legal Conceptions as Applied in Judicial Reasoning' (1913–14) 23 Yale Law Journal 16.

anthropocentric approach, which human rights advocates are used, to a recognition that the environment health and health of human beings are intertwined[153]; thus human rights cannot be seen as superior and without considering the sustainability of the environment. By creating this link, one cannot be said to be "advocating for an anthropocentric approach to environmental protection."[154]

From the foregoing, it can be argued that a purely anthropocentric approach is inimical to environmental management. Consequently, even as we recognize, protect and promote human rights to the environment, we have to keep in mind that the objective must be to improve the quality of the environment and not just to exploit it for human interests. This is the path that the inclusion of the environmental right within the African Charter on Human and Peoples' Rights must lead to in the interpretation and development process as part of an African Union Law development process.[155] In the final analysis, the words of Susanna Borras aptly demonstrate the importance of a human rights approach to environmental protection even as it acknowledges its limits and the need for tempering it in application. She holds as follows:

> a degree of anthropocentrism may be a necessary part of environmental protection not because humanity is at the centre of the biosphere, but because humanity is the only species which possesses the conscious-ness to recognize and respect the morality of rights. In this view, the interests and duties of humanity are inseparable from environmental protection. In any event, a critical stance towards the idea of rights makes for a conceptual point of departure for the consideration of alternative rights-based approaches to environ-mental protection.[156]

---

[153] UNEP, *Global Environment Outlook, GEO-6: Healthy Planet, Healthy People* (Cambridge University Press, 2019) https://wedocs.unep.org/bitstream/handle/20.500.11822/27539/GEO6_2019.pdf?sequence=1&isAllowed=y

[154] Sumudu Attappatu, 'The Right to a Healthy Life or the Life to Die Polluted: The Emergence of a Human Right to a Healthy Life Under International Law' (2002) 16 Tulane Environmental Law Journal 65, 67.

[155] For discussions on development of an African Union Law, see generally, Olufemi Amao, *African Union Law: The Emergence of a Sui Generis Legal Order,* (Routledge Press, 2018).

[156] Borras, (n120) 128.

References

Amao O, African Union Law: The Emergence of a Sui Generis Legal Order, (Routledge Press, 2018)

Attappatu S, 'The Right to a Healthy Life or the Life to Die Polluted: The Emergence of a Human Right to a Healthy Life Under International Law' (2002) 16 Tulane Environmental Law Journal 65

AU, ADB and ECA, Framework and Guidelines on Land Policy in Africa (Addis Ababa, Ethiopia, 2010 https://www.uneca.org/sites/default/files/PublicationFiles/fg_on_land_policy_eng.pdf

Beyerlin U and Marauhn T, International Environmental Law (Hart Publishing, 2011)

Borras S, 'New Transitions from Human Rights to the Environment to the Rights of Nature' (2016) 5(1) Transnational Environmental Law 113

Bosek JK, 'Implementing Environmental Rights in Kenya's New Constitutional Order: Prospects and Potential Challenges' (2014) 14 African Human Rights Law Journal 489

Boyd DR, The Environmental Rights Revolution: A Global Study of Constitutions, Human Rights, and the Environment (UBC Press 2012)

Boyle A, 'Human Rights and the Environment: Where Next?" (2012) 23(3) European Journal of International Law 613

Bruch C and others, 'Natural Resources, Power Sharing, and Peacebuilding in Post conflict Constitutions' in Erin Daly and James R May (eds), Implementing Environmental Constitutionalism: Current Global Challenges (Cambridge University Press, 2018) 100

Cooman F, "The Ogoni Case Before the African Commission on Human and Peoples' Rights' (2003) 52(2) The International and Comparative Law Quarterly 749

du Plessis A, 'Local Environmental Governance and the Role of Local government in the Realization of Section 24 of the South African Constitution' (2010) 21(2) Stellenbosch Law Review 265

du Plessis A, 'The Promise of "Well-being" in Section 24 of the Constitution of South Africa' (2018) 34(2) South African Journal of Human Rights 191

Egede E, 'Human Rights and the Environment: Is There a legally Enforceable Right of a Clean and Healthy Environment for the "Peoples" of the Niger Delta under the Framework of the 1999 Constitution of the Federal Republic of Nigeria' (2007) 19(1) Sri Lanka Journal of International Law 51

Emejuru CT, 'Human Rights and Environment: Whither Nigeria?' (2014) 30 Journal of Law, Policy and Globalization 19

Environmental law Institute and United Nations Environment Programme, Constitutional Environmental Law: Giving Force to Fundamental Principles in Africa (2nd Edition, Washington, 2007)

Glaweski J, 'The Environment, Human Rights and a New South African Constitution' (1991) 7 South African Journal of Human Rights 167

Hohfeld NW, 'Some Fundamental Legal Conceptions as Applied in Judicial Reasoning' (1913–14) 23 Yale Law Journal 16

Kaniaru D, (2011–2012) 29 'Environmental Courts and Tribunals: The Case of Kenya' Pace Environmental Law Review, 566, 581

KNCHR, The Malindi Inquiry Report 2006: Report of a Public Inquiry into Allegations of Human Rights Violations in Magarini, Malindi, July 2006 http://www.knchr.org/Portals/0/Reports/Malindi_Inquiry.pdf

Kotze LJ and Plessis A, 'Some Brief Observations on Fifteen Years of Environmental Rights Jurisprudence in South Africa' (2010) 3(1) Journal of Court Innovation 157

Lakatos I, 'Thoughts on Universalism versus Cultural Relativism with Special Attention to Women's Rights' (2018) 1 Pécs Journal of International and European Law 6

Lewis B, 'Environmental Rights or the Right to Environment: exploring the Nexus Between Human Rights and Environmental Protection' (2012) 8(1) Macquarie Journal of International and Comparative Environmental Law 36

Odote C, 'Kenya: Constitutional Provisions on the Environment' (2012) 1 IUCN Academy of Environmental Journal, E-journal 136

Odote C, 'Public Interest Litigation and Climate Change-An Example from Kenya', in Oliver C. Ruppel et al., Climate Change: International Law and Global Governance, Volume 1: Legal Responses and Global Responsibility (Nomos, 2013a) 817

Odote C, 'Kenya: The New Environmental and Land Court' (2013b) 4 IUCN Academy of Environmental law E-Journal 171, 175–6

Odote C, 'The Role of the Environmental and Land Court in Governing Natural Resources in Kenya' in Patricia K. Mbote, et al., (eds) Law, Environment Africa (Nomos, 2019) 340

Ojwang JB, laying a Basis for Rights: Towards a Jurisprudence of Development, Inaugural Lecture Delivered Before the University of Nairobi on 9th July 1992, (University of Nairobi Press, 1992)

Okonkwo T, 'Environmental Constitutionalism in the United States and Nigeria: Recognizing Environmental Rights' (2017) 5(7) International Journal of Social Science Studies 66
Peel J and Osofky HM, 'A Rights Turn in Climate Change Litigation?' (2018) 7(1) Transnational environmental Law 37, 59
Rodriguez-Rivera LE, 'Is There a Human Rights to Environment Recognized Under International Law? It Depends on the Source' (2001) 12 Colorado Journal of International Environmental Law and Policy 1
Sands P, et al., Principles of International Environmental Law (4th Edition, Cambridge University Press, 2018) 81
Shelton D, 'What Happened in Rio to Human Rights?' (1992) 3(1) Yearbook International Environmental Law 75
Stone DC, 'Should Trees Have Standing? Toward Legal Rights for Natural Objects?' (1972) 45 Southern California Law Review 450.
Thompson SCG, and Barton AM, 'Ecocentric and Anthropocentric Attitudes Toward the Environment' (1994) 14 Journal of Environmental Philosophy 14
UNEP, Global Environment Outlook GEO-6: Healthy Planet, Healthy People (Cambridge University Press, 2019) https://wedocs.unep.org/bitstream/handle/20.500.11822/27539/GEO6_2019.pdf?sequence=1&isAllowed=y
UNEP, Global Environment Outlook (GEO-6), Regional Assessment for Africa (Nairobi, 2016). http://wedocs.unep.org/bitstream/handle/20.500.11822/7595/GEO_Africa_201611.pdf?sequence=1&isAllowed=y
United Nations, transforming our World: the 2030 Agenda for Sustainable Development, A/Res/70/1 https://www.un.org/ga/search/view_doc.asp?symbol=A/RES/70/1&Lang=E accessed 17 July 2019
Van Linde M, "Considering the Interpretation and Implementation of Article 24 of the African Charter on Human and Peoples' Rights in Light of the SERAC Communication' (2003) 3 African Human Rights Law Journal 167.
Wabunoha R, "Environmental Law of Africa", in Patricia Kameri Mbote and Collins Odote (eds), Blazing the Trail: Professor Charles Okidi's Enduring Legacy in the Development of Environmental Law (School of Law, University of Nairobi, 2019).
Yash Ghai Y, (2000) 21 'Universalism and Relativism: Human Rights as Framework for Negotiating Interethnic Claims' Cardozo L Rev 1095, 1095

Part IV

Business Practices, Human Rights and the Environment

16

Human Rights and Energy in a Neoliberal Southern Africa

Lenin Tinashe Chisaira

1 Introduction

Contemporary neoliberal States place capital and industrial energy needs on a higher pedestal than human and environmental rights. This skewed relationship results in inequalities in development and energy policymaking and fuels environmental injustice in the Global South.[1] The inequality has human rights, environmental justice and energy security implications for the lesser powerful communities in the Southern African States of Zimbabwe, Mozambique and the Democratic Republic of the Congo (DRC) as well as in other parts of the Global South and Sub-Saharan Africa. An illustrative example, for instance, concerning dam constructions in Latin America highlights the ever-increasing instances in the Global South where duty bearers do not bother considering downstream impacts, and resultantly the rights of local affected communities

[1] PM Fearnside, 'Environmental and Social Impacts of Hydroelectric Dams in Brazilian Amazonia: Implications for the Aluminium Industry.' (2016) 77 World Development 57.

L. T. Chisaira (✉)
Advocates4Earth, Harare, Zimbabwe

M. Addaney, A. Oluborode Jegede (eds.), *Human Rights and the Environment under African Union Law*, https://doi.org/10.1007/978-3-030-46523-0_16

are disregarded.[2] Southern Africa, like the rest of the world, has been in the claws of neoliberalism and globalization[3] at least from the early 1980s and 1990s and this has had significant tolls on the promotion and protection of human and environmental rights and the right to development.[4] In the time of the climate crisis and accompanying climate change response measures, usually poor sections of the populace get excluded by energy politics, and their human rights are adversely affected by lack of consultation in the development of hydroelectric mega dams, bio-fuel farms and plants and other energy sourcing constructions. Heinberg aptly defines energy politics as a power issue and refers to the concept in terms of it being 'humanity's most basic source of power, and since politics is a contest over power (albeit social power), it should not be surprising that energy is politically contested.'[5] This brings to the fore the related issue of energy security. Energy Security is mainly concerned with the socio-ecological transformations of society, the development of environmentally sound energy sources and the imperative of ensuring that more impoverished members of the society have access to energy.[6] Energy politics is close to, but not the same as, the concept of energy security. These elements have a hold upon the protection and promotion of human, developmental and environmental rights in Southern Africa.

The neoliberal framework vitiates tangible commitment toward fully functional and ascertainable human rights, environmental justice and the right to development. Hordes of communal, rural and low-income peoples who find themselves excluded from the benefits of energy sharing and energy development policies significantly feel this vitiation. In Southern Africa, communities located in the vicinity of mega-dam projects such as the Cahora Bassa, Tugwi-Mukosi, Mpanda Nkuwa, Inga (and Grand Inga Project) and Kariba Dams often found themselves

[2] ibid.

[3] See RE Mazur, 'Realization or deprivation of the right to development under globalization? Debt, structural adjustment, and poverty reduction programs.' (2004) 60 (1) GeoJournal 61.

[4] See Arjun Sengupta, 'On the Theory and Practice of the Right to Development.' (2002) 24 Human Rights Quarterly 837.

[5] Richard Heinberg, *Afterburn: Society Beyond Fossil* Fuels (New Society Publishers 2015) 47.

[6] See OJ Kuik, MB Lima and J Gupta, 'Energy security in a developing world.' (2011) 2 (4) Wiley Interdisciplinary Reviews: Climate Change 627.

excluded. The planning of these projects had minimal consideration on the environment or the human rights of the rural and communal populaces.[7]

In the context of an uncontested hegemony of neoliberalism, capitalism and the free-market system within the region as a whole and specifically within post-socialist Zimbabwe and Mozambique, the human rights of communities that have neither property rights nor significant political and economic power become vulnerable.[8] The relevance of energy to proponents of neoliberalism and capitalism is solely from the multiplication of profits rather than social and environmental development. A white paper pointing to investment opportunities without exposure to risk which was developed by South African business consultancies, J.C. Rathbone Associates Limited (JCRA) and the Centre for Economics and Business Research (CEBR) is ample evidence of this sentiment among the capitalist class around the world.[9]

The real-life impact of the neoliberal world order is felt in the deliberate exclusions of sections of society from sufficient access to energy generated from the commons or common resources such as rivers, land and minerals. The impact is also felt deeply by local communities living in areas earmarked for extractive or energy investments. Despite evidence of investor-related and neoliberal friendly research on energy and profit, the amount of literature, academic or otherwise, dedicated to a critique of neoliberalism, has become at most negligible and at best ineffectual. The few that are critical highlight, from a comparative perspective, the adverse impact of neoliberal- and capitalist-led developments, mega dam construction and privatised water rights. In a critique of the funding

[7] TM Mashingaidze, 'Beyond the Kariba Dam Induced Displacements: The Zimbabwean Tonga's Struggles for Restitution, 1990s–2000s.' (2013) 20 International Journal on Minority and Group Rights 381.

[8] ibid.

[9] JCRA, *Whitepaper on Renewable energy in sub-Saharan Africa* (Cebr and JCRA 2018) 8. https://jcragroup.com/wp-content/uploads/2018/06/Renewable-energy-in-sub-Saharan-Africa_FINAL.pdf accessed 07 September 2018.

behind dam projects on the Nile, Yohannes argues that 'the involvement of the international financial institutions in the promotion of water regimes is part of the general trend to commodify, privatize, and marketize water resources to serve the purpose of global accumulation.'[10]

As the Southern African region plans to increase its energy generation capacities mostly in order to attract investments,[11] it is incumbent to critique and analyze how the system has traditionally affected those communities who live in the vicinity of rivers pegged for hydroelectric mega dam projects. In the region, the human rights plight of the indigenous riverine communities is profoundly felt within the middle Zambezi River area, that is, the area between the Victoria Falls and Lake Cahora Bassa. These people have historically faced forced displacement and resettlement, lack of consultation in developmental projects right from the colonial era up to the post-colonial times.[12] Furthermore, the riverine communities also face the ongoing threats of flooding and hence require legal and policy protection and institutional support at regional and national levels.[13] The problems highlight the need for a pro-indigenous community, environmental justice and human rights-centric system. National and regional legal, policy and institutional frameworks should reflect this support and solidarity.[14]

[10] Okbazghi Yohannes, 'Hydro-politics in the Nile Basin: in search of theory beyond realism and neo-liberalism,' (2009) 3 (1) Journal of Eastern African Studies 74, 87.

[11] See, for instance, A Moyo, 'Investors target energy sector.' The Sunday Mail (Harare 18 February 2018).

[12] See A Isaacman, 'Displaced People, Displaced Energy, and Displaced Memories: The Case of Cahora Bassa, 1970–2004.' (2005) 38 (2) The International Journal of African Historical Studies 201.

[13] See P Muhonda, 'Analysis of institutional mechanisms that support community response to impacts of floods in the Middle-Zambezi river basin, Zimbabwe.' (2014) 76–78 Physics and Chemistry of the Earth 64.

[14] See SH Ali, 'Water Scarcity and Institutional Reform in Southern Africa.' (1999) 24 (2) Water International 116.

2 Human Rights, Neoliberalism and Energy in the Global South: A Theoretical Overview

A vast body of literature has attempted to highlight the relationship between human and environmental rights and the neoliberal economic system in general and specifically with the issues of energy distribution and politics in the Global South.[15] In terms of definition, neoliberalism has been described as 'a theory of political economic practices that proposes that human well-being can best be advanced by liberating individual entrepreneurial freedoms within an institutional framework.'[16] This definition is, however, a sanitized one since neoliberalism essentially rewards corporates at the expense of social systems and poor societies who cannot monetize at the same speed as sophisticated State and private business entities. Neoliberalism is an unequal economic system based on capitalism. In terms of the system's link with the environment and society, Obeng-Odoom points out that neoliberalism 'refers to the financialization and marketization of society, economy, and environment.'[17] Research has also aptly touched on the impact of an unequal society on the distribution of natural resources such as water. Schulz makes an exciting yet highly relevant point that:

> ...the complexities facing humankind due to the uneven distribution of water resources, climate change, and inequitable development remain serious challenges to those governing water resources.[18]

[15] Franklin Obeng-Odoom aptly captures the concept of neoliberalism and market approaches to the environment in Africa in 'Green neoliberalism: Recycling and sustainable urban development in Sekondi-Takoradi' (2014) 41 Habitat International 129–134; See also B Büscher, 'Anti-Politics as Political Strategy: Neoliberalism and Transfrontier Conservation in Southern Africa.' (2010) 41 (1) Development and Change 29–51;

[16] D Harvey, A *Brief History of Neoliberalism* (Oxford University Press 2007) p. 2.

[17] Obeng-Odoom (n 15) 129.

[18] A Schulz, 'Creating a Legal Framework for Good Transboundary Water Governance in the Zambezi and Incomati River Basins.' (2007) 19 (2) Georgetown International Environmental Law Review 117, at 118.

The human and environmental rights of riverine communities and societies are significantly infringed upon by capital-centric neoliberalism. Hence this chapter tries to infuse concepts of environmental justice, neoliberalism critique and human rights-based approaches. Hoddya and Ensor describe the defense of human rights under neoliberalism as fighting for 'human rights from below.'[19] This research is critical in articulating the human rights and environmental justice plight of indigenous communities along the Zambezi as well as in trying to understand how they have faced such plight at the hands of State and business-centered development planning.

The paradox in the Zambezi River Basin has been that economic and political processes in the riparian States' development have usually been less beneficial to the riverine peoples themselves than they have been to investors and businesses situated in places that are distant from the rivers themselves. The riverine indigenous peoples, especially in Zimbabwe and Mozambique, have suffered some avoidable human, developmental and environmental rights violations such as loss of land, fishing rights, livelihoods, unfair levies and human rights violations. A report by Zimbabwe's Ministry of Labour, Public Service and Social Welfare, titled 'The 1995 Poverty Assessment Survey', affirms this view and indicated how the Tonga peoples' displacements 'received no government compensation, were reduced to dependence on food handouts and have consistently registered the highest levels of poverty in the country.'[20] The effects on the Tonga people were unusually harsh, considering that they have been traditionally environmentally conscious and 'prominent agricultural people.'[21]

Some researchers have linked displacement with fewer hectarage for subsistence farming and adverse effects on the human right to food security.[22] The implementation and protection of a continental and regional

[19] Eric T Hoddya and Jonathan E. Ensorb, 'Brazil's landless movement and rights "from below",' (2018) 63 Journal of Rural Studies 74–82.

[20] J McGregor, 'Crocodile crimes: people versus wildlife and the politics of postcolonial conservation on Lake Kariba, Zimbabwe.' (2005) 36 Geoforum, 353 at 360.

[21] KM Kanene, 'Indigenous practices of environmental sustainability in the Tonga community of southern Zambia.' (2016) 8 (1) Jàmbá: Journal of Disaster Risk Studies a331. https://doi.org/10.4102/jamba.v8i1.331.

[22] L Artur and D Hilhorst, 'Floods, resettlement and land access and use in the Lower Zambezi, Mozambique.' (2014) 36 Land Use Policy 361, at 367.

human rights system in the context of mega dam development and energy politics in Southern Africa is tied up to the dominant global capitalist and neoliberal economic systems. The utilization of dams as sources of hydroelectric energy in Southern Africa has been propagated since the advent of colonialism and general industrialization. This energy push was due to the exploitative nature of the colonial economic structure with its insatiable need both for producing agricultral raw materials and for hydroelectric energy and industrial power for use in processing the same raw materials.[23]

With the end of colonial rule, however, the trend continued, in highly ambitious developing Global South States, to exploit rivers and natural resources as a way of supporting the energy needs of investors in projects that hardly catered for the needs of marginalised human communities and for environmental justice. A research on the Bakun Project in Malaysia highlights that 'centralized energy megaprojects, while ostensibly championed for reasons of economies of scale and the ability to bring about transformational change in the shortest period of time, often fail to address broader development goals such as fighting energy poverty and improving the livelihoods of the populations affected by these projects.'[24]

The desire to respond to climate change has also led to renewed interest in water governance and hydroelectricity, giving rise to the neoliberal concept of the green economy. The green economy is a system that 'aims to use economic rationality and market mechanisms to mute the most ecologically damaging effects of globalized capitalism while reviving economic growth in the global North, fostering development in the South, and decoupling economic growth from environmental decline.'[25] Hence even under the guise of saving the environment, neoliberalism conducts ruthless economic exploitation with minimal to zero regards for actual environmental preservation or public consultation on people's concerns and human rights matters. The critical goal is profit and, as Yohannes

[23] See, for instance, Clapperton Chakanetsa Mavhunga, 'Energy, Industry, and Transport in South-Central Africa's History' in CC Mavhunga and H Trischler (eds), *Energy (and) Colonialism, Energy (In)Dependence: Africa, Europe, Greenland, North America* (RCC Perspectives 2014/5).

[24] BK Sovacool and LC Bulan, 'Behind an ambitious mega project in Asia: The history and implications of the Bakun hydroelectric dam in Borneo.' (2011) 39 Energy Policy, 4587.

[25] K Mcafee, 'Green economy and carbon markets for conservation and development: a critical view' (2016) 16 (3) International Environmental Agreements: Politics, Law and Economics, p. 333.

points out, this can be achieved without public consideration.[26] This situation is a vital characteristic of the capitalist system and the neoliberal environment that facilitates the dominant economic model of governance.

The desire for profit and facilitating the interests of investors is one of the main reasons why development and the distribution of energy and even of dams as energy sources hardly go in the interest of the ordinary peoples or for the alleviation of people's livelihoods. Instead, these developments either pose problems for poor communities' environmental justice or derive immense financial and other benefits for the elite. For instance, the drive for mega dam building by Mozambique hasless consideration for people-and-environmental justice development and more regard for the interests of investors such as the China Export-Import Bank. This has left ten thousand people at the risk of relocation. The following passage aptly captures this trajectory: '...the China Export-Import Bank has helped launch the funding for a new 1,500 megawatt dam on the Zambezi River in Mozambique, the Mpanda Nhuwa. Building the dame would eliminate any hopes of rectifying the impacts of a previously built dam on the Zambezi river delta, which has shrunk by half as sediment ans water have been choked off, crippling fisheries and reducing fertile farmland downstream. That dam, the Cahora Bassa - built in the 1970s - required the relocation of more than 10,000 people'.[27]

The entrenchment of neoliberalism and the complex involvement of Chinese investments to offset its immensely polluting and heavy emission industries[28] are posing an emerging challenge for the ordinary people who will continue to suffer the ongoing environmental injustice that was initiated by the States' shift to neoliberalism in both Mozambique and Zimbabwe. The laws in developing states since the 1990s have exhibited the influence of neoliberal tendencies and the drive toward facilitating the creation of an investor friendly climate at the expense of human, environmental and developmental rights.[29] Zimbabwe and Mozambique

[26] Yohannes (n 10) 77.

[27] D Biello, 'The Dam Building Boom: Right Path to Clean Energy?' Yale Environment 360 (New Haven 23 February 2009).

[28] ibid.

[29] See SA Subedi, *International Investment Law: Reconciling Policy and Principle* (3rd edn, Hart Publishing 2016).

share a seemingly similar political and economic history. The influence of the historical development on the law and policy formulation is apparent. The influence has also had the effect of posing some toll upon the less powerful working and peasant communities. The history of colonial rule followed by periods of socialist-leaning politics and the eventual opening up by the two States to neoliberalism and capitalism has had a noticeable policy and legal impact. Zimbabwe and Mozambique, and indeed like the rest of Africa, passed through colonial rule, and violent liberations struggle and in the aftermath of such struggles sought to create egalitarian states.[30] However, the influence of neocolonialism and lack of commitment by the leadership plus problems associated with the transition from colonial rule curtailed such aspirations.[31] Ngang points out that the efforts of African leaders to achieve specific progressive policies and aspirations like Pan-Africanism, human and people's rights and other policies are plausible in theory but almost non-existent in the eyes of the people.[32]

Within a decade and as a result of worldwide pressure following the end of the Cold War, Zimbabwe, Mozambique and a host of States in the developing or Third World fell to the influence of neoliberalism and capitalist modes of development during the so-called golden age of capitalism. In the context of Mozambique, where the Cahora Bassa and Mphanda Nkuwa Dams are situated the State initially proclaimed an egalitarian Marxist-Leninist state. However, the State also went through years of a brutal civil war that had the effect of closing off large parts of the country from development. In Zimbabwe, there was relative peace, then civil strife and elite-led reconciliation at the end of the liberation struggle.

The development of contemporary dam and energy plans in the two States expose the politics of exclusion associated with inherited neoliberalism around the world. For instance, the construction of the Tugwi Mukosi Dam and the plans to construct other dams south of the Lake Cahora Bassa have been met with controversy and lack of consultation

[30] See R Southall, 'The Poverty of the "Middle Classing" Of Development: Key Problems in Southern Africa.' (2017) 39 (1) Strategic Review for Southern Africa 211.

[31] ibid.

[32] CC Ngang, 'Towards a right-to-development governance in Africa.' (2018) 17 (1) Journal of Human Rights 110.

for the more impoverished communities who will experience the significant impact of these developments. Cahora Bassa and Kariba were highly disadvantageous to the peoples of the Zambezi River Valley. Whole communities were forced to relocate and those who still stay in the areas, mostly the fisherfolk, live in undeveloped districts with severe legal restrictions on fishing and agricultural rights and seemingy less capacity for profit and development unlike commercial fish companies operating in the area.[33] Such situations should pose as warnings and lessons for future developmental projects.

Under neoliberal and free-market law-making processes, the laws and policies have historically and currently, rarely catered for the rights of indigenous peoples. Impoverished communities affected by energy and dam development at the time of the construction of the Kariba Dam and Cahora Bassa were generally excluded from the development discourse. They continued to suffer, even more than during the racist colonial era, as the Global South pushed toward full-scale neoliberalism and economic structural adjustment[34] at the end of the Cold War. Mashingaidze makes a comprehensive and extensive case about the plight of the BaTonga people since their relocation during the construction of Lake Kariba. In his view, the riverine Tonga people constitute part of a marginalized and powerless class of people 'since the Southern Rhodesian' government displaced them from the ecologically rich Zambezi River.'[35] The challenges have worsened with renewed green economic interests in climate change adaptation and mitigation mechanisms in which the push for renewable energy production will likely lead to mega hydroelectric dam construction and corporate demands for land for biofuel production.[36]

[33] Observations by the author in a visit to Binga District, Zimbabwe in 2017.

[34] See Mazur (n 3).

[35] TM Mashingaidze, 'Beyond the Kariba Dam Induced Displacements: The Zimbabwean Tonga's Struggles for Restitution, 1990s–2000s.' (2013) 20 International Journal on Minority and Group Rights 381, at 382.

[36] Carol Hunsberger, Esteve Corbera, Saturnino M. Borras Jr., Jennifer C. Franco, Kevin Woods, Courtney Work, Romulo de la Rosa, Vuthy Eang, Roman Herre, Sai Sam Kham, Clara Park, Seng Sokheng, Max Spoor, Shwe Thein, Kyaw Thu Aung, Ratha Thuon and Chayan Vaddhanaphuti, 'Climate change mitigation, land grabbing and conflict: towards a landscape-based and collaborative action research agenda.' (2017) 38 (3) Canadian Journal of Development Studies 305–324.

3 The Legal and Policy Framework Governing the Zambezi River Basin

The law concerning the Zambezi River Basin is made up of the State laws of each riparian country as well as agreements and policy formulated by the regional economic community bloc, namely the Southern African Development Community (SADC). Some scholars have highlighted that 'the current legal framework governing the Zambezi is a patchwork of bilateral and multilateral treaties.'[37] The increasing shift toward neoliberal policies and the desire to attract investments profoundly inform the nature and impact of national and regional laws and policies upon the environment and upon indigenous peoples with the Zambezi Valley.[38] The patchwork of laws referred to by scholars, like Schulz, present a necessity to study the various strands that make up the law and policies governing or guiding the livelihoods, rights and duties of the indigenous riverine and rural communities within the Zambezi River Basin.

The role of law in neoliberal development planning,[39] and the treatment of indigenous disempowered communities, is often a contested terrain and moreso in Southern Africa. The adoption of market solutions to environmental and economic issues in Southern Africa is a reflection of the acceptance of neoliberalism as the dominant model on post-colonial law-making. Hence scholars suggest that the way particular species and communities are implausibly designated as 'requiring special protection from markets'[40] is a clear reflection of the growing strength of neoliberalism in the legal and policy framework framing. The Zambezi River Basin is subject to many international laws and conventions. The most critical regional framework concerning the Zambezi Valley area and its indigenous riverine communities barely cater to peoples' human, developmental, environmental and community rights. The SADC Protocol on Shared

[37] Schulz (n 18) at 168.

[38] See B Büscher, 'Investing in Irony? Development, Improvement and Dispossession in Southern African Coal Spaces.' (2015) 27 European Journal of Development Research 727.

[39] See M Gwisai, 'Theory and Practice of Liberal Democracy in the Post-Colonial State in Africa: The Zimbabwe Experience.' (1991–1992) 9–10 The Zimbabwe Law Review 110.

[40] P Anthias and SA Radcliffe, 'The ethno-environmental fix and its limits: Indigenous land titling and the production of not-quite-neoliberal natures in Bolivia.' (2015) 64 Geoforum, 257 at 260.

Watercourses[41] is the principal regional instruments on shared watercourses within the Zambezi River Basin and guides the relationships among the riparian SADC States. The protocol provides that Member States 'shall maintain a proper balance between resource development for a higher standard of living for their peoples and conservation and enhancement of the environment to promote sustainable development.'[42] This provision is telling on the need to embark on development planning for the peoples of the Zambezi River Basin. Unfortunately, what is contained in the protocol is not matched by lived realities on the ground.

The SADC Treaty[43] was formulated with the objectives of building solidarity within the SADC region during the eras of colonialism, apartheid and in the contemporary post-colonial era. These solidarity ideals originated from the original objective of front-line (SADC) States - encouraging minimal dependence on then apartheid South Africa. Solidarity translates even within the areas of environmental and economic justice.[44] The quality of life and human rights situations within the communities of indigenous riverine peoples in the Zambezi Valley highlights some of the lowest forms of poverty, disenfranchisement and disempowerment that have not been alleviated by economic structural adjustment. In the post-colonial era, most Southern African states adopted neoliberal free-market policies because of the fall of the Soviet Union.[45] The effect of the triumph of neoliberalism and capitalism around the world in the 1990s was intially met with disdain as is reflected in the way some of the SADC States, for instance, Zambia tried unsuccessfully to resist falling into the trap of the global International Financial Institutions (IFIs)-led neoliberal economics.[46] In the end, the neoliberal

---

[41] Revised Protocol on Shared Watercourses in the Southern African Development Community (SADC) (adopted 7 August 2000).

[42] ibid., art 2 (3).

[43] Consolidated Text of the Treaty of the Southern African Development Community (adopted 17 August 1992, entered into force 05 October 1992) (SADC Treaty).

[44] ibid., art 5 (1) (a).

[45] SA Subedi, *International Investment Law: Reconciling Policy and Principle* (3rd Ed) (Oxford and Portland, Oregon: Hart Publishing, 2016), p. 46.

[46] M Hanson and JJ Hentz, 'Neocolonialism and Neoliberalism in South Africa and Zambia' (1999) 114 (3) Political Science Quarterly 479, p. 501.

shift in most of the post-colonial Global South also affected the economic, environmental and social development and laws and policies of the riparian SADC states such as Zimbabwe and Mozambique.

Water-based industries such as fishing also constitute a major part of the contemporary economic and cultural livelihoods of the riverine peoples. The SADC Protocol on Fisheries clearly outlines the need for the protection of the rights of indigenous and small-scale riverine fisherfolk.[47] However these local involvements and enterprises are adversely affected by ecosystem damage from mega dam projects as well as high fishing permit fees, for instance, and this is an area that highlights lack of political will by some of the Zambezi River Basin states to adhere to the letter and spirit of the SADC Protocol on Fisheries.

Besides the SADC Treaty and its protocols, another relevant multilateral treaty is the Zambezi Action Plan (ZACPLAN)[48] adopted in 1987. To its downside, the action plan rarely mentions the rights, aspirations or the need to cater to the riverine communities. The action plan mentions pollution control, soil erosion and other ecological issues but is glaringly lacking on the actual human development and human rights protection elements.[49] This disregard of people's rights in favor of industrial and raw material promotion is one of the critical characteristics of a region leaning toward further neoliberal policy formulation. The ZACPLAN also disempowers regional-led initiatives such as development centric research and projects in favor of national-level initiatives.[50] National-level initiatives pay homage to the customary international law principle of upholding the right to State sovereignty. Sovereignty, as the basis of international law and relations, is essential but can also limit the potential for regional planning, solidarity and cooperation.

---

[47] See ibid., art 12 (1).

[48] Action Plan for the Environmentally Sound Management of the Common Zambezi River System.

[49] See Annex I. To the agreement on the action plan for the environmentally sound management of the Common Zambezi River system, art 14.

[50] Annex II. To the agreement on the action plan for the environmentally sound management of the common Zambezi river system, art 12.

State Laws and Pronounced Policy

The laws of two Zambezi Valley Basin affected States of Mozambique and Zimbabwe highlight a quagmire between anti-colonial history and ever-emboldened neoliberalism. It is essential to study how this background has shaped the emergence and transition to wholescale neoliberalism within these riparian States. Such a background is necessary to understand the context and the plight of indigenous riverine peoples under the throes of neoliberalism-induced human and environmental rights injustices along the middle Zambezi River Valley of the Zambezi River Basin.

Mozambique

The then Peoples' Republic of Mozambique attained independence from Portuguese colonial rule after a liberation struggle fronted by the Marxist-Leninist Frente de Libertação de Moçambique[51] (FRELIMO). The 1974 Carnation Revolution in Portugal catalyzed the victory of the struggle for independence and the resultant end of colonial rule. The revolution in the colonial State Portugal, ended the Caetano regime and recommended the independence of the colonies. Mozambique was a Marxist-Leninist state at the time of its independence in 1975. In the wake of the expansion of neoliberalism in the1990s, the State 'abandoned many features of its state-centred economy in favour of a free-market, capitalist economy.'[52] The global shift to neoliberalism in 1990, with the fall of the Soviet Union mentioned above, affected Mozambique's people. In the 1990s, the formulation of laws affecting the indigenous communities was becoming less progressive and more investor-conscious. Scholars have described the State's Land Law of 1997 as seeking to 'achieve a balance between safeguarding the interests of communities and facilitating investors' access to land.'[53]

---

[51] Front for the Liberation of Mozambique.

[52] E Lunstrum, 'Mozambique, Neoliberal Land Reform, and the Limpopo National Park.' (2008) 98 (3) The Geographical Review 339.

[53] RJE van Den Brink, 'Land Reform in Mozambique.' (2008) 43 Agriculture & Rural Development Notes: Land Policy and Administration https://siteresources.worldbank.org/EXTARD/

To its credit, albeit in theory, the Constitution of Mozambique even after the official dropping of Marxism-Leninism still contains pro-people provisions.[54] The pro-people provisions sometimes are not matched by the actual treatment of the people on the ground. Concerning land rights, for instance, the Constitution[55] states that 'as a universal means for the creation of wealth and of social well-being, the use and enjoyment of land shall be the right of all the Mozambican people.'[56] This provision represents some lingering influence of progressive and historical socialist policies of pre-neoliberal Mozambique. The reality on the ground, however, following the abandonment of socialism, showcases indigenous communities facing severe challenges to their land rights a result of mega dam building, coal mining, natural gas explorations, bio-fuels and other transnational investments in the agriculture sectors. The effect of the Zambezi River's Cahora Bassa Dam on the human rights of the Mozambican peoples is telling of the influence of the State's ideological shift.[57] Mozambique's laws and policies now resemble a wholesale shift toward ultra-investor friendly policies. The Constitution of Mozambique lays out the shift to neoliberalism in the Fundamental Principles Section:

> The economic and social order of the Republic of Mozambique shall aim to satisfy the basic needs of the people and to promote social wellbeing, and shall be based on the following fundamental principles:
> a) on the value of labour;
> b) on market forces;
> c) on the initiatives of economic operators.[58]

This shift to neoliberalism is inevitably incompatible with the human and environmental rights and developmental expectations of the

Resources/336681-1295878311276/WB_ARD_Mzmbq_Note43_web.pdf accessed 20 June 2018.

[54] See, for instance, Articles 2 (1), 96 (1) and 101 (1) of the Constitution of the Republic of Mozambique.

[55] Mozambique's Constitution of 2004 with Amendments through 2007.

[56] Ibid., art 109.

[57] AF Isaacman BS and Isaacman, *Dams, Displacement, and the Delusion of Development: Cahora Bassa and Its Legacies in Mozambique, 1965–2007* (Athens, Ohio University Press, 2013), p. 4.

[58] Constitution of the Republic of Mozambique, Art 97.

indigenous peoples, and especially those whose economic, social and cultural livelihoods depend on natural resources, communal lands, public water rights and affirmative access to riverine resources.

Zimbabwe

Like most post-colonial States, the Republic of Zimbabwe has traveled through legal and policy developments that started with a Marxist and socialist grounding and then transformed into neoliberalism during the 1990s. The transition from white settler rule to black majority rule failed to deliver economic justice for the majority.[59] The treatment of ordinary indigenous and less powerful people and their livelihoods within the Zambezi Valley, especially the Tonga people, continued to be on the downside with continued dispossession and lessened space to claim recognition of economic and environmental rights for the indigenous peoples. The legal framework has failed to tranform the livelihoods of disempowered minoroties. Zimbabwe's laws governing rivers and the rights of indigenous and riverine communities consist of the National Constitution, general law and customary law as well as international commitments resulting from the ratification of international treaties and conventions. Zimbabwe has had a progressive Constitution coming from a 2013 referendum.

Even though the plight of the Tonga people on the shores of Lake Kariba and the Zambezi River is still dire, the 2013 Constitution of Zimbabwe did come up with some progressive provisions on the right of ordinary communal people. The Constitution provides for 'fair representation of all Zimbabwe's regions in institutions and agencies of government.'[60] In addition to the Constitution, Zimbabwe has good statutes that cater to human and environmental rights such as the Environmental Management Act,[61] the Civil Protection Act[62] and the

[59] See M Gwisai (n 39).

[60] Constitution of Zimbabwe Amendment (No 20) Act 2013, sec 18.

[61] [Chapter 20:27].

[62] [Chapter 10:06].

Parks and Wildlife Act.[63] However, even with the comprehensive available legal framework, there is a lack of administrative and political will and no substantive mechanisms for enforcing and ensuring the equitable and affirmative representation of indigenous riverine peoples in institutions that must deal with Zambezi River issues. The continued challenges for the riverine communities, most of whom rely on the Zambezi for economic and cultural activities and fishing, include the scourge of natural hazards such as seasonal floods that turn into disasters due to inadequate infrastructure and lack of development. Muhonda et al. (2014) argue that 'the frequency with which floods occur and, also, the distribution of rainfall within the season, highlights the need for strong institutional mechanisms to mitigate the impacts of naturally induced factors on livelihoods.'[64] The legal and policy framework must, therefore, be responsive to these challenges. Hence national pieces of legislation such as the Civil Protection Act and the Environmental Management Act must be accompanied by the sufficient political will to cater to the challenges faced by the riverine indigenous peoples whose local resources usually traslate into energy sources for others.

4 The Historical Impacts of Law and Policy on Indigenous Zambezi Riverine People

The plight of the indigenous peoples within the Zambezi River Basin is interlinked with the legacy effects of historical events such as colonialism on them. In the middle Zambezi region, the colonial and settler powers' quest for energy to power industry[65] resulted in the displacement of the local indigenous peoples to pave the way for the construction of the Kariba Dam on the Zambezi. The dam construction also separated and split families into the two colonial States. The purpose of the dam was the

63 [Chapter 20:14].

64 Muhonda (n 13) at 71.

65 GJ Williams, 'The changing electrical power industry of the middle Zambezi Valley.' (1984) 69 (3) Geography 257.

powering of the colonial industrial States that made up a significant chunk of the short-lived Federation of Rhodesia and Nyasaland.[66]

In addition to disgruntlement over the construction of the Kariba Dam, there were also strong anti-dam views in Mozambique, especially from communities along the Zambezi River in the State. In Mozambique, at the time of the construction of the Cahora Bassa Dam, the anti-colonial and socialist-oriented FRELIMO viewed the project as 'part of a military and economic alliance between Portugal and South Africa designed to provide cheap energy to the apartheid regime and to perpetuate white rule in the region.'[67] However, the same FRELIMO has shifted stance after it officially dropped Marxism-Leninism.

The independence years across developing States in Africa, Asia and Latin America was also the period of the Cold War, during which an ideological struggle ensued between the capitalist West and the socialist East. However, this period was not sustainable because of international economic pressure on the then Third World States, especially following the fall of the Soviet Union. The governance emphasis, therefore, shifted to pro-investment and pro-business policies. The interests of neoliberal and liberal governance models are focused on the market forces and private sector-led economy at the expense of genuine and people-centric economic development. Scholars have highlighted that the rising influence of neoliberalism is because of the cleverness of the system to place many efforts at managing societal tensions while doing very little, if not nothing toward empowering the social status of their peoples.[68] The departure from socialism to neoliberalism had an unfortunate effect upon most of the poor citizens, as well as rural communities and small-scale farmers.[69] Mozambique continues to face challenges with local peoples'

---

[66] The federation consisted of Northern Rhodesia (now Zambia), Southern Rhodesia (now Zimbabwe) and Nyasaland (now Malawi).

[67] A Isaacman, 'Displaced People, Displaced Energy, and Displaced Memories: The Case of Cahora Bassa, 1970–2004.' (2005) 38 (2) The International Journal of African Historical Studies 201, pp. 201–202.

[68] F Eguren, 'Socialism in the Twenty-First Century and Neo-liberalism: Diverse Ideological Options Do Not Always Generate Different Effects' in G Carbonnier et al. (eds), *Alternative Pathways to Sustainable Development: Lessons from Latin America* (Brill, 2017), p. 124.

[69] A Arnall, 'A climate of control: flooding, displacement and planned resettlement in the Lower Zambezi River valley, Mozambique.' (2014) 180 (2) The Geographical Journal 141, at 146.

acceptance of displacement for further planned energy mega dam construction projects on the Zambezi.[70] These problems have their roots, as mentioned above, in the local community's displacement and disgruntlement following the construction of the Cahora Bassa Dam which was meant to power the factories, cities and economies of colonial Mozambique and apartheid South Africa.[71]

Several factors ensured the entrenchment and character of neoliberal economic and political governance across the developing world. In Southern Africa and especially in the States around the Zambezi River Basin, as the chapter highlights, the adoption of neoliberalism and colonial economic systems came at the expense of the human rights, economic and environmental justice aspirations of the disenfranchised African communities. Of the latter, the indigenous riverine communities suffered the most and continue to suffer from energy politics and economics, mega dam construction and water and land grabbing.[72]

5 Conclusions

Mega Dams and Energy Politics' Impact on Human Rights

The more disproportionate the adverse impacts of energy policies and dams are on the poorer, peasant and working-class sectors of society, the worst the situation for minority groups such as smaller ethnic groups, the disabled and women. As highlighted in the chapter, energy has historically had—and continues to have—gendered dimensions on contemporary society. In the Global South and other less-affluent societies, evidence of the gendered aspects of energy are apparent and continue to be

---

[70] AF Isaacman and D Morton, 'Harnessing the Zambezi: How Mozambique's Planned Mphanda Nkuwa Dam Perpetuates the Colonial Past.' (2012) 45 (2) The International Journal of African Historical Studies 157.

[71] Isaacman (n 57).

[72] See J Franco et al., 'The Global Politics of Water Grabbing' (2013) 34 (9) Third World Quarterly 1651.

highlighted in diverse ways.[73] The growth and shift toward energy-related mega developmental projects deeply entrenched with neoliberal frameworks have a great hold on poverty, inequality and environmental injustice within Southern Africa.

The Case for Alternatives to Contemporary Neoliberalism's Impacts on Environmental and Human Rights

Equitable energy planning has the potential to uplift the livelihoods of the poorest and excluded sectors of Southern Africans, or more specifically of the rural and peasant classes in Zimbabwe and Mozambique. However, this can only be achieved in a legal, policy-making and economic environment that is devoid of unchecked neoliberalism. Scholars have pointed out that '[t]he goal of protecting the social, economic, and development rights of the poor and vulnerable can and should become the basic objective of economic adjustment.'[74] The contemporary examples indicate that energy planning and sourcing for private industry and capital in Southern Africa and other Global South States such as Brazil take precedence over people's rights, human rights and environmental justice rights. Fearnside's research notes that;

> Global dam-building activity is increasingly focused on tropical areas in Africa, Southeast Asia and Latin America. National decisions on promoting and subsidizing dams and electro-intensive exports have multiple perverse effects on political processes in developing countries through the "resource curse" and other mechanisms. Decisions on export priorities and energy policies give little weight to the heavy environmental and social costs of dam projects, as is evident in the example of Brazil.[75]

[73] See Beth Osnes, 'Engaging women's voices through theatre for energy development.' (2013) 49 Renewable Energy 185.

[74] Mazur (n 3) 69.

[75] Fearnside (n 1) 60.

In Southern Africa, the concerns and observations around the conduct, indifferent law making and policy management around the Cahora Bassa, Inga, Tugwi-Mukosi, Mphanda Nkuwa and Kariba Dams create lessons for the future. Contemporary neoliberalism is a system of economic and social exclusion, gendered injustices and the drawbacks of community and developmental rights for communities living near energy and dam projects.

Neoliberal governance systems harm human, developmental, energy and environmental rights. In that regard, the Global South and on particular Sub-Saharan African States can take several measures in order to cushion their populace from unchecked neoliberalism. Regional transparency, cooperation and experience sharing on environmental justice and human rights law and policy development between States such as Mozambique, Zimbabwe, the Democratice Republic of the Congo and others is essential. From an empowerment model, there is a need for capacity building and affirmative enfranchisement of riverine communities to participate in regional and local energy governance as well as human rights and envirometal justice-related law making. There is also a need for promoting and facilitating the public participation of indigenous riverine communities in regional and national river management initiatives, especially within the middle Zambezi River Valley.

References

Ali SH, 'Water Scarcity and Institutional Reform in Southern Africa.' (1999) 24 (2) Water International 116.

Anthias P and Radcliffe SA, 'The ethno-environmental fix and its limits: Indigenous land titling and the production of not-quite-neoliberal natures in Bolivia.' (2015) 64 Geoforum, 257 at 260.

Arjun Sengupta A, 'On the Theory and Practice of the Right to Development.' (2002) 24 Human Rights Quarterly 837.

Arnall A, 'A climate of control: flooding, displacement and planned resettlement in the Lower Zambezi River valley, Mozambique.' (2014) 180 (2) The Geographical Journal 141.

Artur L and Hilhorst D, 'Floods, resettlement and land access and use in the Lower Zambezi, Mozambique.' (2014) 36 Land Use Policy 361.

Biello D, 'The Dam Building Boom: Right Path to Clean Energy?' Yale Environment 360 (New Haven 23 February 2009).

Büscher B, 'Anti-Politics as Political Strategy: Neoliberalism and Transfrontier Conservation in Southern Africa.' (2010) 41 (1) Development and Change 29.

Büscher B, 'Investing in Irony? Development, Improvement and Dispossession in Southern African Coal Spaces.' (2015) 27 European Journal of Development Research 727.

Carbonnier G et al. (eds), *Alternative Pathways to Sustainable Development: Lessons from Latin America* (Brill, 2017).

Fearnside PM, 'Environmental and Social Impacts of Hydroelectric Dams in Brazilian Amazonia: Implications for the Aluminium Industry.' (2016) 77 World Development 57.

Franco J et al., 'The Global Politics of Water Grabbing' (2013) 34 (9) Third World Quarterly 1651.

Gwisai M, 'Theory and Practice of Liberal Democracy in the Post-Colonial State in Africa: The Zimbabwe Experience.' (1991–1992) 9–10 The Zimbabwe Law Review 110.

Hanson M and Hentz JJ, 'Neocolonialism and Neoliberalism in South Africa and Zambia' (1999) 114 (3) Political Science Quarterly 479.

Harvey D, *A Brief History of Neoliberalism* (Oxford University Press 2007).

Hoddya ET and Ensorb JE, 'Brazil's landless movement and rights 'from below',' (2018) 63 Journal of Rural Studies 74.

Hunsberger C, Corbera E, Borras SM and others, 'Climate change mitigation, land grabbing and conflict: towards a landscape-based and collaborative action research agenda.' (2017) 38 (3) Canadian Journal of Development Studies 305.

Isaacman A, 'Displaced People, Displaced Energy, and Displaced Memories: The Case of Cahora Bassa, 1970–2004.' (2005) 38 (2) The International Journal of African Historical Studies 201.

Isaacman AF and Isaacman BS, *Dams, Displacement, and the Delusion of Development: Cahora Bassa and Its Legacies in Mozambique, 1965–2007* (Athens, Ohio University Press, 2013).

Isaacman AF and Morton D, 'Harnessing the Zambezi: How Mozambique's Planned Mphanda Nkuwa Dam Perpetuates the Colonial Past.' (2012) 45 (2) The International Journal of African Historical Studies 157.

JCRA, *Whitepaper on Renewable energy in sub-Saharan Africa* (Cebr and JCRA 2018) 8. https://jcragroup.com/wp-content/uploads/2018/06/Renewable-energy-in-sub-Saharan-Africa_FINAL.pdf accessed 07 September 2018.

Kanene KM, 'Indigenous practices of environmental sustainability in the Tonga community of southern Zambia.' (2016) 8 (1) Jàmbá: Journal of Disaster Risk Studies a331. doi:https://doi.org/10.4102/jamba.v8i1.331.

Kuik OJ, Lima MB and Gupta J, 'Energy security in a developing world.' (2011) 2 (4) Wiley Interdisciplinary Reviews: Climate Change 627.

Lunstrum E, 'Mozambique, Neoliberal Land Reform, And the Limpopo National Park.' (2008) 98 (3) The Geographical Review 339.

Mashingaidze TM, 'Beyond the Kariba Dam Induced Displacements: The Zimbabwean Tonga's Struggles for Restitution, 1990s–2000s.' (2013) 20 International Journal on Minority and Group Rights 381, at 382.

Mazur RE, 'Realization or deprivation of the right to development under globalization? Debt, structural adjustment, and poverty reduction programs.' (2004) 60 (1) GeoJournal 61.

McGregor J, 'Crocodile crimes: people versus wildlife and the politics of postcolonial conservation on Lake Kariba, Zimbabwe.' (2005) 36 Geoforum, 353.

Moyo A, 'Investors target energy sector.' The Sunday Mail (Harare 18 February 2018).

Muhonda P, 'Analysis of institutional mechanisms that support community response to impacts of floods in the Middle-Zambezi river basin, Zimbabwe.' (2014) 76–78 Physics and Chemistry of the Earth 64.

Ngang CC, 'Towards a right-to-development governance in Africa.' (2018) 17 (1) Journal of Human Rights 110.

Obeng-Odoom F, 'Green neoliberalism: Recycling and sustainable urban development in Sekondi-Takoradi' (2014) 41 Habitat International 129.

Osnes B, 'Engaging women's voices through theatre for energy development.' (2013) 49 Renewable Energy 185.

Richard Heinberg R, *Afterburn: Society beyond Fossil* Fuels (New Society Publishers 2015) 47.

Schulz A, 'Creating a Legal Framework for Good Transboundary Water Governance in the Zambezi and Incomati River Basins.' (2007) 19 (2) Georgetown International Environmental Law Review 117.

Southall R, 'The Poverty of the 'Middle Classing' Of Development: Key Problems in Southern Africa.' (2017) 39 (1) Strategic Review for Southern Africa 211.

Sovacool BK and Bulan LC, 'Behind an ambitious mega project in Asia: The history and implications of the Bakun hydroelectric dam in Borneo.' (2011) 39 Energy Policy 4587.

Subedi SA, *International Investment Law: Reconciling Policy and Principle* (3rd edn, Hart Publishing 2016).

van Den Brink RJE, 'Land Reform in Mozambique.' (2008) 43 Agriculture & Rural Development Notes: Land Policy and Administration https://siteresources.worldbank.org/EXTARD/Resources/336681-1295878311276/WB_ARD_Mzmbq_Note43_web.pdf accessed 20 June 2018.

Williams GJ, 'The changing electrical power industry of the middle Zambezi Valley.' (1984) 69 (3) Geography 257.

Yohannes O, 'Hydro-politics in the Nile Basin: in search of theory beyond realism and neo-liberalism,' (2009) 3 (1) Journal of Eastern African Studies 74.

Mavhunga CC and Trischler H (eds), Energy (and) Colonialism, Energy (In) Dependence: Africa, Europe, Greenland, North America (RCC Perspectives 2014/5).

Mcafee K, 'Green economy and carbon markets for conservation and development: a critical view' (2016) 16 (3) International Environmental Agreements: Politics, Law and Economics, p 333.

17

Trade, Human Rights and Environmental Sustainability in Africa with Special Reference to the Extractive Sector

Omphemetse S. Sibanda Snr

1 Introduction

This chapter explores the promotion and protection of human rights and environmental sustainability in the extractive industry, particularly against the backdrop of the recently established Africa Continental Free Trade Agreement[1] (AfCFTA) and the renewed focus of free trade maximization by the African Union (AU). AfCFTA is intended to unlock intra-continental trade and generally boost trade relationships among African countries through trade liberalization.[2] This resonates well with the commitment made by several African nations to implement plans pursuant to the Action Plan for Boosting Intra-Africa Trade (BIAT), and

[1] See T Disenyana, 'The AFCFTA – lessons from the free trade area of the Americas' (Business Day, 27 March 2019) <https://www.businesslive.co.za/bd/opinion/2019-03-27-the-AfCFTA%2D%2Dlessons-from-the-free-trade-area-of-the-americas/> Accessed 27 March 2019.

[2] Generally, see TRALAC, *The African Continental Free Trade Area: A Tralac Guide* (2016).

O. S. Sibanda Snr (✉)
University of Limpopo, Polokwane, South Africa
e-mail: Omphemetse.sibanda@ul.ac.za

M. Addaney, A. Oluborode Jegede (eds.), *Human Rights and the Environment under African Union Law*, https://doi.org/10.1007/978-3-030-46523-0_17

open their markets and embrace liberalization in all industries. AfCFTA is expected to lead to the exponential increase of industrial activities in the continent. This development comes on the backdrop concerns raised about the effect of trade liberalization and industrialization,[3] particularly in the case of the extractive industries sector (mining, quarrying, dredging, extraction, fracking, exploration).[4]

The extractive industry sector is important for AfCFTA to succeed as a framework for a significant contribution to economic growth. Van Dyk succinctly reports that the Southern African Development Community (SADC) member countries are heavily dependent on mining exports of minerals such as diamonds, uranium, platinum, coal, copper and oil.[5] However, it is important that there is synergy and harmless co-existence between extractive industry activities and the rights of communities in and around these activities. The relationship must promote environmental sustainability and respect for the freedoms and rights of the mining-affected communities, particularly rural communities.[6] The distortion of the economy in natural assets trade, for example, has been highlighted to be one of the drawbacks in iron ore exporting countries of Zambia and Papua New Guinea.[7] The discovery of oil in Kenya, for example, sparked fears of "resource curse" accompanied by possible human rights abuses and operations that are not transparent and not accountable.[8] The risks

[3] A Julian, R Bullard, and B Evans (eds), Just Sustainabilities: Development in an Unequal World (Boston, MA: The MIT Press 2003).

[4] See B Meyersfeld, 'Empty Promises and the Myth of Mining: Does Mining Lead to Pro-Poor Development?' (2017) 2 Business and Human Rights Journal 31–53.

[5] D Van Dyk, 'Globalisation and the Mineral Industry: A South African Case Study' in I Masters and E Kisiangani (eds), Natural Resources Governance in Southern Africa (Institute for Global Dialogues: Braamfontein, South Africa 2017) 9.

[6] R César, Human Rights in Minefields: Extractive Economies, Environmental Conflicts, and Social Justice in the Global South (Bogotá, D.C.: Dejusticia 2015). See also A Whitmore (ed), *Pitfalls and Pipelines: Indigenous Peoples and Extractive Industries* (Tebtebba Foundation: Philippines 2012)

[7] See R Aunty, *Sustaining Development in Mineral Economies: The Resource Curse Thesis* (London: Routledge 2006).

[8] See NM Michieka and D Blankenship, 'Avoiding the Resource Curse: Applying the Guiding Principles in Kenya' in Jena Martin and Karen E. Bravo (eds), *The Business and Human Rights Landscape: Moving Forward, Looking Back* (Cambridge University Press 2015). See also L Diamond and J Mosbacher, 'Africa's Coming Resource Curse—and How to Avoid It' (2013) 92 Foreign Affairs 86.

identified by the Working Group on Extractive Industries, Environment and Human Rights[9] (WGEI) of the African Commission on Human and Peoples' Rights (African Commission) include "dis-possession of land and displacement of communities; weak or poorly beneficial terms of concession; environmental degradation and poor labour rights protection; lack of transparency in respect of royalties paid and profits made and avoidance of taxes".[10] In view of the possible risks, environmental sustainability in Africa must not only be legalized, but it must also be considered a human right from constitutional, environmental law and earth jurisprudence perspectives.[11]

There is evidence of some egregious violations of human rights and threats to environmental sustainability linked to activities of the extractive industries and multinational corporations (MNCs) in Africa. The AU, unfortunately, is yet to have a harmonized framework to "regulate" the activities of MNCs.[12] However, there are several AU instruments that may provide some much-needed general guidelines on how to deal with the possible ramifications to MNCs and extractive industries activities. One main instrument to be considered is the African Charter on Human and People's Rights[13] (ACHPR, and commonly referred to as the Banjul Charter). The Malabo Protocol, adopted in 2014, also contains key provisions that can be used mutatis mutandis to address challenges in the extractive industry. The Malabo Protocol is intended to strengthen the jurisdiction of the African Court of Justice and Human Rights (ACJHR) and to cover international crimes committed in Africa including those by

---

[9] The Working Group on Extractive Industries, Environment and Human Rights (WGEI) was established pursuant to resolution 48 (XLVI) 09 adopted at the 46th Ordinary Session of the African Commission on Human and Peoples' Rights held in Banjul, The Gambia, from 11 to 25 November 2009.

[10] African Commission on Human and Peoples' Rights. Extractive Industries & Environment in Africa, October 2018, Issue 1, 2.

[11] See also KSA Ebeku, 'The right to a satisfactory environment and the African Commission' (2003) 3 African Human Rights Law Journal 149.

[12] Generally, see EO Ekhator, 'Regulating the Activities of Multinational Corporations in Nigeria: A Case for the African Union?' (2018) 20 International Community Law Review 30–68.

[13] *African [Banjul] Charter on Human and Peoples' Rights*, African Commission on Human and People's Rights, OAU Doc. CAB/LEG/67/3 rev. 5, 21 I.L.M. 58 (1982), adopted 27 June 1981, entered into force 21 October 1986.

MNCs among others.[14] Also important to consider is the growing jurisprudence of the AU Commission, which has issued some resolutions providing recourse to extractive industry-affected communities. Notable, for example, is AU Commission, Communication *155/96: The Social and Economic Rights Action Centre (SERAC) and the Centre for Economic and Social Rights v Nigeria resolution.*[15]

In this Chapter, a qualitative research approach is employed through an interpretive, critical and descriptive paradigm. The use of an interpretive paradigm assists in constructively understanding certain realities of communities in the extractive industries sector and the regulatory approaches adopted by the AU. Through a descriptive paradigm, the chapter evaluates the challenges and efficacies of socio-economic development in Africa driven by extractive industries.

2 Trade, Human Rights and the Environment: Theoretical and Conceptual Considerations

The Office of the United Nations High Commissioner for Human Rights describes "[a] human rights-based approach as a conceptual framework for the process of human development that is normatively based on international human rights standards and operationally directed to promoting and protecting human rights".[16] The importance of using a human rights-based approach (HRBA) in this Chapter is that it identifies who the holders and/or custodians of the rights to the environment are, and how to redress must be undertaken. Through HRBA corresponding duties and obligations on the AU and member countries would ensure that

[14] Generally, see M V S Sirleaf, 'The African Justice Cascade and the Malabo Protocol, Legal Studies Research Paper Series Working Paper No. 2017-01 January 2016.

[15] AU Commission, Communication *155/96: The Social and Economic Rights Action Centre (SERAC) and the Centre for Economic and Social Rights v Nigeria*, 15th Annual Activity Report [in Decisions 2002–2007, IHRDA, Banjul 2008, 277–293], paragraph 52.

[16] UNICEF, Human Rights-based Approach to Programming <https://www.unicef.org/policyanalysis/rights/index_62012.html#1> Accessed 21 January 2020.

trade liberalization and industrialization[17] do not impact negatively communities in the extractive industry sector. Equally important is developmental transformation theory,[18] whose relevance emanates from the economic integration and development objectives of AfCFTA. The principle of sustainable development will provide a golden threat to the discussions and arguments relating to environmental rights and protection.

The African philosophy of *uBuntu* will also be used as a theoretical basis to determine if, and how, Africa is achieving environmental justice in Africa.[19] Therefore, in this chapter reference is made to *uBuntu* Theory of Environmental Justice and Sustainable Development. *uBuntu* is used when addressing corporate social responsibility (CSR)[20] or what is now termed corporate social and environmental responsibility (CSER).[21] When translated in simple English "*uBuntu*" means "a person is a person because of or through others".[22] Values and principles contained in the African philosophy of *uBuntu* include the need to show compassion reciprocity, dignity, humanity and mutuality in the interests of building and maintaining communities with justice and mutual caring.[23]

[17] Generally, see M Xu, JM David, and SH Kim, 'The Fourth Industrial Revolution: Opportunities and Challenges' (2018)9 International Journal of Financial Research 90.

[18] Generally, see, AR Bakare, 'African Union and the Developmental Transformation of Africa: Challenges, Achievement and Prospects' (2014) 3 European Journal of Sustainable Development 67–86.

[19] See M Sheridan, 'Environmental Conservation through *uBuntu* and other Emerging Perspectives' (2015) 32 African Journal of Range & Forage Science 67.

[20] See, for example, K Mofuoa, 'Applying *uBuntu*-Botho African ethics to stakeholder corporate social responsibility' (2014)12 The Journal of the Iberoamerican Academy of Management 222.

[21] V Olusegun and M Chima (note 21) 590.

[22] Generally, see A West, '*uBuntu* and Business Ethics: Problems, Perspectives and Prospects' (2013) Journal of Business Ethics 121.

[23] See V Olusegun and M Chima, 'Determinants of Nigerian managers' environmental attitude: Africa's *uBuntu* ethics versus global capitalism' (2018) 60 International Business Review 577.

3 AfCFTA: A Brief Overview

AfCFTA Objective, and Ramifications on Extractive Industries and Mining Communities

Listed as general purposes in Article 3 of AfCFTA include promoting industrial development through diversification and regional value chain development, agricultural development and food security. In terms of Article 4 of AfCFTA, the general objectives in Article 3 will be achieved through a number of specific interventions and measures including, among others, elimination of trade barriers of trade in goods, and progressive liberalization of trade-in services. Article 3 of the Protocol on Trade in Goods, in furtherance of the general objectives of AfCFTA, addresses explicitly the enhancement of socio-economic development, diversification and industrialization.[24]

The AfCFTA protocols provide recourse to parties when addressing the relationship between trade and the environment. Article 25(c) of the Protocol on Trade in Goods read with Article 25(g), for instance, can be used by State Parties as justification to implement measures that enforce transparency and accountability in mining operations, and in doing so prevent human rights violations going unnoticed. Furthermore, Article 25(b) can be used by a State Party to prohibit exploration or mineral extraction in a particular community when it is of the view that such a measure is necessary to protect human, animal or plant life or health. Also, Article 25(g) can be argued when putting an embargo on the exploration or extraction of exhaustible minerals. In fact, the provisions of Article 26 of the Protocol on Trade in Services can also apply mutatis mutandis in the extractive sector, particularly with regard to illegal mining taking place in disused and abandoned mines where it threatens the security of the country. A typical example would be when traded goods are by-products of illegal mining in part or in whole, and the continuation of such illegal mining activities relating to fissionable materials such as uranium, for example, threaten the security of the country and the communities in and around the illegal mining environment.[25]

[24] Protocol on Trade in Goods, Article 3(f).

[25] See Protocol on Trade in Goods, Article 26(b)(i).

Challenges and the Weaknesses Related to Operationalization of AfCFTA

The operationalization of AfCFTA will be constrained by some weaknesses and challenges. First, for example, the large membership and the varying development levels of the AU member states are considered to be one of the weakest links to AfCFTA.[26] These levels of development may trigger contentions and imbalance of interests of member countries. Second, the AU recognized eight regional economic communities (RECs) as supplementary to AfCFTA, but not all of these are effective.[27] Out of the recognized eight RECs only Common Market for Eastern and Southern Africa (COMESA), East African Community (EAC), Economic Community of West African States (ECOWAS) and SADC have, to date, been the most successful in terms of meeting certain of their objectives.[28] Only ECOWAS and SADC have been able to achieve above-average overall integration. This is problematic because these RECs are now the extension of AfCFTA, which has as its ultimate goal of achieving continental economic liberalization and integration.[29] The third point, and a serious weakness of AfCFTA, is the fact that no impact assessment was conducted throughout the process that led to the establishment of AfCFTA.[30] The welfare effects of AfCFTA may further deepen the development gap among member countries.[31]

---

[26] See generally, T Disenyana (note 1).

[27] A Geda and EH Seid, 'The potential for internal trade and regional integration in Africa' (2019) Journal of African Trade 19, 21.

[28] Generally, see DL Sparks, 'The Sustainable Development Goals and Agenda 2063: Implications for Economic Integration in Africa' (2016) 8 Research in Applied Economics 45.

[29] Generally, see CC Nshimbi, 'Issues in African Informality: What is the Relevance for Regional or Continental Integration? (2018) 48 Africa Insight 41.

[30] See UN Report. (2017), 'The Continental Free Trade Area (CFTA) in Africa – A Human Rights Perspective' <https://www.ohchr.org/Documents/Issues/Globalization/TheCFTA_A_HR_ImpactAssessment.pdf.> accessed 21 January 2019; MG Plummer, D Cheong and S Hamanaka, 'Methodology for impact assessment of free trade agreements' (Mandaluyong City, Philippines: Asian Development Bank 2010).

[31] See A Mandal, 'Welfare Effect of Free Trade Agreements: A Theoretical Note' (2019)52 Foreign Trade Review 115 on the welfare effects of FTAs.

4 Sample of Relevant Continental and International Human Rights Framework, and Normative Regulatory Regimes

The Banjul Charter

According to Article 24 of the Banjul Charter "[a]ll peoples shall have the right to a general satisfactory environment favourable to their development".[32] The preamble to the Banjul Charter further commits to the principles of international human rights under the auspice of the United Nations human rights system. It is therefore important that the extractive industries sector and State Parties take environmental sustainability into account in their operations and decision-making. The industries must also take into consideration the best human rights interest of the mining-affected communities as they use opportunities presented by the new AfCFTA regime. Moreover, the Banjul Charter requires that the right to cultural practices and development, and the socio-economic conditions of these communities should never be compromised by extractive and exploration operations.[33] The obligation of State Parties is covered in the provisions of the different articles of the Banjul Charter. For example, Article 24 of the Charter read with Article 22 states that the realization of the right to development is achieved through promoting the right to a generally satisfactory environment.

---

[32] See generally, AO Enabulele, 'Incompatibility of national law with the African Charter on Human and Peoples' Rights: Does the African Court on Human and Peoples' Rights have the final say?' (2016) 16 African Human Rights Law Journal 1.

[33] See Banjul Charter, Art 22.

The African Commission and the ACHPR

ACHPR

The main African human rights protection system is constituted by the AU Commission and the ACHPR.[34] The relationship between the two is complimentary. The *African Commission on Human and Peoples' Rights v. Republic of Kenya*,[35] also referred to as the Ogiek case, was the first indigenous rights case by the ACHPR and an important jurisprudence in environmental sustainability. The case was referred to the ACHPR by the AU Commission on the basis of grave, mass human rights violations. The matter in question of the Ogiek Community of the Mau Forest involved indigenous minority ethnic group in Kenya inhabiting the greater Mau Forest Complex. The Ogiek were in October 2009, through the Kenya Forestry Service, served with a 30-day eviction notice by the Kenyan government.[36] The government, among others, claimed that the Ogiek community used the Mau Forest in an unsustainable manner. This was denied in the submission of the merits of the applicant.[37] Consequently, the ACHPR held that the Kenyan government violated Article 22 of the Banjul Charter.[38] In terms of Article 22(1), "[a]ll peoples shall have the right to their economic, social and cultural development with due regard to their freedom and identity and in the equal enjoyment of the common heritage of mankind". Further in Article 22(2), the Charter provides that "States shall have the duty, individually or collectively, to ensure the exercise of the right to development". The ACHPR found that the Ogieks have not "been actively involved in developing and determining health, housing and other economic and social programmes affecting them".[39]

[34] See W Fahmy, 'From the Establishment of the Court of Justice of the African Union to Malabo Protocol: The Defies to the Regional Judicial Mode of Protection of Human Rights' (2019) 7 Russian Law Journal 165, 174.

[35] African Commission on Human and Peoples' Rights v. Republic of Kenya Application No. 006/2012, Judgment 26 May 2017 (*Ogiek* case).

[36] Ogiek case, para 6–7.

[37] *Ogiek* case, para 43C.

[38] Ogiek case, para 211.

[39] Ogiek case, para 210.

It is commendable that the *Ogiek* judgment highlighted the importance of consultation with all affected stakeholders. Furthermore, the judgment located environmental issues within HRBA. Having regard to the South African case of *Koyabe and Others v Minister for Home Affairs and Others* (Lawyers for Human Rights as Amicus Curiae),[40] which associated *uBuntu* with a general obligation to treat people with respect and dignity, the Ogiek judgment in my view embodied the dictates of the African philosophy of *uBuntu*. It should be noted that one unique characteristic of the ACHPR is that it has a broad mandate. Its jurisdictional competence is not matched by its counterparts in the Americas or Europe. Its mandate pursuant to Article 3 of the Protocol involves adjudicating the Banjul Charter, its Protocol and other human rights instruments ratified by the Member State that approached it.[41] This broad jurisdiction should be welcomed breaking the limitations and/or barriers in addressing human rights and environmental rights violations in Africa.

African Commission

The African Commission has to date pronounced on some important issues affecting human rights, consequent to the planning and operations of MNCs and extractive industries. In the case of *Social and Economic Rights Action Center & Another v Nigeria* (hereinafter SERAC case)[42] the African Commission ruled that the complainants' right to health and the right to a generally satisfactory environment had been violated by Nigeria linked to the development-related activities and conducts in the Niger Delta.[43] The SERAC case resolution is critical as a reference point for requiring extractive industries to ensure safe water for communities. It is

[40] See 2010 4 SA 327 (CC).

[41] YJ Reventlow and R Curling, 'The Unique Jurisdiction of the African Court on Human and Peoples' Rights: Protection of Human Rights Beyond the African Charter' (2019) 33 Emory International Law Journal 203, 206.

[42] AU Commission, *Communication No 155/96: The Social and Economic Rights Action Centre and the Centre for Economic and Social Rights v Nigeria*, 30th sess (13–27 October 2001) 8–9.

[43] AU Commission, Communication *155/96: The Social and Economic Rights Action Centre (SERAC) and the Centre for Economic and Social Rights v Nigeria*, 15th Annual Activity Report [in Decisions 2002–2007, IHRDA, Banjul 2008, 277–293], paragraph 52.

imperative from trade and human rights position that mining and exploration corporations observe the fundamental human rights to clean and safe drinking water to communities in and around their activities. These fundamental rights derive also from global instruments such as the International Covenant on Economic, Social and Cultural Rights (ICESCR). The CESCR General Comment 15 in 2002 explains that the right to drinking water "entitles everyone to sufficient, safe, acceptable, physically accessible and affordable water for personal and domestic uses".[44] Environmental sustainability in Africa must not only be legalized, but it must also be considered human rights from constitutional, environmental law and earth jurisprudence perspectives.

In another case involving Nigeria, the *Centre for Minority Rights Development (Kenya) and Minority Rights Group International on behalf of Endorois Welfare Council v Kenya* (hereinafter Endorois case),[45] the African Commission addressed among others the displacement of Endorois and restricting their access to Lake Bogoria. The Commission concluded that the displacement violated their rights to religious practice and culture, to property, to free disposition of natural resources and to development.[46]

The resolutions of the African Commission and relevant provisions of the Banjul Charter, as discussed above, put African in line with the jurisprudence of the UN, which serves as an authority for an obligation toward the protection of communities from abuse by MNCs and extractive industries. For instance, the UN Declaration on the Rights of Indigenous Peoples[47] (UNDRIP), adopted by the General Assembly on 13 September 2007, is a key international human rights law instrument for use by indigenous peoples to exercise their rights in relation to extrac-

---

[44] On 28 July 2010 the General Assembly Resolution 64/292 adopted the "Human Right to Water and Sanitation" and "recognizes the right to safe and clean drinking water and sanitation as a human right that is essential for the full enjoyment of life and all human rights".

[45] Communication 276/2003, *Centre for Minority Rights Development (Kenya) and Minority Rights Group International on behalf of Endorois Welfare Council v Kenya.*

[46] African Commission, 'Case 276/2003: *Centre for Minority Rights Development (Kenya) and Minority Rights Group International on behalf of Endorois Welfare Council* versus *Kenya*', February 2010, 1.

[47] UN General Assembly, 'United Nations Declaration on the Rights of Indigenous Peoples: resolution/adopted by the General Assembly' 2 October 2007, A/RES/61/295, <https://www.refworld.org/docid/471355a82.html> Accessed 8 July 2019.

tive industries. In terms of the Declaration, "Indigenous peoples have the right to the lands, territories, and resources which they have traditionally owned, occupied or otherwise used or acquired", including the right to "own, use, develop and control the lands, territories and resources that they possess".[48] Obligations are placed on States, among others, to respect the fact that indigenous communities also have the right not to be relocated without compensation,[49] the right to redress,[50] the right to development[51] and the right to land and resources thereto.[52] In 2002 the UN Permanent Forum on Indigenous Issues (UNPFII) was created as a platform for indigenous peoples to report how extractive industries corporations cause and/or threaten to cause environmental degradation, cultural ethnocide and gross human rights violations. Alongside the UNPFII is the UN Special Rapporteur on the Rights of Indigenous Peoples and the UN Expert Mechanism on the Rights of Indigenous Peoples which, among others, have been involved in addressing problems caused for indigenous peoples by the extractive industries corporations.[53] These include human rights violations such as arbitrary detention, persecution, coercion, torture, extra-judicial killings of community representatives, destruction of property and rape.

Extractive Business and the Mandate of WGEI Resolution 48 of 2009

WGEI Resolution 48 (XLVI) 09 adopted in 2009 at the 46th Ordinary Session of the African Commission sets a broad mandate for African Commission Working Group on Extractive Industries, Environment and Human Rights Violations[54] (WGEI). The mandate includes the respon-

[48] Ibid. Art 26.

[49] Ibid. Art 10.

[50] Ibid. Art 28.

[51] Ibid. Art 23 read with art 32.

[52] Ibid. Art 26 read with Art 29

[53] See Whitmore (n3) xxv.

[54] See Working Group on Extractive Industries, 'Environment and Human Rights Violations' <https://www.achpr.org/specialmechanisms/detailmech?id=13> accessed 25 November 2019.

sibility to "(a) Examine the impact of extractive industries in Africa within the context of the African Charter on Human and Peoples' Rights"; and "(f) Formulate recommendations and proposals on appropriate measures and activities for the prevention and reparation of violations of human and peoples' rights by extractive industries in Africa". As part of its strategic priorities, WGEI is required to "strengthen the regional responses to issues of human rights in the extractive industries in Africa".[55] Complementing the WGEI mandate is a set of Strategic Priorities (SPs), namely: SP1: Developing a Monitoring & Emergency Response Mechanism; SP2: Knowledge production & Development of Normative Frameworks/tools; SP3: Visibility & Reconnecting with Wider Stakeholders; and SP4: Sustainability of the Work and the Activities of the WGEI.

Cumulatively the SPs address several matters relevant to the operations of the extractive industry and the protection of the environment. SP4, for instance, involves developing a coordinated approach to the work of the WGEI and the [AU Commission's] response to issues of extractive industries, human rights and the environment; development of the internal rules and procedures for the WGEI; having a clear and coordinated approach to the provision of support for the work of the WGEI, and obtaining sufficient funding to carry out activities and achieve strategic objectives.[56] Pursuant to Strategic Priority 3, for example, the discourse of WGEI on extractive industries and environmental protection is shared with the public through various avenues, activities and consultations.[57] For example, the *Draft Study of the African Commission on Human and Peoples' Rights on Extractive Industries, Environment and Human Rights in Africa*[58] is a product of a public consultation process that took place during the 65th Ordinary Session of the Commission.

The African Commission is aware of the intensified expansion of extractive industries sector on the African continent, and the need to put

---

[55] See ibid.

[56] See ibid.

[57] See, for example, A du Preez, 'The Marikana Massacre: Seeing it All' (2015)16 Safundi 419.

[58] Draft Study of the African Commission on Human and Peoples' Rights on Extractive Industries, 'Environment and Human Rights in Africa' <https://www.achpr.org/announcement/detail?id=75> accessed 23 November 2019.

in place oversight and regulatory mechanisms to avoid abuses as those experienced in the Niger Delta.[59] According to the Commission, the fact that African "[mineral] resources attract the interests of various international actors including major powers and companies in the extractive industries sector",[60] which must be kept under check. The African Commission has observed and appreciated the exploitative practices of MNCs and extractive industries.[61]

Another important activity of WGEI was the National Dialogue on the Rights of Indigenous Peoples and Extractive Industries, held in Nairobi, Kenya from 7 to 8 October 2019.[62] Part of the resolution, for instance, included requesting the Government of Kenya to ratify the ILO Convention 169 and endorse and domesticate the UN Declaration on the Rights for Indigenous Peoples (UNDRIP), to oversee the implementation of the African Commission ruling on the Ogiek case.[63] Also, for the Government of Kenya "[t]ogether with extractive industries develop and implement national public participation models for the sector taking into account all citizens of the country including full participation of Indigenous Populations".[64]

The Malabo Protocol

In 2014, the AU and its Heads of State and Government meeting in Malabo, Equatorial Guinea, adopted the Protocol on Amendments to the Protocol of the ACHPR Statute, otherwise known as the Malabo

---

[59] See JP Eaton, 'The Nigerian Tragedy, Environmental Regulation of Transnational Corporations, and the Human Right to a Healthy Environment' (1997) 15 BUILJ 261, 264–271.

[60] Generally see 'Background Study on the Operations of the Extractive Industries Sector in Africa and its Impacts on the Realisation of Human and Peoples' Rights under the African Charter on Human and Peoples' Rights' <https://www.achpr.org/public/Document/file/English/Background%20Study%20on%20the%20Operations%20of%20the%20Extractive%20Industries%20Sector%20in%20Africa_ENG.pdf> 3 accessed 25 November 2019.

[61] Ibid. 5.

[62] The National Dialogue on the Rights of Indigenous Peoples and Extractive Industries, held in Nairobi, Kenya from 7 to 8 October 2019, Nairobi, Kenya <https://www.achpr.org/news/viewdetail?id=203> accessed 20 November 2019.

[63] Ibid.

[64] See Ibid.

Protocol.[65] The Malabo Protocol has the potential to play a very critical role in consequence management and regulation of liabilities of MNCs and extractive industry in general. The Malabo Protocol extends the jurisdictional competence of the proposed African Court of Justice and Human Rights (African Criminal Court) to include corporate criminal responsibility. It thus will enable the African Criminal Court to play a complementary and supplementary regional role to the International Criminal Court.[66] The Protocol establishes several redress avenues such as an independent office of a public defender. Thus, communities affected by operations of MNCs and the extractive industry will be able to approach the office of the public defender for intervention, including criminal prosecution of the violators of their environmental rights and privileges.[67]

Another important feature of Malabo Protocol is that it creates a Trust Fund "for legal aid and assistance and for the benefit of victims of crimes or human rights violations and their families".[68] Furthermore, the Malabo Protocol creates a Victims and Witnesses Unit with the responsibility to provide, among others, "protective measures and security arrangements, counselling and other appropriate assistance".[69] It is in light of arrangements such as the Malabo Protocol Trust Fund, that AU must be lauded for putting in place interventions that align to CSR or CSER.[70] The Victims and Witnesses Unit, taking into account the values of *uBuntu*, creates a *sui generis* governance regime.

[65] Protocol on Amendments to the Protocol on the Statute of the African Court of Justice and Human Rights <https://au.int/sites/default/files/treaties/36398-treaty-0045_-protocol_on_amendments_to_the_protocol_on_the_statute_of_the_african_court_of_justice_and_human_rights_e.pdf> accessed 19 November 2019.

[66] See S Nimigan, 'The Malabo Protocol, the ICC, and the Idea of 'Regional Complementarity' (2019)10 Journal of International Criminal Justice 1078.

[67] Generally, see, A Oyeniyi and A Ordor, 'Addressing Human Rights Concerns in the Extractive Resource Industry in Sub-Saharan Africa using the Lens of Article 46 (C) of the Malabo Protocol' (2018)11 Law and Development Review 843.

[68] Art 46 M.

[69] Art 22B.

[70] Generally, see OA Jacob, *Global commerce and human rights: towards an African legal framework for corporate human rights responsibility and accountability* (PhD: University of Witwatersrand 2015).

African communities suffer injustices perpetrated by MNCs, and not only at the hands of natural persons. This has been recognized as needing attention in the Malabo Protocol, which provides that "the Court shall have jurisdiction over legal persons, with the exception of States".[71] It is commendable that the Malabo Protocol addresses corporate criminal responsibility. Oyeniyi and Ordor[72] argue that the Malabo Protocol will promote corporate respect for human rights and access to judicial remedies by victims of corporate human rights abuse through the provisions of Article 46 (C) of the Malabo Protocol. In particular, that expanded jurisdiction provided by the Malabo Protocol is consistent with the commitment of African countries to implement Pillar II of the United Nations Guiding Principles on Business and Human Rights on respect for human rights by corporations.[73] There is a growing consensus that the African philosophy of *uBuntu* is not only an important guiding ethical principle in the business environment, *uBuntu* generally plays a significant role in the broader corporate governance framework.[74] Finally, the Malabo Protocol is also consistent with Pillar III of the Guiding Principles on the provision of remedies for human rights violations.[75]

---

[71] Art 46C.

[72] A Oyeniyi and A Ordor (note 76) 843–878.

[73] See International Trade Union Confederation, The United Nations "Protect, Respect, Remedy" Framework for Business and Human Rights and the United Nations Guiding Principles for Business and Human Rights, May 2012 <https://www.ituc-csi.org/IMG/pdf/12-04-23_ruggie_background_fd.pdf> accessed 20 October 2019.

[74] A West, '*uBuntu* and Business Ethics: Problems, Perspectives and Prospects' (2013) Journal of Business Ethics 121. See also S H Kayuni and R Tambulasi, '*uBuntu* and corporate social responsibility: The case of selected Malawian organizations' (2012)3 African Journal of Economic and Management Studies 64.

[75] Generally, see, L Lizarazo-Rodríguez, 'The UN 'Guiding Principles on Business and Human Rights': Methodological Challenges to Assessing the Third Pillar: Access to Effective Remedy' (2018)36 Nordic Journal of Human Rights 353–370 discussing Pillar III.

5 Trade, Development and the Impact of Extractive Industries Sector in Africa

Abuses, Human Suffering and Human Rights Violations

In *The New Colonialism: Britain's Scramble for Africa's Energy and Mineral Resources*, Curtis reports that many British companies control Africa's mineral resources through trade and investment projects, an occurrence which in my view will be heightened by the new drive for liberalized African trade. What the report also highlighted is the alleged complicity of these foreign mining companies in environmental degradation in Africa, and in other social ills, such unfair or forced resettlement programs, unfair labor conditions and abuses, and their disputed association with killings in or near mine areas.[76]

A case study of the Karonga district in northern Malawi by Human Rights Watch revealed the negative impact of the bourgeoning mining industry on mining-affected communities, for example. At the heart of the challenges experienced by these communities has been a lack of proper regulatory and governance structures. Concerning is the apparent inadequate normative standards and safeguards securing the balance between the mining-affected community and the mining industry, resulting in gross abuse of the former in several respects.[77] Still ringing loud among abuses is the 10 November 1995 execution of nine leaders of the Movement for the Survival of the Ogoni People (MOSOP) following a judgment by a military tribunal in defiance of the recommendation for

[76] See J Ruggie, 'Report of the Special Representative of the Secretary-General on the Issue of Human Rights and Transnational Corporations and Other Business Enterprises, Addendum, Corporations and Human Rights: A Survey of the Scope and Patterns of Alleged Corporate-Related Human Rights Abuse' UN Doc A/HRC/8/5/Add.2 (23 May 2003). See also B Meyersfeld, 'Empty Promises and the Myth of Mining: Does Mining Lead to Pro-Poor Development?' (2017) 2 Business and Human Rights Journal 31.

[77] Human Rights Watch, 'They Destroyed Everything: Mining and Human Rights in Malawi' 2016 <https://www.hrw.org/sites/default/files/report_pdf/malawi09_brochure_web.pdf> accessed 20 June 2019.

the stay of execution by the ACHPR.[78] This infamous execution was linked in part to the oil extraction operations in the Niger Delta where Shell had operations, and the military's troubling of the indigenous people of the Niger Delta, especially the Ogoni people.[79]

Part of the complaints before the African Commission in the *SERAC* case was that the oil consortium exploiting the oil reserves in the Ogoniland showed no regard to the health or environment of the communities. The consortium contaminated water streams, soil and air causing serious "short-term and long-term health impacts, including skin infections, gastrointestinal and respiratory ailments, and increased risk of cancers, and neurological and reproductive problems".[80] It was also alleged that the Nigerian government did its best to protect the oil extractive industry, even going so far as to destroying and threatening the food security of the Ogoni people. In *African Commission on Human and Peoples' Right v. Republic of Kenya*[81] the ACHPR gave a judgment confirming that the government of Kenya violated rights of the Ogiek community through various conducts, including making it impossible for them to secure ownership and use of their ancestral land on which they depend for their social, economic and cultural existence. The judgment is important toward building a jurisprudence of protecting the rights of vulnerable indigenous and/or ethnic communities against activities that seek to displace them. It also gives traction to the interpretation and development of the normative content of Article 24 of the Banjul Charter.[82]

[78] See *International Pen and Others (On Behalf of Saro-Wiwa) v Nigeria* (2000) AHRLR 212 (ACHPR 1998) at par:103.

[79] See, KSA Ebeku, 'The right to a satisfactory environment and the African Commission' (2003) 3 African Human Rights Law Journal 159–160.

[80] Ebeku ibid. 161. See, also, AY Emmanuel, CS Jerry and DA Dzigbodi, 'Review of Environmental and Health Impacts of Mining in Ghana' (2018)17 Journal of Health and Pollution 43, 43.

[81] *African Commission on Human and Peoples' Right v. Republic of Kenya* (Application 006/2012) Judgment of the Court 26 May 2017.

[82] See generally M Van Der Linde and M Louw, 'Considering the Interpretation and Implementation of Article 24 of the African Charter on Human and Peoples' Rights in Light of the SERAC Communication' (2003) 3 African Human Rights Law Journal 167.

Iniquitous Beneficiation and Exploitation

A recent Mining Royalties Research Report[83] by Corruption Watch, relating to research conducted in the mining communities of the provinces of Limpopo and the North West, revealed how mining corporations denied beneficiation to mining-affected communities.[84] This investigation drew deserved attention to the challenges experienced by mining-affected communities and the human rights violations against these communities. The Corruption Watch Report resonates with some of the findings in the 2014 study by Chikwanha[85] reporting on corruption in the extractive industry in Africa. Countries like Botswana were identified as one of the few that has been able to put in place industry governance and management system that helps ferret out corruption.

A national hearing by the South African Human Rights Commission (SAHRC) revealed, for example, "socio-economic challenges experienced by mining-affected Communities".[86] Communities are not provided with adequate and equitable benefit sharing or distribution. The SAHRC also expressed concern, for example, on companies not "giving effect to the internationally recognised precautionary principle in matters dealing with environmental protection"[87] and focused on "gaining immediate economic benefit of mining activities over the maintenance and protection of the environment, particularly in those areas that are crucial for sustaining ecological biodiversity, natural heritage, cultural significance and life".[88] Also noted by the SAHRC was the Government's "inability to provide certain information about the monitoring of mining activities in

---

[83] Corruption Watch, 'Mining Royalties Report 2018' <https://www.corruptionwatch.org.za/wp-content/uploads/2019/03/Mining-royalties-research-report-final1.pdf> accessed 20 September 2019.

[84] See generally E Morgera, 'Fair and Equitable Benefit-sharing' in L Kramer and E Orlando (eds), *Principles of Environmental Law* (Edward Elgar 2018) 323–327.

[85] AB Chikwanha, *Combating corruption in the extractive industry in Africa* (Swedish International Development Cooperation Agency 2016).

[86] South African Human Rights Commission (SAHRC) Report. *National Hearing on the Underlying Socio-economic Challenges of Mining-affected Communities in South Africa* (Braamfontein, South Africa 2016).

[87] Ibid. 1.

[88] Ibid. 3.

protected areas".[89] Extractive industries seem to take little responsibility and accountability. They have been allowed to gain profits and amassed wealth at the expense of fundamental rights of the poorly paid and exploited miners.[90]

Extractive Industry Sector, Socio-Economic and Environmental Sustainability

Addressing a sustainable and socially acceptable trade and development in the context of the extractive industry sector has become increasingly important.[91] The recognition of sustainability is not only occasioned by the need to ensure "development that meets the needs of the present without compromising the ability of future generations to meet their own needs".[92] It is also about intra-generational justice and the "interface of human ecology and political ecology, linking environmental issues to economic constraints, land rights, social justice, and values".[93] Closely linked to environmental sustainability is humanity. The requirement for sustainability can thus be justified by the South African indigenous concept of *uBuntu,* "a holistic view of human identity as interconnected with the environment and with other persons",[94] which is now used as part of the South African public law, private law and constitutional interpretation. What then to do next as the way forward in ensuring that the AfCFTA regimes do not comprise environmental sustainability and lead to the violation of human rights of mining-affected communities?

[89] Idem.

[90] See PM Makua and KO Odeku, 'Harmful mining activities, environmental impacts and effects in the mining communities in South Africa: a critical perspective' (2017) Environmental Economics 14.

[91] See J Dubiński, 'Sustainable development of mining mineral resources' (2013) 12 J. Sust. Min 1.

[92] World Commission on Environment and Development, *Our common future* (Oxford University Press, Oxford 1978) 8.

[93] B Tor, R Rohde, E Sjaastad, P Wisborg, and T Lebert, 'Land Reform, Range Ecology, and Carrying Capacities in Namaqualand, South Africa' (2006) 96 Annals of the Association of American Geographers 524.

[94] Carmen Gonzalez, 'Environmental Justice, Human Rights, and the Global South' (2015)13 Santa Clara J. Int'l L 191, making reference to Jacqueline Church, 'Sustainable Development and the Culture of *uBuntu*' (2012) 45 De Jure 511, 524–31.

6 Conclusion

Harmonized Human Rights and Environmental Sustainability Framework

Africa needs National Extractive Industry Programmes of Action (NEIPA). NEIPA must embody certain core principles on the advancement of human rights and protection of the environment. In particular, the *uBuntu* approach to environmental management and sustainability must be at the core of NEIPAs in order to properly respond to African conditions and circumstances. *uBuntu* should not only be confined to human relations, but it must also include an appreciation of environmental sustainability.[95] According to Chibvongodze, *uBuntu* serves as an important foundational basis "in the preservation and conservation of the environmental".[96] Furthermore, NEIPAs must be founded on HRBA to extractive industry regulation. In its *Resolution on Climate Change and Human Rights and the Need to Study its Impact on Africa* of 2009[97] the AU Commission acknowledged a shortfall of built-in human right protection safeguard in several instruments. Although the Resolution specifically related to the study of the impact of climate change on human rights in Africa,[98] reference to it herein remains relevant because activities in the extractive industry sector may have influenced the occurrence of climate change and in turn negatively affect the sustainability of the environment and human rights.

---

[95] See DT Chibvongodze, '*uBuntu* is Not Only about the Human! An Analysis of the Role of African Philosophy and Ethics in Environment Management' (2016) 53 J Hum Eco*l* 157.

[96] Idem.

[97] Resolution 153 adopted at the 46th Ordinary Session of the ACHPR held from 11 to 25 November 2009 in Banjul, The Gambia.

[98] Paragraph 4 of the Resolution.

Extractive Industries Transparency and Accountability Guidelines

AU countries must be held to a set of Extractive Industries Transparency and Accountability Guidelines (EITAG). EITAG will help foster the prevailing efforts by the Extractive Industries Transparency Initiative (EITI). Some of the provisions of EITAG may be developed from the provisions of the Malabo Protocol, particularly section 46C on criminal liability of corporations. Interestingly, countries like Nigeria that have a history of abuses in the extractive industry environment have joined the EITI and set up its Nigerian Extractive Industries Transparency Initiative (NEITI),[99] while on the other hand, a country like Botswana with an exemplary record on how to handle the balance between the mining industry and the mining communities took a decision not to join EITI. Botswana has a well-developed and effective extractive industry management and oversight regime, which is perceived as a good model.[100]

Partnerships, Public Participation and Open Governance

It is important to have in place a critical partnership with non-governmental organizations and institutions and to allow them to fully participate in policy and governance discourses relating to environmental protection as may be affected by the extractive industries sector. This recommendation resonates with the *Rio Declaration on Environment and Development*,[101] which in Principle 10 for instance, calls for "Environmental

[99] See *Nigeria's Struggle with Corruption: The Success of Nigeria's Extractive Industries Transparency Initiative,* Testimony of David Goldwyn Before the House International Relations Committee Subcommittee on Africa, May 18th, 2006.

[100] M Silibasizo, 'The Private-Public Partnership Model: The Debswana Case' (2006) 4, OPEN SPACE. Open Society Institute for Southern Africa.

[101] *Rio Declaration on Environment and Development*, General Assembly 19th plenary meeting, A/CONF.151/26 (Vol. I), 14 June 1992

issues are best handled with the participation of all concerned citizens, at the relevant level" and for State Parties to enable "appropriate access to information concerning the environment that is held by public authorities, including information on hazardous materials and activities in their communities, and the opportunity to participate in decision-making processes". Public participation is also demanded by the United Nations' *Guidelines for the Development of National Legislation on Access to Information, Public Participation and Access to Justice in Environmental Matters*, United Nations Environment Programme[102] in recognition of Principle 10 of the Rio Declaration of 1992. The African Convention on the Conservation of Nature and Natural Resources of 1993, article 17(3) requires State Parties to "take the necessary measures necessary to enable active participation by the local communities in the process of planning and management of natural resources upon which such communities depend with a view to creating local incentives for the conservation and sustainable use of such resources".

Continental Governance and Management Framework

There is a need for an overarching continental framework to address the interface among trade, human rights and environmental sustainability in the context of extractive industries to close the existing governance gap.[103] Therefore, the main proposition in this study is that there is a need to work toward a *Continental Governance and Management Framework for Extractive Industries Sector*. The framework must cover relevant principles, such as corporate civil and criminal liability, environmental ethics and preservation education.[104]

---

[102] *Guidelines for the Development of National Legislation on Access to Information, Public Participation and Access to Justice in Environmental Matters*, United Nations Environment Programme, adopted by the Governing Council of the United Nations Environment Programme in decision SS.XI/5, part A of 26 February 2010.

[103] Generally, see Penelope Simons and A Macklin, *The Governance Gap: Extractive Industries, Human Rights, and the Home State Advantage* (Routledge 2014).

[104] See CA Bowers, 'The case against John Dewey as an environmental and eco-justice philosopher' (2003) 25 Environmental Ethics 25.

State Oversight and Extractive Industry Sector Accountability and Responsibility

Challenges faced by African countries in overseeing their mining and extractive studies are in part attributable to the lack of effective oversight and accountability regime.[105] It is important that governments play a more involved role as part of ensuring a "viable oversight mechanisms to monitor the collection and use of revenues from extractive industry revenues, and in ensuring that the interests of civil society and citizens are taken into account, from community-level environmental concerns to the allocation and disbursement of revenues collected by central governments".[106] Environmental sustainability is a human rights issue, which needs protection from a number of activities including trade-related activities in the extractive industries sector. The Ogoni people in Nigeria, for example, framed their environmental justice claims against the Government and extractive industries in the language of human rights.[107] Any failure by extractive industry-intensive trade and development economy to protect the environment is an affront to human rights.

References

African Commission on Human and Peoples' Rights v. Republic of Kenya Application No. 006/2012, Judgment 26 May 2017 (Ogiek case).

African [Banjul] Charter on Human and Peoples' Rights, African Commission on Human and People's Rights, OAU Doc. CAB/LEG/67/3 rev. 5, 21 I.L.M. 58 (1982), adopted 27 June 1981, entered into force 21 October 1986.

African Commission on Human and Peoples' Rights. Extractive Industries & Environment in Africa, October 2018, Issue 1.

---

[105] S Bryan and B Hofmann (ed), *Transparency and Accountability in Africa's Extractive Industries: The Role of the Legislature* (National Democratic Institute for National Affairs 2007) 8.

[106] Idem.

[107] See generally Carmen Gonzalez. Environmental Justice, Human Rights, and the Global South, (2015)13 *Santa Clara J. Int'l L* 151.

African Commission, 'Case 276/2003: Centre for Minority Rights Development (Kenya) and Minority Rights Group International on behalf of Endorois Welfare Council versus Kenya', February 2010, 1.

AU Commission, Communication No 155/96: The Social and Economic Rights Action Centre and the Centre for Economic and Social Rights v Nigeria, 30th sess (13–27 October 2001).

African Commission on Human and Peoples' Right v. Republic of Kenya (Application 006/2012) Judgment of the Court 26 May 2017.

Aunty R, Sustaining Development in Mineral Economies: The Resource Curse Thesis (London: Routledge 2006).

'Background Study on the Operations of the Extractive Industries Sector in Africa and its Impacts on the Realisation of Human and Peoples' Rights under the African Charter on Human and Peoples' Rights' <https://www.achpr.org/public/Document/file/English/Background%20Study%20on%20the%20Operations%20of%20the%20Extractive%20Industries%20Sector%20in%20Africa_ENG.pdf> 3 accessed 25 November 2019.

Bakare AR, 'African Union and the Developmental Transformation of Africa: Challenges, Achievement and Prospects' (2014) 3 European Journal of Sustainable Development 67.

Bowers CA, 'The case against John Dewey as an environmental and eco-justice philosopher' (2003) 25 Environmental Ethics 25.

Bryan S and Hofmann B (ed), Transparency and Accountability in Africa's Extractive Industries: The Role of the Legislature (National Democratic Institute for National Affairs 2007) 8.

César R, Human Rights in Minefields: Extractive Economies, Environmental Conflicts, and Social Justice in the Global South (Bogotá, D.C.: Dejusticia 2015).

Chibvongodze DT, '*uBuntu* is Not Only about the Human! An Analysis of the Role of African Philosophy and Ethics in Environment Management' (2016) 53 J Hum Ecol 157.

Chikwanha AB, Combating corruption in the extractive industry in Africa (Swedish International Development Cooperation Agency 2016).

Communication 276/2003, Centre for Minority Rights Development (Kenya) and Minority Rights Group International on behalf of Endorois Welfare Council v Kenya.

Corruption Watch, 'Mining Royalties Report 2018' <https://www.corruption-watch.org.za/wp-content/uploads/2019/03/Mining-royalties-research-report-final1.pdf> accessed 20 September 2019.

Diamond L and Mosbacher J, 'Africa's Coming Resource Curse—and How to Avoid It' (2013) 92 Foreign Affairs 86.

Disenyana, T 'The AFCFTA – lessons from the free trade area of the Americas' (Business Day, 27 March 2019) <https://www.businesslive.co.za/bd/opinion/2019-03-27-the-AfCFTA%2D%2Dlessons-from-the-free-trade-area-of-the-americas/> Accessed 27 March 2019.

Dubiński J, 'Sustainable development of mining mineral resources' (2013) 12 J. Sust. Min 1.

du Preez A, 'The Marikana Massacre: Seeing it All' (2015)16 Safundi 419.

du Plessis M, 'Shambolic, shameful and symbolic: Implications of the African Union's immunity for African leaders' (2014) 4 ISS Paper 278.

Draft Study of the African Commission on Human and Peoples' Rights on Extractive Industries, 'Environment and Human Rights in Africa' <https://www.achpr.org/announcement/detail?id=75> accessed 23 November 2019.

Eaton JP, 'The Nigerian Tragedy, Environmental Regulation of Transnational Corporations, and the Human Right to a Healthy Environment' (1997) 15 BUILJ 261.

Ebeku KSA, 'The right to a satisfactory environment and the African Commission' (2003) 3 African Human Rights Law Journal 149.

Ekhator EO, 'Regulating the Activities of Multinational Corporations in Nigeria: A Case for the African Union?' (2018) 20 International Community Law Review 30.

Emmanuel AY, Jerry CS and Dzigbodi DA, 'Review of Environmental and Health Impacts of Mining in Ghana' (2018)17 Journal of Health and Pollution 43.

Enabulele AO, 'Incompatibility of national law with the African Charter on Human and Peoples' Rights: Does the African Court on Human and Peoples' Rights have the final say?' (2016) 16 African Human Rights Law Journal 1.

Fahmy W, 'From the Establishment of the Court of Justice of the African Union to Malabo Protocol: The Defies to the Regional Judicial Mode of Protection of Human Rights' (2019) 7 Russian Law Journal 165.

Geda A and EH Seid, 'The potential for internal trade and regional integration in Africa' (2019) Journal of African Trade 19, 21.

Gonzalez C, 'Environmental Justice, Human Rights, and the Global South' (2015)13 Santa Clara J. Int'l L 191, making reference to Jacqueline Church, 'Sustainable Development and the Culture of *uBuntu*' (2012) 45 De Jure Law Journal 511.

Guidelines for the Development of National Legislation on Access to Information, Public Participation and Access to Justice in Environmental Matters, United Nations Environment Programme, adopted by the Governing Council of the United Nations Environment Programme in decision SS. XI/5, part A of 26 February 2010.

Human Rights Watch, 'They Destroyed Everything: Mining and Human Rights in Malawi' 2016 <https://www.hrw.org/sites/default/files/report_pdf/malawi09_brochure_web.pdf> accessed 20 June 2019.

Inman D, Smis A, Cirimwami EA and Bahalaokwibuye CB, 'The (in)willingness to implement the recommendations of the African Commission on Human and Peoples' Rights: Revisiting the Endorois and the Mamboleo decisions' (2018) 2 African Human Rights Yearbook 400.

International Pen and Others (On Behalf of Saro-Wiwa) v Nigeria (2000) AHRLR 212 (ACHPR 1998.

International Trade Union Confederation, The United Nations "Protect, Respect, Remedy" Framework for Business and Human Rights and the United Nations Guiding Principles for Business and Human Rights, May 2012 <https://www.ituc-csi.org/IMG/pdf/12-04-23_ruggie_background_fd.pdf > accessed 20 October 2019.

Jacob OA, Global commerce and human rights: towards an African legal framework for corporate human rights responsibility and accountability (PhD: University of Witwatersrand 2015).

Julian A, Bullard R, and Evans B (eds), Just Sustainabilities: Development in an Unequal World (Boston, MA: The MIT Press 2003).

Kayuni H and Tambulasi R, '*uBuntu* and corporate social responsibility: The case of selected Malawian organizations' (2012)3 African Journal of Economic and Management Studies 64.

Lizarazo-Rodríguez M, 'The UN 'Guiding Principles on Business and Human Rights': Methodological Challenges to Assessing the Third Pillar: Access to Effective Remedy' (2018)36 Nordic Journal of Human Rights 353.

Makua PM and Odeku KO, 'Harmful mining activities, environmental impacts and effects in the mining communities in South Africa: a critical perspective' (2017) Environmental Economics 14.

Mandal A, 'Welfare Effect of Free Trade Agreements: A Theoretical Note' (2019)52 Foreign Trade Review 115 on the welfare effects of FTAs.

Meyersfeld B, 'Empty Promises and the Myth of Mining: Does Mining Lead to Pro-Poor Development?' (2017) 2 Business and Human Rights Journal 31.

Michieka NM and Blankenship D, 'Avoiding the Resource Curse: Applying the Guiding Principles in Kenya' in Jena Martin and Karen E. Bravo (eds), The Business and Human Rights Landscape: Moving Forward, Looking Back (Cambridge University Press 2015).

Mofuoa K, 'Applying *uBuntu*-Botho African ethics to stakeholder corporate social responsibility' (2014)12 The Journal of the Iberoamerican Academy of Management 222.

Morgera E, 'Fair and Equitable Benefit-sharing' in L Kramer and E Orlando (eds), Principles of Environmental Law (Edward Elgar 2018) 323.

National Dialogue on the Rights of Indigenous Peoples and Extractive Industries, held in Nairobi, Kenya from 7 to 8 October 2019, Nairobi, Kenya <https://www.achpr.org/news/viewdetail?id=203> accessed 20 November 2019.

Nigeria's Struggle with Corruption: The Success of Nigeria's Extractive Industries Transparency Initiative, Testimony of David Goldwyn Before the House International Relations Committee Subcommittee on Africa, May 18th, 2006.

Nimigan S, 'The Malabo Protocol, the ICC, and the Idea of 'Regional Complementarity' (2019)10 Journal of International Criminal Justice 1078.

Nshimbi CC, 'Issues in African Informality: What is the Relevance for Regional or Continental Integration? (2018) 48 Africa Insight 41.

Olusegun V and Chima M, 'Determinants of Nigerian managers' environmental attitude: Africa's *uBuntu* ethics versus global capitalism' (2018) 60 International Business Review 577.

Oyeniyi A and Ordor A, 'Addressing Human Rights Concerns in the Extractive Resource Industry in Sub-Saharan Africa using the Lens of Article 46 (C) of the Malabo Protocol' (2018)11 Law and Development Review 843.

Protocol on Amendments to the Protocol on the Statute of the African Court of Justice and Human Rights <https://au.int/sites/default/files/treaties/36398-treaty-0045_-protocol_on_amendments_to_the_protocol_on_the_statute_of_the_african_court_of_justice_and_human_rights_e.pdf> accessed 19 November 2019.

Resolution 153 adopted at the 46th Ordinary Session of the ACHPR held from 11 to 25 November 2009 in Banjul, The Gambia.

Reventlow YJ and Curling R, 'The Unique Jurisdiction of the African Court on Human and Peoples' Rights: Protection of Human Rights Beyond the African Charter' (2019) 33 Emory International Law Journal 203, 206.

Rio Declaration on Environment and Development, General Assembly 19th plenary meeting, A/CONF.151/26 (Vol. I), 14 June 1992.

Ruggie R, 'Report of the Special Representative of the Secretary-General on the Issue of Human Rights and Transnational Corporations and Other Business Enterprises, Addendum, Corporations and Human Rights: A Survey of the Scope and Patterns of Alleged Corporate-Related Human Rights Abuse' UN Doc A/HRC/8/5/Add.2 (23 May 2003).

Simons P and Macklin A, The Governance Gap: Extractive Industries, Human Rights, and the Home State Advantage (Routledge 2014).

Silibasizo M, 'The Private-Public Partnership Model: The Debswana Case' (2006) 4, OPEN SPACE. Open Society Institute for Southern Africa.

Sirleaf MVS, 'The African Justice Cascade and the Malabo Protocol, Legal Studies Research Paper Series Working Paper No. 2017-01 January 2016.

Sheridan M, 'Environmental Conservation through *uBuntu* and other Emerging Perspectives' (2015) 32 African Journal of Range & Forage Science 67.

South African Human Rights Commission (SAHRC) Report. National Hearing on the Underlying Socio-economic Challenges of Mining-affected Communities in South Africa (Braamfontein, South Africa 2016).

Sparks DL, 'The Sustainable Development Goals and Agenda 2063: Implications for Economic Integration in Africa' (2016) 8 Research in Applied Economics 45.

'The Natural Resource Paradox: Resource Wealth Amid Human Poverty' In Africa Progress Report 2013 – Equity in Extractives: Stewarding Africa's Natural Resources for All. Africa Progress Panel, 2013.

Tor B, Rohde R, Sjaastad E, Wisborg P, and Lebert T, 'Land Reform, Range Ecology, and Carrying Capacities in Namaqualand, South Africa' (2006) 96 Annals of the Association of American Geographers 524.

TRALAC, The African Continental Free Trade Area: A Tralac Guide (2016).

UNICEF, Human Rights-based Approach to Programming <https://www.unicef.org/policyanalysis/rights/index_62012.html#1> Accessed 21 January 2020.

UN General Assembly, 'United Nations Declaration on the Rights of Indigenous Peoples: resolution / adopted by the General Assembly' 2 October 2007, A/RES/61/295, < https://www.refworld.org/docid/471355a82.html> Accessed 8 July 2019.

UN Report. (2017), 'The Continental Free Trade Area (CFTA) in Africa – A Human Rights Perspective' <https://www.ohchr.org/Documents/Issues/Globalization/TheCFTA_A_HR_ImpactAssessment.pdf.> accessed 21 January 2019; MG Plummer, D Cheong and S Hamanaka, 'Methodology for impact assessment of free trade agreements' (Mandaluyong City, Philippines: Asian Development Bank 2010).

Van Der Linde M and Louw M, 'Considering the Interpretation and Implementation of Article 24 of the African Charter on Human and Peoples' Rights in Light of the SERAC Communication' (2003) 3 African Human Rights Law Journal 167.

Van Dyk D, 'Globalisation and the Mineral Industry: A South African Case Study' in I Masters and E Kisiangani (eds), Natural Resources Governance in Southern Africa (Institute for Global Dialogues: Braamfontein, South Africa 2017).

West A, '*uBuntu* and Business Ethics: Problems, Perspectives and Prospects' (2013) Journal of Business Ethics 121.

Whitmore A (ed), Pitfalls and Pipelines: Indigenous Peoples and Extractive Industries (Tebtebba Foundation: Philippines 2012).

Working Group on Extractive Industries, 'Environment and Human Rights Violations' <https://www.achpr.org/specialmechanisms/detailmech?id=13> accessed 25 November 2019.

World Commission on Environment and Development, Our common future (Oxford University Press, Oxford 1978)

Xu M, David JM, and Kim SH, 'The Fourth Industrial Revolution: Opportunities and Challenges' (2018)9 International Journal of Financial Research 90.

Index[1]

[1] Note: Page numbers followed by 'n' refer to notes.

M. Addaney, A. O. Jegede (eds.), *Human Rights and the Environment under African Union Law*, https://doi.org/10.1007/978-3-030-46523-0

The manufacturer's authorised representative in the EU is Springer Nature Customer Service Centre GmbH, Europaplatz 3, 69115 Heidelberg, Germany. If you have any concerns regarding our products, please contact ProductSafety@springernature.com

Printed and bound by CPI Group (UK) Ltd, Croydon, CR0 4YY

23/09/2026

02221534-0001